Analysis and Geometry of Metric Measure Spaces

Volume 56

CRM PROCEEDINGS & LECTURE NOTES

Centre de Recherches Mathématiques
Montréal

Analysis and Geometry of Metric Measure Spaces

Lecture Notes of the 50th Séminaire de Mathématiques Supérieures (SMS), Montréal, 2011

Galia Dafni
Robert John McCann
Alina Stancu
Editors

The Centre de Recherches Mathématiques (CRM) of the Université de Montréal was created in 1968 to promote research in pure and applied mathematics and related disciplines. Among its activities are special theme years, summer schools, workshops, postdoctoral programs, and publishing. The CRM is supported by the Université de Montréal, the Province of Québec (FQRNT), and the Natural Sciences and Engineering Research Council of Canada. It is affiliated with the Institut des Sciences Mathématiques (ISM) of Montréal. The CRM may be reached on the Web at www.crm.math.ca.

American Mathematical Society
Providence, Rhode Island USA

The production of this volume was supported in part by the Fonds pour la Formation de Chercheurs et l'Aide à la Recherche (Fonds FCAR) and the Natural Sciences and Engineering Research Council of Canada (NSERC).

2000 *Mathematics Subject Classification.* Primary 53C23; Secondary 35-06, 53-06, 58-06, 60-06, 49Q15, 49N60.

Group photo courtesy of Suzette Paradis

Library of Congress Cataloging-in-Publication Data

ISBN-13 978-0-8218-9418-7

Library of Congress Cataloging-in-Publication Data has been applied for by the AMS. See www.loc.gov/publish/cip/.

Visit the AMS home page at `http://www.ams.org/`

10 9 8 7 6 5 4 3 2 1   18 17 16 15 14 13

Contents

Foreword

The 50th edition of the Séminaire de Mathématiques Supérieures took place from June 27 to July 8, 2011 on the campus of the Université de Montréal. This university has supported the summer school, which has acquired international renown, for over 50 years, often in collaboration with NATO. The 2011 SMS, on the topic of "Metric measure spaces: geometric and analytic aspects" was the start of a new funding structure in which the school was supported by the Canadian mathematical community, as well as at the North-American level. We are pleased to acknowledge the institutional partners and sponsors who supported this endeavor: the Centre de recherche mathématiques de Montréal (CRM — principal partner), the Fields Institute of Mathematical Sciences, the Pacific Institute for Mathematical Sciences (PIMS), the Mathematical Sciences Research Institute (MSRI), the Institut de sciences mathématiques de Montréal (ISM), the Département de mathématiques et de statistique de l'Université de Montréal (UdeM), Concordia University, and the Canadian Mathematical Society (CMS).

In recent decades, metric-measure spaces have emerged as a fruitful source of mathematical questions in their own right, and as indispensable tools for addressing classical problems in geometry, topology, dynamical systems, and partial differential equations. The summer school was designed to lead young scientists to the research frontier concerning the analysis and geometry of metric-measure spaces, by exposing them to a series of minicourses featuring leading researchers who highlighted both the state-of-the-art and some of the exciting challenges which remain. Attended by over 60 people, mostly graduate students, coming from 13 countries in Africa, Asia, Europe and North America, this 50th edition of the Séminaire de Mathématiques Supérieures was a resounding success.

There were thirteen four-hour minicourses, revealing the breadth and depth of recent work in this area. Versions of eight of these appear in this volume. The notes come under three general themes. The first and perhaps most prevalent was the use of techniques from optimal transportation to define geometric notions on metric measure spaces. In particular the notion of lower curvature bounds, as well as an actual generalization of such a notion to arbitrary metric measure spaces are presented. A shift toward more analytic aspects follows by focussing on the notion of weak gradient and Sobolev spaces of functions in metric measure spaces. As a natural continuation, the final theme consists of functional and geometric inequalities ranging from concentration of measure, isoperimetry and minimizers, to heat kernel estimates on non-compact manifolds, and respectively Sobolev inequalities in the Sierpinsky carpet.

Thus the reader will find in Yann Ollivier's notes a definition of Ricci curvature on discrete metric spaces, based on the idea of comparing the (average transportation) distance between balls to the distance between their centers. Far-reaching consequences of this idea are explored in various geometries. The alternative definition of Lott – Villani and Sturm in path-metric spaces, based on McCann's notion of displacement convexity, and its consequences are also recalled. This is complemented by Robert McCann and Nestor Guillen's introductory notes on optimal transportation, which offer insight into the techniques and applications starting with the classical transportation problem of Monge and Kantorovich and leading to the recent theorems concerning smoothness of optimal maps by Ma, Trudinger, Wang and Loeper. Along the way they are excursions into two-player zero sum games, convex duality and linear programming, fully nonlinear partial differential equations, the economics of optimal pricing, and connections with minimal Lagrangian submanifolds in semi-Riemannian geometry. Minimal submanifolds are the subject of the course of Guy David. Starting with the famous Plateau problem in higher dimensions (not yet solved) as an illustration, David's notes discuss the structure of minimizers to functionals on currents with a given lower-dimensional rectifiable current as boundary. The focus is on lack of smoothness, and minimizers in the sense of Almgren, concluding with the proof of Jean Taylor's theorem which characterizes them. Regularity questions for optimal maps are taken up once again in the lecture notes of Young-Heon Kim. Focusing his attention on the Riemannian setting, Kim outlines the state-of-the art through a series of examples and counterexamples, building up to his recent joint work on Hölder continuity of optimal maps between not-necessarily-smooth distributions of mass.

The second theme of the summer school, roughly coming under the "analytic aspects" of metric spaces, concerned notions of differentiability on metric measure spaces and the corresponding function spaces. The two themes come together in the lecture notes by Luigi Ambrosio, which review and compare various notions of weak gradients and Sobolev spaces in metric measure spaces, such as upper gradients, absolute continuity on lines, and Cheeger's energy, and showed the identification of weak gradients using optimal transportation techniques, without relying on doubling or Poincaré assumptions. Results of Ambrosio with Gigli and Savaré also involve Otto's gradient flow with respect to the Wasserstein distance and the curvature-dimension conditions of Lott – Sturm – Villani.

The final theme could be described as functional and geometric inequalities. The mini-course of Ollivier concludes with results on concentration of measure and joint work with Villani on the Brunn – Minkowski inequality for the discrete hypercube. The notes of Emanuel Milman describe the relations between isoperimetric inequalities, concentration of measure, and functional inequalities such as Poincaré and Sobolev. Based on these, he gives a new proof of a Bobkov's lower bound for the Cheeger constant of a convex domain. For such domains (and more generally, Ricci non-negative metric-measure spaces) the Cheeger constant is proportional to square root of the spectral gap. The inequality in question bounds its reciprocal from above by the product of average distance, and the deviation of this distance, separating an arbitrary point in Euclidean space from the domain.

The relations between various inequalities are also the subject of the course by Thierry Coulhon. Working in the Riemannian context, but this time on a noncompact manifold with volume doubling, Coulhon discusses the implications of various heat kernel estimates to L^p boundedness of Riesz transforms, an important question which relates back to the second theme since it concerns the compatibility of various definitions of weak derivatives and the corresponding Sobolev spaces. An example was given of a fractal-like manifold where local and global bounds differ. Heat kernel estimates, and in particular this type of different local and global behavior, also played a prominent role in the lecture notes by Martin Barlow, where Gaussian bounds were shown to be equivalent to a parabolic Harnack inequality on the one hand, and doubling and Poincaré on the other. These inequalities (or their failure) were discussed in detail for the Sierpinski carpet.

We hope that this volume succeeds in capturing some of the excitement of the lectures themselves, presenting the reader with a glimpse into this active area of research and its connections with other branches of contemporary mathematics. Just as two of the speakers, Robert McCann and Young-Heon Kim, recounted positive memories of past SMS summer schools they attended as students, it is hoped that the 2011 SMS will remain memorable both for its participants and the readers of this volume, some of whom may even emerge as lecturers and authors in a future edition of the SMS.

Galia Dafni
Robert J. McCann
Alina Stancu

List of Speakers

Luigi Ambrosio (Scuola Normale Superiore di Pisa)
Calculus in metric measure spaces with Ricci curvature bounded from below

Martin T. Barlow (University of British Columbia)
Heat equation on some fractal metric spaces

Thierry Coulhon (Université de Cergy-Pontoise)
Heat kernel estimates, Sobolev type inequalities and Riesz transform on non-compact Riemannian manifolds

Guy David (Université Paris-Sud)
Regularity results for minimal sets

Piotr Hajłasz (University of Pittsburgh)
Sobolev mappings into metric spaces

Vitali Kapovitch (University of Toronto)
Fundamental groups of manifolds with lower Ricci curvature bounds

Young-Heon Kim (University of British Columbia)
Ma–Trudinger–Wang curvature and regularity of optimal transport

Pekka Koskela (University of Jyväskylä)
Quasiconformal mappings and function spaces

Robert J. McCann (University of Toronto)
Optimal transportation

Emanuel Milman (Technion—Israel Institute of Technology)
Isoperimetric, functional and concentration inequalities

Yann Ollivier (CNRS, Université Paris-Sud)
Discrete Ricci curvature with applications

Felix Otto (Max-Planck-Institut für Mathematik in den Naturwissenschaften)
Burger's equation as a gradient flow on two-phase Wasserstein space

Karl-Theodor Sturm (Universität Bonn)
Ricci bounds for metric measure spaces and geometric analysis

Photo: Suzette Paradis

List of Participants

Tayeb Aïssiou
Concordia University
Canada

Suliman Albandik
Georg-August-Universität Göttingen
Germany

Luigi Ambrosio
Scuola Normale Superiore di Pisa
Italy

Sebastien Andres
University of British Columbia
Canada

Alex Austin
University of Illinois at Chicago
USA

Martin T. Barlow
University of British Columbia
Canada

Jorge Basilio
The Graduate Center, CUNY
USA

Mohammad Bavarian
University of British Columbia
Canada

Ryan Benty
Concordia University
Canada

Raquel Cabral
Concordia University
Canada

Li Chen
Université Paris-Sud
France

Shibing Chen
University of Toronto
Canada

Jong Keun Choi
Yonsei University
South Korea

Thierry Coulhon
Université de Cergy-Pontoise
France

Galia Dafni
Concordia University
Canada

Guy David
Université Paris-Sud
France

Qintao Deng
Huazhong Normal University
China

Jian Duan
University of Arkansas
USA

Tullia Dymarz
Université Paris-Sud
France

Nathaniel Eldredge
Cornell University
USA

Andrew Fenwick
Concordia University
Canada

Richard Fournier
Université de Montréal
Canada

Paul M. Gauthier
Université de Montréal
Canada

Saman Gharib
University of British Columbia
Canada

Nicola Gigli
Université de Nice - Sophia Antipolis
France

Dorian Goldman
Université Pierre et Marie Curie
France

Jasun Gong
University of Pittsburgh
USA

Heather Griffin
University of Arkansas
USA

Metin Alpen Gür
Indiana University Bloomington
USA

Piotr Hajłasz
University of Pittsburgh
USA

John Harvey
University of Notre-Dame
USA

Erwan Hillion
Université de Neuchâtel
Switzerland

Xueping Huang
Bielefeld University
Germany

François Huard
Bishop's University
Canada

Lionel Jassionnesse
Université de Bourgogne
France

Chanyoung Jun
University of Illinois at
Urbana-Champaign
USA

Ehsan Kamalinejad
University of Toronto
Canada

Vitali Kapovitch
University of Toronto
Canada

Young-Heon Kim
University of British Columbia
Canada

Kyle Kinnenberg
University of California, Los Angeles
USA

Jun Kitagawa
Princeton University
USA

Slawomir Kolasinski
University of Warsaw
Poland

Jonathan Korman
University of Toronto
Canada

Pekka Koskela
University of Jyväskylä
Finland

Artem Kozhevnikov
École Normale Supérieure
France

Eric Larson
Washington State University
USA

Hanjin Lee
McGill University
Canada

Woon Yin (Paul) Lee
University of California, Berkeley
USA

Stephen Lewis
University of Washington
USA

Xiangyu Liang
Université Paris-Sud
France

Janna Lierl
Cornell University
USA

Shiping Liu
Max-Planck-Institut für Mathematik in den Naturwissenschaften
Germany

Wolfgang Loehr
Universität Duisburg-Essen
Germany

Charly Makitu Kivuvu
University of South Africa
South Africa

Robert J. McCann
University of Toronto
Canada

Emanuel Milman
Technion—Israel Institute of Technology
Israel

Frank Morgan
Williams College
USA

Michael Munn
New York City College of Technology, CUNY
USA

Mohammad Najafi Ivaki
Concordia University
Canada

Vincent Nolot
Université de Bourgogne
France

Yann Ollivier
Université Paris-Sud
France

Felix Otto
Max-Planck-Institut für Mathematik in den Naturwissenschaften
Germany

Melanie Pivarski
Roosevelt University, Chicago
USA

Mohammad Rostami
University of Waterloo
Canada

Boutayeb Salahaddine
Université de Cergy-Pontoise
France

Christian Seis
Max-Planck-Institut für Mathematik in den Naturwissenschaften
Germany

Shichang Song
University of Illinois at Urbana-Champaign
USA

Alina Stancu
Concordia University
Canada

Benjamin Steinhurst
Cornell University
USA

Karl-Theodor Sturm
Universität Bonn
Germany

Daniele Valtorta
Università degli Studi di Milano
Italy

Eduardo Santillan Zeron
Cinvestav-IPN
Mexico

Xiangwen Zhang
McGill University
Canada

Yongsheng Zhang
Stony Brook University
USA

Centre de Recherches Mathématiques
CRM Proceedings and Lecture Notes
Volume **56**, 2013

An Overview on Calculus and Heat Flow in Metric Measure Spaces and Spaces with Riemannian Curvature Bounded from Below

Luigi Ambrosio

ABSTRACT. In this paper we provide a short overview of the recent work in collaboration with N. Gigli and G. Savaré on the development of a differential calculus in metric measure spaces based on optimal transportation tools. This leads to a new approach to the theory of Sobolev spaces on metric measure spaces. In the second part we provide an application of these ideas, still obtained in collaboration with N. Gigli and G. Savaré, on the study of a synthetic notion of "Riemannian Ricci bound from below" which enjoys several nice properties including stability with respect to measured Gromov – Hausdorff convergence.

1. Introduction and overview

In these notes I will illustrate the main results contained in the recent two papers [4, 5], written in collaboration with N. Gigli and G. Savaré. I will basically present results and proofs in some detail for [4], and present only the main results of [5]. These notes follow to a large extent the presentation given in Montréal, in July 2011.

1.1. Some by now "classical" results. Let us consider in $\mathbb{R}^n$ the heat equation (with the notation $u_t(x) = u(t,x)$)

$$\frac{\mathrm{d}}{\mathrm{d}t}u_t = \Delta u_t.$$

Classically, this PDE can be viewed as the gradient flow of the energy

$$\mathrm{Dir}(u) := \frac{1}{2}\int_{\mathbb{R}^n} |\nabla u|^2 \,\mathrm{d}x \quad (+\infty \text{ if } u \notin H^1(\mathbb{R}^n))$$

in the Hilbert space $H = L^2(\mathbb{R}^n)$. Indeed, formally $t \mapsto u_t$ solves the ODE $u' = -\nabla \mathrm{Dir}(u)$ in H because

$$\mathrm{Dir} \text{ "differentiable" at } u \iff -\Delta u \in L^2,\ \nabla \mathrm{Dir}(u) = -\Delta u$$

The precise meaning of "differentiability" is provided by convex analysis and it will be specified later on in rigorous terms.

2010 *Mathematics Subject Classification.* 49J52, 49Q20, 58J35, 35K90, 31C25.
Work written within the research project ERC ADG GeMeThNES..
This is the final form of the paper.

In 1998, Jordan, Kinderlehrer and Otto proved [16] that the same equation arises as gradient flow of the *entropy* functional

$$\mathrm{Ent}(\rho\mathscr{L}^n) := \int_{\mathbb{R}^n} \rho \log \rho \, \mathrm{d}x \quad (+\infty \text{ if } \mu \text{ is not a.c. w.r.t. } \mathscr{L}^n)$$

in the space $\mathscr{P}_2(\mathbb{R}^n)$ of Borel probability measures in $\mathbb{R}^n$ with finite quadratic moments, with respect to Wasserstein distance W_2 (I am denoting here by $\mathscr{L}^n$ the Lebesgue measure in $\mathbb{R}^n$). Recall that W_2^2 is defined by the minimum transportation cost, in the Kantorovich formulation, using $c(x,y) = d^2(x,y)$ as cost function, namely

$$W_2^2(\mu,\nu) := \min\left\{\int_{\mathbb{R}^n\times\mathbb{R}^n} |x-y|^2 \, \mathrm{d}\gamma(x,y) : (\pi_1)_\sharp\gamma = \mu, (\pi_2)_\sharp\gamma = \nu\right\}.$$

Here and in the sequel I will adopt the standard push forward notation: any $f\colon X \to Y$ Borel induces a map $f_\sharp\colon \mathscr{P}(X) \to \mathscr{P}(Y)$ given by

$$f_\sharp\mu(B) := \mu\big(f^{-1}(B)\big) \quad \forall B \in \mathscr{B}(Y).$$

The proof of this equivalence, and the reasons for it, can be found at many levels:

(1) By the so-called Otto calculus [22], i.e., formally viewing $\mathscr{P}(X)$ as an infinite-dimensional Riemannian manifold and computing with this structure the gradient flow of Ent.

(2) Proving that the implicit time discretization scheme (the so-called Euler scheme), traditionally used for the time discrete approximation of gradient flows, when done with energy $E = \mathrm{Ent}$ and distance $d = W_2$, does converge to the heat equation. Recall that this scheme involves a time step τ and the recursive minimization of

$$y \mapsto E(y) + \frac{1}{2\tau} d^2(y,x)$$

to provide a discrete (piecewise constant in time) solution to the gradient flow. This is the strategy pursued in [16].

(3) Giving a rigorous meaning to what "gradient flow of Ent in $\mathscr{P}(X)$ w.r.t. W_2" means, and checking that solutions of this gradient flow are solutions to the heat equation. Then, having proved that W_2 gradient flows are contained in L^2 gradient flows, it suffices to apply standard uniqueness results for $(\mathrm{d}/\mathrm{d}t)u_t = \Delta u_t$.

The last strategy, used in [3], is more abstract, but still uses to some extent the differentiable structure of $\mathbf{R}^n$. The question is: are there deeper reasons for this equivalence? This is motivated also by a long series of papers where the JKO result is extended to more general classes of metric spaces: Riemannian manifolds [11], Finsler spaces [20], Wiener spaces [12] (a class of infinite-dimensional Gaussian spaces), Alexandrov spaces [14], etc.

1.2. Metric measure spaces. We consider a *metric measure space* $(X, d, \mathfrak{m})$, with $\mathfrak{m} \in \mathscr{P}(X)$. In this framework it is still possible to define a "Dirichlet energy," that we call Cheeger functional:

$$\mathrm{Ch}(f) := \frac{1}{2}\inf\left\{\liminf_{n\to\infty}\int_X |\nabla f_n|^2 \, \mathrm{d}\mathfrak{m} : f_n \in \mathrm{Lip}(X), \int_X |f_n - f|^2 \, \mathrm{d}\mathfrak{m} \to 0\right\},$$

where

$$|\nabla g|(x) := \limsup_{y\to x} \frac{|g(y)-g(x)|}{d(y,x)}$$

is the *slope* (also called local Lipschitz constant). Our terminology is motivated by Cheeger's seminal paper [8], where a similar relaxation procedure is considered. Cheeger considered arbitrary functions f_n in the approximation procedure, and upper gradients g_n of them in place of the slopes; for this reason our functional is a priori larger than the original functional in [8]. A nontrivial fact, a consequence of the identification theorem of weak gradients (see (8.1)), is that the two functionals coincide.

Also, one can consider the so-called *relative entropy functional* $\mathrm{Ent}_{\mathfrak{m}}: \mathscr{P}(X) \to [0,\infty]$

$$\mathrm{Ent}_{\mathfrak{m}}(\rho\mathfrak{m}) := \int_X \rho \log \rho \, \mathrm{d}\mathfrak{m} \quad (+\infty \text{ if } \mu \text{ is not a.c. w.r.t. } \mathfrak{m}).$$

The basic result is that the equivalence between L^2-gradient flow of Ch and W_2-gradient flow of $\mathrm{Ent}_{\mathfrak{m}}$ *always* holds, if the latter is properly understood. But, without additional assumptions on the space, both objects can be trivial, as the following simple example shows.

Example 1.1 (Triviality of Ch). Let $X = [0,1]$, d the Euclidean distance, $\mathfrak{m} = \sum_{n\geq 1} 2^{-n}\delta_{q_n}$, where $\{q_n\}_{n\geq 1}$ is an enumeration of $[0,1]\cap\mathbb{Q}$. Let $A_n \supset \mathbb{Q}\cap X$ be open sets with $\mathscr{L}^1(A_n) \to 0$ and

$$\chi_n(t) := \int_0^t \left(1 - \chi_{A_n}(s)\right) \mathrm{d}s \quad t \in [0,1].$$

Then $f\circ\chi_n \to f$ in $L^2(X,\mathfrak{m})$ for all $f \in \mathrm{Lip}(X)$ and $f\circ\chi_n$ is locally constant in $\mathbb{Q}\cap X$ hence

$$\mathrm{Ch}(f) = 0 \quad \forall f \in \mathrm{Lip}(X).$$

It follows that $\mathrm{Ch} \equiv 0$ in $L^2(X,\mathfrak{m})$.

1.3. Identification of weak gradients. A closely related question, relevant in particular for the second paper, is the identification of weak gradients. The first one, that we call *relaxed* gradient $|\nabla f|_*$, is the object that provides integral representation to Ch:

$$\mathrm{Ch}(f) = \frac{1}{2}\int_X |\nabla f|_*^2 \, \mathrm{d}\mathfrak{m} \quad \forall f \in D(\mathrm{Ch}).$$

It has all the natural properties (locality, chain rules, etc.) a weak gradient should have, see Theorem 3.2 and (3.1) below. This gradient is useful when doing "vertical" variations $\epsilon \mapsto f + \epsilon g$ (i.e., in the *dependent* variable). On the other hand, when computing variations of the entropy, the "horizontal" variations $\epsilon \to f(\gamma_\epsilon)$ (i.e., in the *independent* variable) are necessary. These are related to another weak gradient $|\nabla f|_w$, defined as follows.

We consider the so-called *weak* upper gradient property by requiring

$$|f(\gamma_1) - f(\gamma_0)| \leq \int_\gamma G$$

along "almost all" curves γ in $AC^2([0,1];X)$. Then, we define $|\nabla f|_w$ as the weak upper gradient G with smallest $L^2(X,\mathfrak{m})$ norm. This definition crucially depends on the notion of null set of curves, that we shall specify later on.

The remarkable fact is that these two gradients *always* coincide (and, of course, maybe both trivial without extra assumptions). The proof of this identification uses ideas from optimal transportation, such as lifting of solutions to the heat flow to probability measures in $AC^2([0,1];X)$ and the energy dissipation rate of $\mathrm{Ent}_{\mathfrak{m}}$ along the L^2 gradient flow of Ch. I think that this identification result, which a priori has nothing to do with entropy and optimal transportation, is a nice illustration of the power of the optimal transport theory.

1.4. Why gradients are not trivial in Lott – Sturm – Villani spaces. In these spaces one imposes convexity along W_2 geodesics of $\mathrm{Ent}_{\mathfrak{m}}$ (the so-called $\mathrm{CD}(0,\infty)$ condition) or of functionals

$$\rho\mathfrak{m} \mapsto -\int_X \rho^{1-1/N}\,\mathrm{d}\mathfrak{m}$$

(the $\mathrm{CD}(0,N)$ condition). In this case the gradient flow of $\mathrm{Ent}_{\mathfrak{m}}$ is not trivial, and since it coincides with the L^2 gradient flow of Ch, also the latter is not trivial. As a consequence, pathological situations as those described in Example 1.1 cannot occur.

Notice that, formally, the energy dissipation rate is

$$\frac{\mathrm{d}}{\mathrm{d}t}\int_X \rho_t\log\rho_t\,\mathrm{d}\mathfrak{m} = \int_X \log\rho_t\Delta\rho_t\,\mathrm{d}\mathfrak{m} = -\int_{\{\rho_t>0\}}\frac{|\nabla\rho_t|^2}{\rho_t}\,\mathrm{d}\mathfrak{m} = -4\int_X |\nabla\sqrt{\rho_t}|^2\,\mathrm{d}\mathfrak{m}.$$

We shall develop a calculus that makes this result meaningful and rigorous.

The standing assumption on the metric measure structure in these notes are that (X,d) is a compact metric space and $\mathfrak{m}\in\mathscr{P}(X)$. Notice that the results in [4] require neither global or local compactness assumptions on (X,d) nor finiteness of $\mathfrak{m}$, and are therefore appropriate for dealing with infinite-dimensional spaces. Those of [5], instead, have been established for the moment under the additional assumption that $\mathfrak{m}\in\mathscr{P}(X)$. Good prerequisites needed for the reading of these notes are the basic facts of optimal transport theory, see [2, 3, 26].

2. Hopf – Lax formula and Hamilton – Jacobi semigroup

Given a function $f\colon X\to\mathbb{R}$ bounded from below, we define

$$Q_tf(x) := \inf_{y\in X}\left(f(y)+\frac{1}{2t}d^2(x,y)\right) \qquad \text{(Hopf – Lax formula)}$$

Theorem 2.1 (Properties of Q_tf). *Assume that f is bounded and lower semicontinuous. It holds:*

(1) $Q_tf(x)\uparrow f(x)$ *as* $t\downarrow 0$;
(2) $Q_t(Q_sf(x))\geq Q_{t+s}f(x)$, *with equality if* (X,d) *is geodesic*;
(3) $(\mathrm{d}^+/\,\mathrm{d}t)Q_tf(x)+\frac{1}{2}|\nabla Q_tf(x)|^2\leq 0$;
(4) $Q_tf(x)$ *restricted to* $(\epsilon,\infty)\times X$ *is Lipschitz for all* $\epsilon>0$.

Sketch of proof. (1) It follows by the lower semicontinuity of f, which ensures also that minimizers do exist.

(2) It follows by

$$\begin{aligned}(2.1)\quad \inf_y\left(\inf_z f(z)+\frac{1}{2s}d^2(z,y)\right)+\frac{1}{2t}d^2(x,y)\\ &= \inf_z\inf_y\left(\frac{1}{2s}d^2(z,y)+\frac{1}{2t}d^2(x,y)\right)+f(z)\\ &\geq \inf_z\frac{1}{2(s+t)}d^2(x,z)+f(z),\end{aligned}$$

noticing that the last inequality is an equality in geodesic spaces.

In order prove (3), we set

$$\begin{cases} D_f^+(x,t) := \max\{d(x,y) : y \text{ minimizer}\}, \\ D_f^-(x,t) := \min\{d(x,y) : y \text{ minimizer}\}. \end{cases}$$

Since any limit of minimizers is a minimizer, D_f^+ is upper semicontinuous, while D_f^- is lower semicontinuous.

In addition $D_f^+(x,s) \geq D_f^-(x,s) \geq D_f^+(x,t)$ if $0 < t < s$. Indeed, if x_t minimizes for $Q_t f(x)$ and x_s minimizes for $Q_s f(x)$, it holds

$$f(x_t) + \frac{d^2(x_t,x)}{2t} \leq f(x_s) + \frac{d^2(x_s,x)}{2t},$$
$$f(x_s) + \frac{d^2(x_s,x)}{2s} \leq f(x_t) + \frac{d^2(x_t,x)}{2s}.$$

Adding up and using $1/t > 1/s$ we deduce that $D_f^-(x,s) \geq D_f^+(x,t)$.

It follows that, given x, $D_f^+(x,t) = D_f^-(x,t)$ with at most countably many exceptions. We prove first that $(\mathrm{d}^{\pm}/\,\mathrm{d}t)Q_t f(x) = -[D_f^{\pm}(x,t)]^2/(2t^2)$. Choosing x_t at maximum distance and x_s at minimum distance yields

$$\begin{aligned} Q_s f(x) - Q_t f(x) &\leq \frac{1}{2s} d^2(x,x_t) + f(x_t) - f(x_t) - \frac{1}{2t} d^2(x,x_t) \\ &= \frac{\left(D_f^+(x,t)\right)^2}{2}\left(\frac{1}{s} - \frac{1}{t}\right) \\ Q_s f(x) - Q_t f(x) &\geq \frac{1}{2s} d^2(x,x_s) + f(x_s) - f(x_s) - \frac{1}{2t} d^2(x,x_s) \\ &= \frac{\left(D_f^-(x,s)\right)^2}{2}\left(\frac{1}{s} - \frac{1}{t}\right). \end{aligned}$$

One can then use semicontinuity of $D_f^{\pm}$ and monotonicity $D_f^+ \geq D_f^-$ to obtain the formula for the right and left time derivatives of $Q_t f$.

To conclude the proof of (3), it suffices to show that $|\nabla Q_t f|(x) \leq D_f^+(x,t)/t$. The same trick used before, now for variations in space, yields:

$$\begin{aligned} Q_t f(x) - Q_t f(y) &\leq \frac{1}{2t} d^2(x,z) + f(z) - f(z) - \frac{1}{2t} d^2(z,y) \\ &\leq d(x,y)\left(\frac{D_f^-(y,t)}{t} + \frac{d(x,y)}{2t}\right) \end{aligned}$$

and we can use the upper semicontinuity of D_f^+ to conclude. □

If y is kept fixed and we let $x \to y$ we obtain the sharper inequality

$$|\nabla^+ Q_t f|(y) \leq \frac{D_f^-(y,t)}{t}, \tag{2.2}$$

where the *ascending slope* $|\nabla^+ f|$ is defined by

$$|\nabla^+ f|(y) := \limsup_{x \to y} \frac{[f(x) - f(y)]^+}{d(x,y)}.$$

2.1. Hamilton – Jacobi and optimal transportation. Why are Hopf – Lax formula and the Hamilton – Jacobi equation relevant in the theory of optimal transport?

c-transform. Given a cost function $c\colon X\times Y\to\mathbb{R}$, the c-transforms $\varphi^c\colon Y\to\mathbb{R}\cup\{-\infty\}$, $\psi^c\colon X\to\mathbb{R}\cup\{-\infty\}$ are defined by

$$\varphi^c(y):=\inf_{x\in X}c(x,y)-\varphi(x),\qquad \psi^c(x):=\inf_{y\in Y}c(x,y)-\psi(y).$$

Notice the analogy with convex analysis: $\psi^c=(-\psi)^*$ if X is Hilbert and $c(x,y)=\langle x,y\rangle$. The relation with the HL formula is also obvious:

$$\psi^c=Q_1(-\psi).$$

Then, we say that $\varphi\colon X\to\mathbb{R}\cup\{-\infty\}$ is *c-concave* if $\varphi=\psi^c$ for some $\psi\colon Y\to\mathbb{R}\cup\{-\infty\}$. As in convex analysis, $\varphi\mapsto\varphi^c$ is an involution in the class of c-concave functions: $(\varphi^c)^c=\varphi$.

Definition 2.2. We say that a c-concave function $\varphi\colon X\to\mathbb{R}\cup\{-\infty\}$ is a Kantorovich potential relative to (μ,ν) if it satisfies

$$\varphi(x)+\varphi^c(y)=c(x,y)\quad\text{for }\gamma\text{-a.e. }(x,y) \tag{2.3}$$

for any optimal plan γ from μ to ν.

Proposition 2.3. *If φ is a Kantorovich potential from μ to ν it holds:*

$$|\nabla^+\varphi|(x)\le d(x,y)\quad\textit{for }\gamma\textit{-a.e. }(x,y)$$

for any optimal plan γ from μ to ν. In particular $\int|\nabla^+\varphi|^2\,\mathrm{d}\mu\le W_2^2(\mu,\nu)$.

PROOF. Since $\varphi=(\varphi^c)^c$ we may write $\varphi=Q_1(-\varphi^c)$. Combining this with the optimality condition (2.3) and with (2.2) gives

$$|\nabla^+\varphi(x)|=|\nabla^+Q_1(-\varphi^c)|(x)\le D^-_{-\varphi^c}(x,1)\le d(x,y)\quad\gamma\text{-a.e. in }X\times X.\qquad\square$$

2.2. The classical Brenier theorem and its metric counterpart. In the Euclidean case $c(x,y)=|x-y|^2/2$, if φ is differentiable at x and

$$\varphi(x)+\varphi^c(y)=\tfrac12|x-y|^2,$$

one can differentiate at x and obtain $\nabla\varphi(x)=(x-y)$, which tells us that y is uniquely determined by x and

$$|\nabla^+\varphi|(x)=d(x,y). \tag{2.4}$$

We say that a *metric* Brenier theorem holds if (2.4) holds γ-a.e. for any optimal plan γ, so that in particular $W_2^2(\mu,\nu)=\int|\nabla^+\varphi|^2\,\mathrm{d}\mu$. The following simple example shows that this equality may fail in general; see Theorem 9.14 for a positive result.

Example 2.4. $X=[0,1]$, $\mu=\delta_0$, $\mu_t=t^{-1}\chi_{[0,t]}\mathscr{L}^1$. In this case

$$\varphi(x)=\frac{x^2}{2}-x,\qquad\int|\nabla^+\varphi|^2\,\mathrm{d}\mu_0=0$$

while $W_2^2(\mu_0,\mu_1)=\frac13$.

2.3. Optimal transport and Kantorovich potentials in geodesic spaces. If (X, d) is Polish (i.e., complete and separable) and geodesic we may formulate the optimal transport problem in terms of *geodesic plans*, namely probability measures $\boldsymbol{\pi}$ concentrated in the Polish space $\mathrm{Geo}(X)$ of constant speed geodesics:

$$\min\left\{\int d^2(\gamma_0, \gamma_1)\,\mathrm{d}\boldsymbol{\pi}(\gamma) : (e_0)_\sharp\boldsymbol{\pi} = \mu, (e_1)_\sharp\boldsymbol{\pi} = \nu\right\}. \tag{2.5}$$

Here $e_t\colon C([0,1]; X) \to X$ are the evaluation maps, namely $e_t(\gamma) = \gamma_t$. The relation with the classical optimal plans $\boldsymbol{\gamma}$ of Kantorovich theory is that if $\boldsymbol{\pi}$ is a minimizer in (2.5), then $(e_0, e_1)_\sharp\boldsymbol{\pi}$ is an optimal plan, and that any optimal $\boldsymbol{\gamma}$ admits a (possibly nonunique) "lifting" $\boldsymbol{\pi}$, i.e., $(e_0, e_1)_\sharp\boldsymbol{\pi} = \boldsymbol{\gamma}$. The nice fact is that constant speed geodesics are in 1-1 correspondence with optimal geodesic plans:

Theorem 2.5. *Any constant speed geodesic μ_t in $\mathscr{P}(X)$ can be represented as $(e_t)_\sharp\boldsymbol{\pi}$ for a suitable optimal geodesic plan $\boldsymbol{\pi}$. Conversely, any optimal geodesic plan $\boldsymbol{\pi}$ induces a constant speed geodesic $(e_t)_\sharp\boldsymbol{\pi}$.*

The following fundamental result provides a deeper connection between geodesics and the Hopf–Lax formula; see the seminal paper [7] and [26] for much more on this subject.

Theorem 2.6. *Let μ_t, $t \in [0,1]$ be a constant speed geodesic and let φ be a Kantorovich potential relative to μ_0, μ_1. Then, for all $t \in (0,1]$, $\varphi_t := Q_t(-\varphi^c)$ is a Kantorovich potential, relative to the scaled cost $c_t := c/t$, from μ_{1-t} to μ_1.*

Sketch of proof. It is obvious that $\varphi_t + \varphi^c \le c_t$. The key implication is

$$\varphi(\gamma_0) + \varphi^c(\gamma_1) = c(\gamma_0, \gamma_1) \quad \text{implies} \quad \varphi_t(\gamma_{1-t}) + \varphi^c(\gamma_1) = c_t(\gamma_{1-t}, \gamma_1). \tag{2.6}$$

Hence, if $\boldsymbol{\pi}$ is an optimal geodesic plan, $\varphi + \varphi^c = c$ $\boldsymbol{\pi}$-a.e. implies $\varphi_t + \varphi^c = c_t$ $\boldsymbol{\pi}_t$-a.e., where

$$\boldsymbol{\pi}_t := (\gamma_{1-t}, \gamma_1)_\sharp\boldsymbol{\pi}$$

is an optimal geodesic plan from μ_{1-t} to μ_1. The implication (2.6) is not difficult to prove, and related to the fact that characteristic lines for the Hamilton–Jacobi equation are geodesics; see also (2.1). □

3. Cheeger's energy and relaxed gradients

Let us recall the definition of Ch we already mentioned in the introduction:

$$\mathrm{Ch}(f) := \frac{1}{2}\inf\left\{\liminf_{h\to\infty}\int_X |\nabla f_h|^2\,\mathrm{d}\mathfrak{m} : f_h \in \mathrm{Lip}(X), \int_X |f_h - f|^2\,\mathrm{d}\mathfrak{m} \to 0\right\}.$$

By construction $\mathrm{Ch}\colon L^2(X, \mathfrak{m}) \to [0, \infty]$ is lower semicontinuous, and it is easily seen to be convex. Can we provide an integral representation to it?

Relaxed slope. $G \in L^2(X, \mathfrak{m})$ is a relaxed slope of f if G bounds from above a function in

$$\{\text{weak } L^2 \text{ limit points of } |\nabla f_n|,\ f_n \in \mathrm{Lip}(X),\ \|f_n - f\|_2 \to 0\}$$

or equivalently in

$$\{\textit{strong } L^2 \text{ limit points of } G_n \ge |\nabla f_n|,\ f_n \in \mathrm{Lip}(X),\ \|f_n - f\|_2 \to 0\}.$$

The equivalence between the two characterizations of relaxed slopes follows by Mazur's lemma: the first characterization is useful to show that f has a relaxed slope iff $\mathrm{Ch}(f) < \infty$, while the second one is useful to perform diagonal arguments and to show that the collection of relaxed slopes is a convex closed set, possibly empty. This motivates the next definition.

Definition 3.1 (Minimal relaxed slope). We call *minimal relaxed slope*, and denote by $|\nabla f|_*$, the function with smallest $L^2(X, \mathfrak{m})$ norm among relaxed slopes.

Theorem 3.2. *Let $f \in D(\mathrm{Ch})$. Then:*
(1) $\mathrm{Ch}(f) = \frac{1}{2}\int |\nabla f|_*^2 \,\mathrm{d}\mathfrak{m}$;
(2) *if G_1, G_2 are relaxed slopes, so is* $\min\{G_1, G_2\}$;
(3) $|\nabla f|_* \leq G$ *$\mathfrak{m}$-a.e. for any relaxed slope G;*
(4) *$g = f$ $\mathfrak{m}$-a.e. on a Borel set B implies $|\nabla f|_* = |\nabla g|_*$ $\mathfrak{m}$-a.e. on B.*

Calculus rules. If $N \subset \mathbb{R}$ is Lebesgue negligible, then $|\nabla f|_* = 0$ a.e. in $f^{-1}(N)$. In addition, we have the (weak) chain rule

$$|\nabla \phi(f)|_* \leq |\phi'(f)|\,|\nabla f|_* \quad \text{with equality if } \phi' \geq 0. \tag{3.1}$$

SKETCH OF PROOF. (1) Any weak limit point of $|\nabla f_n|$ yields a relaxed slope, hence $\mathrm{Ch}(f) \geq \frac{1}{2}\int |\nabla f|_*^2 \,\mathrm{d}\mathfrak{m}$. By definition of $|\nabla f|_*$ there exists $G \leq |\nabla f|_*$ which is the weak limit of $|\nabla g_n|$ with $g_n \to f$ in $L^2(X, \mathfrak{m})$. Now, using Mazur's lemma, we can find convex combinations f_n on g_n such that the corresponding convex combinations G_n of the slopes converge strongly in $L^2(X, \mathfrak{m})$ to G. Since $|\nabla f_n| \leq G_n$ it follows that

$$\int |\nabla f|_*^2 \,\mathrm{d}\mathfrak{m} \geq \int G^2 \,\mathrm{d}\mathfrak{m} \geq \liminf_n \int |\nabla f_n|^2 \,\mathrm{d}\mathfrak{m} \geq 2\,\mathrm{Ch}(f).$$

(2) By approximation, it suffices to show that $\chi_{X\setminus B} G_1 + \chi_B G_2$ is a relaxed slope if B is closed. Set $\rho(x) = \mathrm{dist}(x, B)$, $\chi_r(x) = \min\{1, r^{-1}\rho\}$, so that $\chi_r \uparrow \chi_{X\setminus B}$ as $r \downarrow 0$, and pass to the limit in

$$|\nabla(\chi_r f_{n,1} + (1-\chi_r) f_{n,2})| \leq \chi_r |\nabla f_{n,1}| + (1-\chi_r)|\nabla f_{n,2}| + \mathrm{Lip}(\chi_r)|f_{n,1} - f_{n,2}|.$$

(3) Just take $\widetilde{G} := \min\{|\nabla f|_*, G\}$. Its L^2 norm is strictly smaller than $\||\nabla f|_*\|_2$ if the set $\{|\nabla f|_* > G\}$ has positive $\mathfrak{m}$-measure. □

3.1. Heat flow and Laplacian. Let's start with some reminders on the classical theory of gradient flows of convex and l.s.c. functionals $F\colon H \to \mathbb{R} \cup \{+\infty\}$ in a Hilbert space H.

Subdifferential ∂F. It is the multivalued map defined by

$$\partial F(x) := \{p \in H : F(x) + \langle p, y - x\rangle \leq F(y)\ \forall y \in H\}$$

for all $x \in D(F) := \{F < \infty\}$. The set $\partial F(x)$ is closed and convex. The *gradient* $\nabla F(x)$ is the element with minimal norm in $\partial F(x)$.

Definition 3.3 (Gradient flow). It is a locally absolutely continuous map $x\colon (0, \infty) \to H$ satisfying

$$-x'(t) \in \partial F\big(x(t)\big) \qquad \text{for a.e. } t > 0.$$

In addition, we say that $x(t)$ starts from $\bar{x}$ if $\lim_{t\downarrow 0} x(t) = \bar{x}$.

Theorem 3.4 (Existence and uniqueness). *For all $\bar{x} \in \overline{D(F)}$ there exists a unique gradient flow starting from $\bar{x}$ and the induced semigroup*

$$S_t\colon [0,\infty) \times \overline{D(F)} \to \overline{D(F)}$$

is contractive. In addition, we have the regularizing effects:

(1) $S_t\bar{x} \in D(\partial F) \subset D(F)$ *for all* $t > 0$ *and*

$$F(S_t\bar{x}) \le \inf_{v\in D(F)} \left(F(v) + \frac{1}{2t} d^2(v,\bar{x}) \right);$$

(2) $(\mathrm{d}^+/\,\mathrm{d}t) S_t\bar{x} = -\nabla F(S_t\bar{x})$ *for all* $t > 0$;

(3) $t \mapsto |\nabla F|^2(S_t\bar{x})$ *is nonincreasing, so that* $S_t\bar{x}$ *is Lipschitz in* (ϵ,∞) *for all* $\epsilon > 0$;

(4) $(\mathrm{d}^+/\,\mathrm{d}t) F(S_t\bar{x}) = -|\nabla F|^2(S_t\bar{x}) = -|(\mathrm{d}^+/\,\mathrm{d}t) S_t\bar{x}|^2$ *for all* $t > 0$.

According to these results, we may choose $H = L^2(X,\mathfrak{m})$ and $F = \mathrm{Ch}$ and define

$$-\Delta f := \text{the element with minimal } L^2\text{-norm of } \partial\,\mathrm{Ch}(f)$$

so that (by the density of $D(\mathrm{Ch}) \supset \mathrm{Lip}(X)$ in $L^2(X,\mathfrak{m})$) we obtain a L^2 *heat flow* $\mathbf{h}_t f$ solving (the derivative being understood in $L^2(X,\mathfrak{m})$)

$$\frac{\mathrm{d}}{\mathrm{d}t}\mathbf{h}_t\bar{f} = \Delta \mathbf{h}_t\bar{f}$$

starting from any initial condition $\bar{f} \in L^2(X,\mathfrak{m})$.

Remarks. (1) $\Delta = \Delta_{d,\mathfrak{m}}$. Even in the classical situations, say on a Riemannian manifold (M,g) endowed with the Riemannian distance and the volume measure, $\Delta f = \mathrm{div}(\nabla f)$, where ∇f depends on the metric g (to associate a vector ∇f to df) while div depends on the volume form $\mathfrak{m}$, via the adjoint formula

$$\int g\,\mathrm{div}\,F\,\mathrm{d}\mathfrak{m} = -\int \langle \nabla g, F\rangle\,\mathrm{d}\mathfrak{m}.$$

(2) Δ need not to be linear in this context! Take $X = \mathbb{R}^2$ with the L^∞ norm, to get

$$\mathrm{Ch}(f) = \frac{1}{2}\int \left(\left|\frac{\partial f}{\partial x}\right| + \left|\frac{\partial f}{\partial y}\right| \right)^2 \mathrm{d}x\,\mathrm{d}y.$$

Notwithstanding this potential lack of linearity, a reasonable calculus can be developed:

$$-\int g\Delta f\,\mathrm{d}\mathfrak{m} \le \int |\nabla f|_*|\nabla g|_*\,\mathrm{d}\mathfrak{m}, \tag{3.2}$$

$$-\int \phi(f)\Delta f\,\mathrm{d}\mathfrak{m} = \int \phi'(f)|\nabla f|_*^2\,\mathrm{d}\mathfrak{m}. \tag{3.3}$$

The inequality (3.2) follows by

$$\mathrm{Ch}(f) - \epsilon\int g\Delta f\,\mathrm{d}\mathfrak{m} \le \frac{1}{2}\int |\nabla(f+\epsilon g)|_*^2\,\mathrm{d}\mathfrak{m},$$

noticing that $|\nabla(f+\epsilon g)|_* \le |\nabla f|_* + \epsilon|\nabla g|_*$ for $\epsilon > 0$, so that

$$-\epsilon\int g\Delta f\,\mathrm{d}\mathfrak{m} \le \epsilon\int |\nabla f|_*|\nabla g|_*\,\mathrm{d}\mathfrak{m} + \frac{\epsilon^2}{2}\int |\nabla g|_*^2\,\mathrm{d}\mathfrak{m}.$$

The proof of (3.3) is based on the chain rule (3.1).

Proposition 3.5 (Properties of the heat flow). *The following properties hold:*

(1) (Homogeneity) $\mathbf{h}_t(\lambda f) = \lambda \mathbf{h}_t f$ $\forall \lambda \in \mathbb{R}$;
(2) (Comparison principle) *if* $f \le g$, *then* $\mathbf{h}_t f \le \mathbf{h}_t g$ *for all* $t \ge 0$;
(3) (Energy dissipation) *if* $J \subset \mathbb{R}$ *is an interval,* $f_t \colon X \to J$ *and* $\mathrm{e} \colon J \to \mathbb{R}$ *is convex and locally* $C^{1,1}$, *then*

$$\int \mathrm{e}(\mathbf{h}_t f)\,\mathrm{d}\mathfrak{m} = \int \mathrm{e}(f)\,\mathrm{d}\mathfrak{m} - \int_0^t \int \mathrm{e}''(\mathbf{h}_s f)|\nabla \mathbf{h}_s f|_*^2 \,\mathrm{d}\mathfrak{m}\,\mathrm{d}s.$$

(4) (Mass preservation) $\int \mathbf{h}_t f\,\mathrm{d}\mathfrak{m} = \int f\,\mathrm{d}\mathfrak{m}$ *for all* $t \ge 0$.

Strictly speaking, (3) does not cover the most interesting case, the case of the entropy $\mathrm{e}(z) = z \log z$ when $\mathbf{h}_t f \ge 0$ and $J = [0, \infty)$:

$$\int \mathbf{h}_t f \log \mathbf{h}_t f\,\mathrm{d}\mathfrak{m} = \int f \log f\,\mathrm{d}\mathfrak{m} - \int_0^t \int_{\{\mathbf{h}_s f > 0\}} \frac{|\nabla \mathbf{h}_s f|_*^2}{\mathbf{h}_s f}\,\mathrm{d}\mathfrak{m}\,\mathrm{d}s. \tag{3.4}$$

It can be recovered by the approximation $f \mapsto \max\{f, \epsilon\}$, $\epsilon \downarrow 0$ (this is possible thanks to the fact that $\mathfrak{m}(X) < \infty$); the general case is much more delicate, see [4]).

Sketch of proof. (1) Since Ch is 2-homogeneous, one can prove that $\partial\,\mathrm{Ch}(\lambda f) = \lambda \partial\,\mathrm{Ch}(f)$, hence f_t is a gradient flow iff λf_t is. Uniqueness then gives the identity.

(2) Since Δ is not linear, the standard argument (take $f_t - g_t$ and use that it is a gradient flow) does not apply. We appeal to the Euler scheme and prove that $f \le g$ and

$$f_\tau \text{ minimizer of } \frac{1}{2\tau}\|\cdot - f\|_2^2 + \mathrm{Ch}(\cdot), \qquad g_\tau \text{ minimizer of } \frac{1}{2\tau}\|\cdot - g\|_2^2 + \mathrm{Ch}(\cdot),$$

(here τ is the time step) implies $f_\tau \le g_\tau$.

(3) By monotone approximation, $\mathrm{e} \in C^{1,1}(\mathbb{R})$. Then

$$\frac{\mathrm{d}}{\mathrm{d}t}\int \mathrm{e}(\mathbf{h}_t f)\,\mathrm{d}\mathfrak{m} = \int \mathrm{e}'(\mathbf{h}_t f)\Delta \mathbf{h}_t f\,\mathrm{d}\mathfrak{m} = -\int \mathrm{e}''(\mathbf{h}_t f)|\nabla \mathbf{h}_t f|_*^2\,\mathrm{d}\mathfrak{m}.$$

(4) Suffices to use the function identically equal to 1 in the PDE. □

3.2. Absolutely continuous functions and metric speed. A curve $\gamma \colon [0,1] \to X$ is said to be absolutely continuous if

$$d(\gamma_t, \gamma_s) \le \int_t^s f(r)\,\mathrm{d}r \quad \forall [t,s] \subset [0,1] \tag{3.5}$$

for some $f \in L^1(0,1)$.

If γ is absolutely continuous, the *metric speed* (also called metric derivative) $|\dot\gamma| \colon [0,1] \to [0,\infty]$ is defined by

$$|\dot\gamma_t| := \lim_{h \to 0} \frac{d(\gamma_{t+h}, \gamma_t)}{|h|}.$$

It is possible to prove that the limit exists for a.e. t, that $|\dot\gamma| \in L^1(0,1)$, and that it is the minimal L^1 function for which the bound (3.5) holds.

3.3. Kuwada's lemma. This lemma, taken from [14], provides one of the two key connections between the "Eulerian" or "vertical" viewpoint implicit in the theory of relaxed gradients and the "Lagrangian" of "horizontal" viewpoint of the theory of optimal transportation.

Lemma 3.6. *Let $f_0 \in L^2(X, \mathfrak{m})$ be a probability density, $f_t = \mathbf{h}_t f_0$. Then the curve $\mu_t := f_t \mathfrak{m}$ is absolutely continuous in $\mathscr{P}(X)$ and*

$$|\dot{\mu}_t|^2 \leq \int_{\{f_t>0\}} \frac{|\nabla f_t|_*^2}{f_t} \, \mathrm{d}\mathfrak{m} \quad \textit{for a.e. } t > 0.$$

It is also convenient to introduce the Fisher information functional, defined on $\{\rho : \rho \geq 0, \sqrt{\rho} \in D(\mathrm{Ch})\}$, as follows:

$$F(\rho) := 4 \int |\nabla \sqrt{\rho}|_*^2 \, \mathrm{d}\mathfrak{m} = \int_{\{\rho>0\}} \frac{|\nabla \rho|_*^2}{\rho} \, \mathrm{d}\mathfrak{m}$$

(the last equality follows by chain rule).

Sketch of proof of Kuwada lemma. We prove an integral version of the lemma, namely

$$W_2^2(\mu_t, \mu_s) \leq \ell \int_t^s F(f_r) \, \mathrm{d}r$$

with $0 \leq s < t < \infty$ and $\ell := (s - t)$. By Kantorovich's duality formula, it suffices to show

$$\int -\varphi \, \mathrm{d}\mu_t + \int Q_1 \varphi \, \mathrm{d}\mu_s \leq \frac{\ell}{2} \int_t^s F(f_r) \, \mathrm{d}r,$$

where φ runs in the class of bounded continuous functions. Replacing φ by $Q_\epsilon \varphi$ and letting $\epsilon \downarrow 0$ we can assume that $Q_t \varphi$ is Lipschitz in $[0, 1] \times X$.

Now we set $g(r) := \int Q_r \varphi \, \mathrm{d}\mu_{t+\ell r}$, so that $\int \varphi \, \mathrm{d}\mu_t = g(0)$ and $\int Q_1 \varphi \, \mathrm{d}\mu_s = g(1)$, and we write the inequality as

$$\int_0^1 g'(r) \, \mathrm{d}r \leq \frac{\ell}{2} \int_t^s F(f_r) \, \mathrm{d}r.$$

Using the HJ subsolution property of $Q_r \varphi$ and the "integration by parts" we get

$$\begin{aligned} g'(r) &= \int (\frac{\mathrm{d}}{\mathrm{d}r} Q_r \varphi) f_{t+\ell r} \, \mathrm{d}\mathfrak{m} + \ell \int Q_r \varphi \Delta f_{t+\ell r} \, \mathrm{d}\mathfrak{m} \\ &\leq -\frac{1}{2} \int |\nabla Q_r \varphi|_*^2 f_{t+\ell r} \, \mathrm{d}\mathfrak{m} + \ell \int |\nabla Q_r \varphi|_* \sqrt{f_{t+\ell r}} \frac{|\nabla f_{t+\ell r}|_*}{\sqrt{f_{t+\ell r}}} \, \mathrm{d}\mathfrak{m}. \end{aligned}$$

Eventually the Young inequality gives

$$g'(r) \leq \frac{\ell^2}{2} F(f_{t+\ell r})$$

and an integration in $(0, 1)$ with respect to r gives the result. □

4. W_2-gradient flow of $\mathrm{Ent}_\mathfrak{m}$

Since the ambient space $\mathscr{P}(X)$ is not linear (at least if we take the viewpoint of optimal transportation), what do we mean by gradient flow?

Key idea (De Giorgi). Encode the *system* $x'(t) = -\nabla F(x(t))$ in a *single* differential inequality, by looking at the rate of energy dissipation:

$$\text{(DG)} \qquad \frac{\mathrm{d}}{\mathrm{d}t}F(x(t)) \leq -\frac{1}{2}|\nabla F|^2(x(t)) - \frac{1}{2}|x'(t)|^2.$$

Indeed, in a sufficiently smooth setting, along *any* curve $y(t)$, we have

$$\begin{aligned}\frac{\mathrm{d}}{\mathrm{d}t}F(y(t)) &= \langle \nabla F(y(t)), y'(t)\rangle \\ &\geq -|\nabla F(y(t))||y'(t)| \qquad (= \text{ iff } -y'(t) \text{ is parallel to } \nabla F(y(t))) \\ &\geq -\frac{1}{2}|\nabla F|^2(y(t)) - \frac{1}{2}|y'(t)|^2 \quad (= \text{ iff } |\nabla F|(y(t)) = |y'(t)|).\end{aligned}$$

All terms in (DG) make sense in a metric space (X, d): $|x'|$ can be replaced by the metric derivative and $|\nabla F|$ by the *descending slope* $|\nabla^- F|$, so that the speed is 0 at minimum points. By looking at integral versions of this optimal dissipation rate we can write down an energy dissipation inequality and an energy dissipation identity:

$$\text{(EDI)} \qquad F(x(t)) + \int_0^t \left(\frac{1}{2}|x'(r)|^2 + \frac{1}{2}|\nabla^- F|^2(x(r))\right) \mathrm{d}r \leq F(x(0)) \qquad \forall t \geq 0.$$

$$\text{(EDE)} \qquad F(x(t)) + \int_0^t \left(\frac{1}{2}|x'(r)|^2 + \frac{1}{2}|\nabla^- F|^2(x(r))\right) \mathrm{d}r = F(x(0)) \qquad \forall t \geq 0.$$

5. Properties of the slope of K-convex functions

The following lemma provides very useful properties of the descending slope of K-convex and l.s.c. functions.

Lemma 5.1. *If F is K-convex, l.s.c. in a geodesic metric space, we have the upper gradient property*

$$\text{(5.1)} \qquad F(y(0)) \leq F(y(t)) + \int_0^t |y'_r|\,|\nabla^- F|(y(r))\,\mathrm{d}r$$

along any absolutely continuous curve $y\colon [0,t] \to X$. As a consequence, (EDE) *and* (EDI) *are equivalent for F, $|x'(t)| = |\nabla^- F|(x(t))$ for a.e. $t > 0$, $t \mapsto F(x(t))$ is locally a.c. in $(0,\infty)$, with derivative equal to $-|\nabla^- F|^2(x(t))$.*

Sketch of proof. For simplicity, assume $K = 0$, see [3] for the general case. In this simpler case, using monotonicity of difference quotients one has the *duality formula for the descending slope*:

$$\text{(5.2)} \qquad |\nabla^- F|(x) = \sup_{y \neq x} \frac{[F(x) - F(y)]^+}{d(x,y)}.$$

It implies at once that $x \mapsto |\nabla^- F|(x)$ is l.s.c. in X, and provides the *one-sided* (because no modulus is present) and *local* (because the factor in front of the distance is not constant) Lipschitz property:

$$\text{(5.3)} \qquad F(x) - F(y) \leq |\nabla^- F|(x) d(x,y) \qquad \forall x \in D(F).$$

A real analysis lemma (see [3, Lemma 1.2.6]) then shows that (5.3) implies the upper gradient property.

Now, if we have an (EDI) solution, we can bound $F\big(x(0)\big)$ from above using (5.1) to get

$$\int_0^t \frac{1}{2}|x'(r)|^2 + \frac{1}{2}|\nabla^- F|^2\big(x(r)\big)\,\mathrm{d}r \le \int_0^t |x'(r)|\,|\nabla^- F|\big(x(r)\big)\,\mathrm{d}r.$$

Since t is arbitrary, this implies that $|x'| = |\nabla^- F|(x)$ a.e. in $(0,\infty)$. □

These results apply of course to the $\mathrm{Ent}_{\mathfrak{m}}$ in $\mathscr{P}(X)$, under the $\mathrm{CD}(K,\infty)$ assumption, and provide lower semicontinuity of $|\nabla^- \mathrm{Ent}_{\mathfrak{m}}|$, the upper gradient property of $|\nabla^- \mathrm{Ent}_{\mathfrak{m}}|$ and the equivalence of the (EDE) and (EDI) formulations.

6. Fisher bounds squared slope from above

We have seen that the energy dissipation rate of $\mathrm{Ent}_{\mathfrak{m}}$ along the L^2 heat flow is given by the Fisher information functional. It is natural to relate this functional to energy dissipation of $\mathrm{Ent}_{\mathfrak{m}}$ seen from the Wasserstein viewpoint.

Proposition 6.1. *In a* $\mathrm{CD}(K,\infty)$ *space* $(X,d,\mathfrak{m})$, *assume that* $\rho \in L^1(X,\mathfrak{m})$ *is a probability density with* $\sqrt{\rho} \in D(\mathrm{Ch})$. *Then*

$$|\nabla^- \mathrm{Ent}_{\mathfrak{m}}|^2(\rho\mathfrak{m}) \le \int_{\{\rho>0\}} \frac{|\nabla\rho|_*^2}{\rho}\,\mathrm{d}\mathfrak{m} \quad \Big(= 4\int |\nabla\sqrt{\rho}|_*^2\,\mathrm{d}\mathfrak{m}\Big).$$

Notice that it is precisely this inequality that prevents, in $\mathrm{CD}(K,\infty)$ spaces, triviality of the theory!

Sketch of proof. By approximation (recall that Ch is defined by approximation with Lipschitz functions and that $|\nabla^- \mathrm{Ent}_{\mathfrak{m}}|$ is l.s.c.) we can assume that $\sqrt{\rho} \in \mathrm{Lip}(X)$. By truncation, we can also assume that $c^{-1} \ge \sqrt{\rho} \ge c > 0$, so that $\log\rho \in \mathrm{Lip}(X)$.

Let us consider another density η and an optimal plan $\boldsymbol{\pi}_\eta$ from ρ to η. Then, following [26, Theorem 20.1], we can estimate:

$$\begin{aligned}\mathrm{Ent}_{\mathfrak{m}}(\rho\mathfrak{m}) - \mathrm{Ent}_{\mathfrak{m}}(\eta\mathfrak{m}) &\le \int \log\rho(\rho-\eta)\,\mathrm{d}\mathfrak{m} = \int \log\rho(x) - \log\rho(y)\,\mathrm{d}\boldsymbol{\pi}_\eta \\ &\le \int \big(|\nabla^- \log\rho|(x) + \omega_x(y)\big) d(x,y)\,\mathrm{d}\boldsymbol{\pi}_\eta(x,y) \\ &\le W_2(\eta\mathfrak{m},\rho\mathfrak{m})\bigg(\int \big(|\nabla^- \log\rho|(x) + \omega_x(y)\big)^2\,\mathrm{d}\boldsymbol{\pi}_\eta\bigg)^{1/2}\end{aligned}$$

where $\omega_x(y)$ is a uniformly bounded modulus of continuity with $\omega_x(x) = 0$. Dividing both sides by $W_2(\eta\mathfrak{m},\rho\mathfrak{m})$ and letting $\eta\mathfrak{m} \to \rho\mathfrak{m}$ gives the result, by the weak convergence of $\boldsymbol{\pi}_\eta$ to the identity plan $\boldsymbol{\pi}_\rho$, concentrated on the diagonal (since the first marginal is fixed, the limit works even though $|\nabla^- \log\rho|$ is discontinuous). □

7. Identification of gradient flows

Coming back to the notation used in the introduction, where we dealt with the Jordan – Kinderleherer – Otto [16] result for entropy and the heat equation, the conventional strategy goes as follows:

$$\begin{cases}\{\text{gradient flow of Ent}\} \subset \{\text{gradient flow of Dir}\} \\ \text{Uniqueness of gradient flow of Dir}\end{cases} \quad \Longrightarrow \quad = \text{holds}.$$

Namely, proves that gradient flows of the entropy are included in gradient flows of the Dirichlet energy and one can conclude that they coincide using uniqueness in the larger class. The new strategy, initiated in [14], proves instead

$$\begin{cases}\{\text{gradient flow of Ch}\} \subset \{\text{gradient flowof } \mathrm{Ent}_{\mathfrak{m}}\} \\ \text{Uniqueness of gradient flow of } \mathrm{Ent}_{\mathfrak{m}}\end{cases} \Longrightarrow \qquad = \text{holds.}$$

The new strategy is feasible thanks to the recent uniqueness result proved by Gigli in [13] for the W_2 gradient flow of $\mathrm{Ent}_{\mathfrak{m}}$. This result is surprising, because *no* contractivity property of W_2 can be expected at this level of generality [21], not even in Finsler (non-Riemannian) spaces.

We want to show that any L^2 heat flow $f_t := \mathbf{h}_t f_0$ (with f_0 probability density) is a W_2-gradient flow with $\mu_t := f_t\mathfrak{m}$, i.e.,

$$\int f_t \log f_t \,\mathrm{d}\mathfrak{m} + \int_0^t \frac{1}{2}|\dot{\mu}_r|^2 + \frac{1}{2}|\nabla^- \mathrm{Ent}_{\mathfrak{m}}|^2(\mu_r)\,\mathrm{d}r \le \int f_0 \log f_0 \,\mathrm{d}\mathfrak{m}.$$

Indeed, Kuwada's Lemma 3.6 and Proposition 6.1 give for almost every $t > 0$

$$(7.1) \qquad |\dot{\mu}_t|^2 \le \int_{\{f_t>0\}} \frac{|\nabla f_t|_*^2}{f_t}\,\mathrm{d}\mathfrak{m}, \quad |\nabla^- \mathrm{Ent}_{\mathfrak{m}}|^2(f_t\mathfrak{m}) \le \int_{\{f_t>0\}} \frac{|\nabla f_t|_*^2}{f_t}\,\mathrm{d}\mathfrak{m},$$

while the Hilbertian energy dissipation gives

$$\frac{\mathrm{d}}{\mathrm{d}t}\int f_t \log f_t \,\mathrm{d}\mathfrak{m} = -\int_{\{f_t>0\}} \frac{|\nabla f_t|_*^2}{f_t}\,\mathrm{d}\mathfrak{m} \quad \text{for a.e. } t > 0.$$

Coming now to uniqueness, the key contribution of [13] is the proof of convexity (in the usual sense) of $\mu \mapsto |\nabla^- \mathrm{Ent}_{\mathfrak{m}}|^2(\mu)$. Once we know this, if we have two EDI solutions μ_t^1, μ_t^2, both starting from $\bar{\mu}$, namely

$$\mathrm{Ent}_{\mathfrak{m}}(\mu_t^i) + \int_0^t \frac{1}{2}|\dot{\mu}_s^i|^2 + \frac{1}{2}|\nabla^- \mathrm{Ent}_{\mathfrak{m}}|^2(\mu_s^i)\,\mathrm{d}s \le \mathrm{Ent}_{\mathfrak{m}}(\bar{\mu}) \quad i = 1, 2$$

we can combine them into $\mu_t := (\mu_t^1 + \mu_t^2)/2$ to get (using convexity of $\mathrm{Ent}_{\mathfrak{m}}$ and of the squared metric derivative as well)

$$\mathrm{Ent}_{\mathfrak{m}}(\mu_t) + \int_0^t \frac{1}{2}|\dot{\mu}_s|^2 + \frac{1}{2}|\nabla^- \mathrm{Ent}_{\mathfrak{m}}|^2(\mu_s)\,\mathrm{d}s \le \mathrm{Ent}_{\mathfrak{m}}(\bar{\mu}).$$

By the upper gradient property the inequality has to be an equality, and this can happen only if $2\,\mathrm{Ent}_{\mathfrak{m}}(\mu_t) = \mathrm{Ent}_{\mathfrak{m}}(\mu_t^1) + \mathrm{Ent}_{\mathfrak{m}}(\mu_t^2)$. Strict convexity of the entropy then gives the result.

Another byproduct of the inclusion of L^2 gradient flows of Ch into W_2-gradient flows of $\mathrm{Ent}_{\mathfrak{m}}$ is that all inequalities (see (7.1) in particular) should be equalities, so that the energy dissipation rates are equal a.e. in $(0, \infty)$:

$$|\nabla^- \mathrm{Ent}_{\mathfrak{m}}|^2(f_t\mathfrak{m}) = \int_{\{f_t>0\}} \frac{|\nabla f_t|_*^2}{f_t}\,\mathrm{d}\mathfrak{m} \quad \text{for a.e. } t > 0.$$

By letting $t \downarrow 0$, this can be used to show that Fisher *coincides* with slope:

$$|\nabla^- \mathrm{Ent}_{\mathfrak{m}}|^2(f\mathfrak{m}) = \int_{\{f>0\}} \frac{|\nabla f|_*^2}{f}\,\mathrm{d}\mathfrak{m}.$$

8. Weak gradients and their identification

Let's start from the Euclidean case. We discuss only the case $W^{1,2}$, although all $W^{1,p}$ spaces $1 < p < \infty$ (and even the $W^{1,1}$ and BV spaces) could be treated, see [6]. The two standard definitions of Sobolev spaces are of W type (weak derivatives)

$$W^{1,2}(\mathbb{R}^n) := \left\{ u \in L^2(\mathbb{R}^n) : \frac{\partial u}{\partial x_i} \in L^2(\mathbb{R}^n), 1 \leq i \leq n \right\}$$

and of H type (strong derivatives)

$$H^{1,2}(\mathbb{R}^n) := \{\text{completion of } C^\infty \cap W^{1,2} \text{ for the } W^{1,2} \text{ norm}\}.$$

The celebrated "$H = W$" theorem by Meyers – Serrin in 1960 provides equivalence of the two definitions, even in any open domain. Another less known approach goes back to a paper by B. Levi [17] in 1906. B. Levi was looking for a function space where the minimization of the Dirichlet energy in a planar domain with given boundary conditions could find a solution, I will adopt his definition to n space dimensions, denoting for any $i = 1, \ldots, n$ by x_i the ith variable and by x_i' the block of the remaining $(n-1)$ variables.

Definition 8.1 (Beppo Levi space). Let $u\colon \mathbb{R}^n \to \mathbb{R}$. We say that $u \in \mathrm{BL}^{1,2}(\mathbb{R}^n)$ if:

(a) for $i = 1, \ldots, n$ and for $\mathscr{L}^{n-1}$-a.e. $x_i' \in \mathbb{R}^{n-1}$, $u(x_i', \cdot)$ is absolutely continuous in $\mathbb{R}$;

(b) $\sum_{i=1}^n \int_{\mathbb{R}^{n-1}} \int_{\mathbb{R}} |\partial u/\partial x_i|^2 \, \mathrm{d}x_i \, \mathrm{d}x_i' < \infty$.

Using Fubini's theorem we can still define a gradient in BL spaces. However, a drawback of this approach is that it is not clear whether the BL property is frame-indifferent or not. It is actually true, but to realize this one has to pass through the equivalence with the H and W definitions, that we are now going to discuss. Actually, another way to bypass this objection would be to consider in some sense all or, more precisely, almost all curves (not only geodesics). This is exactly the strategy pursued in the metric theory. For a proof of this result, see for instance [1] or [15].

Theorem 8.2. $\mathrm{BL}^{1,2}(\mathbb{R}^n) \subset W^{1,2}(\mathbb{R}^n)$. *In addition, any* $u \in W^{1,2}(\mathbb{R}^n)$ *has a version (for instance* $\limsup_\epsilon u * \rho_\epsilon$*) in* $\mathrm{BL}^{1,2}(\mathbb{R}^n)$.

In metric spaces, the W definition seems to be difficult to adapt. On the other hand, the H definition roughly corresponds to Cheeger's energy Ch (with $\mathrm{Lip}(X)$ playing the role of C^∞), while Levi's definition corresponds to Shanmugalingam's notion [23] of the Newtonian space $N^{1,2}(X, d, \mathfrak{m})$, that I am now going to illustrate. Actually in [4] a different notion of gradient is used, a priori smaller than the gradient of [23]; however, to simplify the exposition, I will just confine myself to the relaxed gradient and the gradient of [23].

Definition 8.3. Given $\Gamma \subset AC([0,1]; X)$ we define

$$\mathrm{Mod}_2(\Gamma) := \inf\left\{ \int g^2 \, \mathrm{d}\mathfrak{m} : g\colon X \to [0,\infty] \text{ Borel}, \int_\gamma g \geq 1 \text{ for all } \gamma \in \Gamma \right\}.$$

Here the curvilinear integral is defined, as in the theory of upper gradients, using the metric derivative, namely $\int_\gamma g := \int_0^1 g(\gamma_s)|\dot\gamma_s| \, \mathrm{d}s$.

We now define $N^{1,2}(X,d,\mathfrak{m})$ by

$$\left\{f\colon X\to\mathbb{R} : \left|\int_{\partial\gamma} f\right| \le \int_\gamma G \text{ for Mod}_2\text{-a.e. } \gamma, \text{ for some } G\in L^2(X,\mathfrak{m})\right\}$$

and $|\nabla f|_S \in L^2(X,\mathfrak{m})$ as the function G with smallest L^2 norm.

Lemma 8.4 (Absolute continuity lemma). *Any* $f\in N^{1,2}(X,d,\mathfrak{m})$ *is absolutely continuous along* Mod_2*-a.e. curve* γ.

PROOF. Let Γ be the set of curves γ where the u.g. property with $|\nabla f|_S$ does not hold,

$$\Gamma_1 := \{\gamma : \gamma\supset\gamma'\in\Gamma\},$$

and

$$\Gamma_2 := \left\{\gamma : \int_\gamma |\nabla f|_S = \infty\right\}.$$

Now, Γ_1 is Mod_2-negligible because Γ is (any g admissible for Γ is admissible for Γ_1), Γ_2 is Mod_2-negligible by the "Markov" inequality

$$\mathrm{Mod}_2\left(\Gamma_2\cap\left\{\int_\gamma |\nabla f|_S\ge n\right\}\right)\le\frac{1}{n^2}\int|\nabla f|_S^2\,d\mathfrak{m}\to 0.$$

If $\gamma\notin(\Gamma_1\cup\Gamma_2)$ we have

$$\left|\int_{\partial\gamma'} f\right|\le\int_{\gamma'}|\nabla f|_S\le\int_\gamma|\nabla f|_S<\infty\qquad\forall\gamma'\subset\gamma$$

which yields immediately the absolute continuity property of $t\mapsto f(\gamma_t)$.

The gradient $|\nabla f|_S$ has pointwise minimality properties analogous to $|\nabla f|_*$ (see Theorem 3.2(3)), in particular if G satisfies the weak upper gradient property

$$\left|\int_{\partial\gamma} f\right|\le\int_\gamma G\quad\text{for Mod}_2\text{-a.e. curve }\gamma$$

then $|\nabla f|_S\le G$ $\mathfrak{m}$-a.e. in X. □

Are the gradients $|\nabla f|_*$, $|\nabla f|_S$ equal? While the first gradient is relevant in connection with the L^2 heat flow and the "vertical" derivative, the second one is relevant in connection with the derivative of $\mathrm{Ent}_\mathfrak{m}$ and the "horizontal" derivative.

If we assume doubling & Poincaré (with the S-gradient in the right-hand side), then we can approximate any $f\in N^{1,2}(X,d,\mathfrak{m})$, see for instance [8], in the strong norm and even in the Lusin sense by Lipschitz maps f_n. This leads to the equality of gradients.

The strategy is to consider the maximal function

$$M(x) := \sup_{r>0}\frac{\int_{B_r(x)}|\nabla f|_S(y)\,d\mathfrak{m}(y)}{\mathfrak{m}(B_r(x))}$$

and to prove that $f|_{\{M\le n\}}$ is Cn-Lipschitz. Defining f_n as a Lipschitz extension of $f|_{\{M\le n\}}$, locality of gradients and

$$\mathfrak{m}(\{M>n\}) = o\left(\frac{1}{n^2}\right)$$

provide the result.

With "optimal transportation tools" we can provide in [4], see also [6], the equivalence of gradients and the density in energy of Lipschitz maps without doubling & Poincaré. This requires an approximation by Lipschitz functions f_n in "energy," namely

$$\limsup_{n\to\infty}\int|\nabla f_n|^2\,\mathrm{d}\mathfrak{m}\le\int|\nabla f|_S^2\,\mathrm{d}\mathfrak{m},\qquad \int|f_n-f|^2\,\mathrm{d}\mathfrak{m}\to 0. \tag{8.1}$$

By uniform convexity, this provides also

$$\lim_{n\to\infty}\int\big||\nabla f_n|-|\nabla f|_S\big|^2\,\mathrm{d}\mathfrak{m}=0.$$

Notice that, as soon as we know that the Sobolev spaces are reflexive, we can use Mazur's lemma (i.e., take convex combinations) to improve the approximation from weak to strong (while, without doubling & Poincaré, the Lusin approximation seems really to be out of reach).

In order to prove (8.1) we need, besides Kuwada's lemma, three more auxiliary results. In the sequel $\mathrm{AC}^2([0,1];X)$ denotes the class of absolutely continuous functions with square integrable metric derivative.

Lemma 8.5. *If $\boldsymbol{\eta}\in\mathscr{P}(C([0,1];X))$ concentrated on $AC^2([0,1];X)$ has uniformly bounded time marginals, i.e., $(e_t)_\sharp\boldsymbol{\eta}\le C(\boldsymbol{\eta})\mathfrak{m}$ for all $t\in[0,1]$, then*

$$[\boldsymbol{\eta}(\Gamma)]^2\le C(\boldsymbol{\eta})\Big(\int\int_0^1|\dot\gamma_s|^2\,\mathrm{d}s\,\mathrm{d}\boldsymbol{\eta}(\gamma)\Big)\,\mathrm{Mod}_2(\Gamma)\quad\forall\Gamma\subset\mathrm{AC}^2([0,1];X).$$

In particular $\mathrm{Mod}_2(\Gamma)=0$ *implies* $\boldsymbol{\eta}(\Gamma)=0$.

PROOF. If g is admissible for Γ we have

$$[\boldsymbol{\eta}(\Gamma)]^2\le\Big(\int_0^1\int g(\gamma_s)|\dot\gamma_s|\,\mathrm{d}s\,\mathrm{d}\boldsymbol{\eta}(\gamma)\Big)^2.$$

Then, it suffices to apply Hölder and to minimize w.r.t. g. □

Proposition 8.6 (Superposition principle [3,18]**).** *Let $(\mu_t)_{t\in[0,T]}\subset\mathscr{P}(X)$ be absolutely continuous with L^2-integrable metric derivative. Then there exists $\boldsymbol{\eta}\in\mathscr{P}(C([0,T];X))$ concentrated on $\mathrm{AC}^2([0,T];X)$ and satisfying*

(1) $\mu_t=(e_t)_\sharp\boldsymbol{\eta}$ *for all* $t\in[0,T]$;
(2) $|\dot\mu_t|^2=\int|\dot\gamma_t|^2\,\mathrm{d}\eta(\gamma)$ *for a.e.* $t\in(0,T)$.

Lemma 8.7 (Stability of weak upper gradients [15,23]**).** *If $f_n\to f$ in $L^2(X,\mathfrak{m})$, $G_n\to G$ weakly in $L^2(X,\mathfrak{m})$ and*

$$\Big|\int_{\partial\gamma}f_n\Big|\le\int_\gamma G_n\quad\text{for }\mathrm{Mod}_2\text{-a.e. curve }\gamma,$$

then there is a version $\tilde f$ of f satisfying

$$\Big|\int_{\partial\gamma}\tilde f\Big|\le\int_\gamma G\quad\text{for }\mathrm{Mod}_2\text{-a.e. curve }\gamma.$$

Using this lemma with f_n equal to the optimal sequence in the definition of Ch and $G_n=|\nabla f_n|$, weakly convergent to $|\nabla f|_*$, we obtain

$$|\nabla\tilde f|_S\le G=|\nabla f|_*\quad\mathfrak{m}\text{-a.e. in }X.$$

The proof of the converse inequality is constructive: we need Lipschitz functions f_n satisfying $f_n \to f$ in $L^2(X, \mathfrak{m})$ and

$$\limsup_{n\to\infty} \int |\nabla f_n|^2 \, d\mathfrak{m} \le \int |\nabla f|_S^2 \, d\mathfrak{m}.$$

By a diagonal argument, it suffices to find $f_n \in D(\mathrm{Ch})$ satisfying

$$\limsup_{n} \int |\nabla f_n|_*^2 \, d\mathfrak{m} \le \int |\nabla f|_S^2 \, d\mathfrak{m}.$$

By a truncation argument we assume that $0 < c \le f \le c^{-1} < \infty$ and by homogeneity $\int f^2 \, d\mathfrak{m} = 1$. We set $k = f^2$, $k_t = \mathbf{h}_t k$, $\mu_t = k_t \mathfrak{m} \in \mathscr{P}(X)$, $\boldsymbol{\eta}$ given by the superposition principle. Then we argue as in Proposition 6.1, this time using the S-gradient in place of the relaxed slope:

$$\begin{aligned}
\int k \log k - k_t \log k_t \, d\mathfrak{m} &\le \int \log k (k - k_t) \, d\mathfrak{m} \\
&= \int \log k(\gamma_0) - \log k(\gamma_t) \, d\boldsymbol{\eta}(\gamma) \\
&\le \int \int_0^t |\nabla \log k|_S(\gamma_s) |\dot{\gamma}_s| \, ds \, d\boldsymbol{\eta}(\gamma) \\
&\le \left(\int_0^t \int |\nabla \log k|_S^2(\gamma_s) \, d\eta \, ds \right)^{1/2} \left(\int_0^t \int |\dot{\gamma}_s|^2 \, d\boldsymbol{\eta}(\gamma) \, ds \right)^{1/2} \\
&\le \frac{1}{2} \int_0^t \int |\nabla \log k|_S^2 k_s \, d\mathfrak{m} \, ds + \frac{1}{2} \int |\dot{\mu}_s|^2 \, ds.
\end{aligned}$$

By the Kuwada lemma we get

$$\int k \log k - k_t \log k_t \, d\mathfrak{m} \le \frac{1}{2} \int_0^t \int |\nabla \log k|_S^2 k_s \, d\mathfrak{m} \, ds + \frac{1}{2} \int_0^t \int_{\{k_s > 0\}} \frac{|\nabla k_s|_*^2}{k_s} \, d\mathfrak{m} \, ds.$$

The entropy dissipation formula (3.4) then gives

$$\int_0^t \int_{\{k_s > 0\}} \frac{|\nabla k_s|_*^2}{k_s} \, d\mathfrak{m} \, ds \le \int_0^t \int |\nabla \log k|_S^2 k_s \, d\mathfrak{m} \, ds,$$

so that from the identity $|\nabla \log k|_S = |\nabla k|_S / k = 2 |\nabla f|_S / f$ we get

$$\frac{4}{t} \int_0^t \mathrm{Ch}(\sqrt{k_s}) \, ds \le \frac{4}{t} \int_0^t \int \frac{|\nabla f|_S^2}{f^2} k_s \, d\mathfrak{m} \, ds.$$

Letting $t \downarrow 0$ and using the w^*-convergence in $L^\infty(X, \mathfrak{m})$ of k_s to $k = f^2$ gives the result.

9. Riemannian Ricci lower bounds

As noticed by Cordero Erausquin, Sturm and Villani, all Minkowski spaces ($\mathbb{R}^n$ endowed with the Lebesgue measure and any norm $\|\cdot\|$) satisfy the $\mathrm{CD}(0, n)$, and therefore the $\mathrm{CD}(0, \infty)$ condition. On the other hand, Cheeger and Colding ruled out in [9] the possibility of obtaining these spaces as limits of Riemannian manifolds with uniform lower bounds on Ricci curvature and uniform upper bounds on volume.

In [5] we tried to give an answer to the following question: Is there a more restrictive notion, still *stable* and *strongly consistent* with the Riemannian case, that rules out Minkowski (non-Hilbert) spaces? Recall that the $\mathrm{CD}(K,\infty)$ condition is stable [19, 24, 25], meaning that measured Gromov – Hausdorff limits of $\mathrm{CD}(K,\infty)$ spaces are $\mathrm{CD}(K,\infty)$, and strongly consistent, meaning that a Riemannian manifold M endowed with Riemannian distance $d = d_M$ and the volume measure $\mathfrak{m} = \mathrm{vol}_M$ is $\mathrm{CD}(K,\infty)$ iff $\mathrm{Ric}_M \geq KI$. So, a positive answer to this question would provide more insight, among other things, on the closure of Riemannian manifolds under uniform Ricci lower bounds.

This led us to the definition of spaces with *Riemannian* Ricci lower bounds. We have 3 equivalent definitions (and their equivalence is far from being trivial), summarized below, and this class of spaces provides a positive answer to the question I raised.

Definition 9.1 ($\mathbf{RCD}(K,\infty)$ spaces). We say that $(X,d,\mathfrak{m})$ has *Riemannian Ricci curvature* bounded from below by $K \in \mathbb{R}$, and write $\mathrm{RCD}(K,\infty)$, if one of the following equivalent conditions hold:

(i) $(X,d,\mathfrak{m})$ is a strong $\mathrm{CD}(K,\infty)$ space and the L^2 gradient flow $\mathbf{h}_t$ of Ch is linear;
(ii) $(X,d,\mathfrak{m})$ is a strong $\mathrm{CD}(K,\infty)$ space and the W_2 gradient flow $\mathbf{H}_t$ of $\mathrm{Ent}_{\mathfrak{m}}$ is additive (i.e., convex and concave) on $\mathscr{P}(X)$;
(iii) for all $\mu \in \mathscr{P}(X)$ with $\operatorname{supp}\mu \subset \operatorname{supp}\mathfrak{m}$, $\mathbf{H}_t\mu$ is a gradient flow in the EVI_K sense.

I will illustrate later on what strong $\mathrm{CD}(K,\infty)$ and EVI_K mean; in the next subsections I will instead list some properties of this class of spaces.

9.1. Stability under Gromov – Hausdorff limits of $\mathbf{RCD}(K,\infty)$ spaces. We say that two metric measure spaces $(X,d_X,\mathfrak{m}_X)$ and $(Y,d_Y,\mathfrak{m}_Y)$ are *isomorphic* if there exists a bijective isometry $f\colon \operatorname{supp}\mathfrak{m}_X \to \operatorname{supp}\mathfrak{m}_Y$ such that $f_\sharp\mathfrak{m}_X = \mathfrak{m}_Y$. We will denote by $\mathbb{X}$ the set of isomorphism classes of metric measure spaces that we will consider (as we said at the beginning we confine ourselves to compact metric spaces (X,d) and probability reference measures $\mathfrak{m}$):

$$\mathbb{X} := \{(X,d,\mathfrak{m}) : (X,d) \text{ is compact and } \mathfrak{m} \in \mathscr{P}(X)\}.$$

Definition 9.2. Given two metric measure spaces $(X,d_X,\mathfrak{m}_X)$, $(Y,d_Y,\mathfrak{m}_Y)$, we consider the product space $(X\times Y, d_{XY})$, where

$$d_{XY}\big((x_1,y_1),(x_2,y_2)\big) := \sqrt{d_X^2(x_1,x_2)+d_Y^2(y_1,y_2)}.$$

We say that $(\mathbf{d},\gamma)$ is an admissible coupling if:

(a) $\mathbf{d}$ is a pseudodistance on $X \sqcup Y$ which coincides with d_X (resp. d_Y) when restricted to $\operatorname{supp}\mathfrak{m}_X \times \operatorname{supp}\mathfrak{m}_X$ (resp. $\operatorname{supp}\mathfrak{m}_Y \times \operatorname{supp}\mathfrak{m}_Y$) and $\mathbf{d}|_{X\times Y}\colon X\times Y \to [0,\infty)$ is Borel.
(b) γ is a Borel measure on $X\times Y$, $\pi^X_\sharp\gamma = \mathfrak{m}_X$ and $\pi^Y_\sharp\gamma = \mathfrak{m}_Y$.

It is not hard to see that the set of admissible couplings is always nonempty. The *cost* $C(\mathbf{d},\gamma)$ of a coupling is given by

$$C(\mathbf{d},\gamma) := \int \mathbf{d}^2(x,y)\,\mathrm{d}\gamma(x,y).$$

In analogy to the definition of W_2, Sturm's distance $\mathbb{D}$ is then defined as

$$\mathbb{D}^2\big((X,d_X,\mathfrak{m}_X),(Y,d_Y,\mathfrak{m}_Y)\big) := \inf C(\mathbf{d},\gamma),$$

the infimum being taken among all couplings $(\mathbf{d},\gamma)$ of $(X,d_X,\mathfrak{m}_X)$ and $(Y,d_Y,\mathfrak{m}_Y)$.

Theorem 9.3. *Let* $(X_n, d_n, \mathfrak{m}_n) \in \mathbb{X}$ *be* $\mathrm{RCD}(K,\infty)$ *spaces. If*

$$\mathbb{D}\big((X_n, d_n, \mathfrak{m}_n), (X, d, \mathfrak{m})\big) \to 0,$$

then $(X, d, \mathfrak{m})$ *is a* $\mathrm{RCD}(K,\infty)$ *space as well.*

In the proof of this result it is the EVI_K formulation that plays a decisive role.

9.2. Tensorization of $\mathrm{RCD}(K,\infty)$ spaces. Remember [25] that the product of two nonbranching $\mathrm{CD}(K,\infty)$ spaces is still $\mathrm{CD}(K,\infty)$, but if the nonbranching assumption is removed the validity of the implication is still an open problem. Here nonbranching means that the map $(e_0, e_t) : \mathrm{Geo}(X) \to X^2$ is injective for all $t > 0$; in other words, geodesics can't split. The following result proves that the same property persists for the stronger $\mathrm{RCD}(K,\infty)$ notion.

Theorem 9.4. *If* $(X, d_X, \mathfrak{m}_X)$ *and* $(Y, d_Y, \mathfrak{m}_Y)$ *are* $\mathrm{RCD}(K,\infty)$ *and nonbranching, so is* $(X \times Y, \sqrt{d_X^2 + d_Y^2}, \mathfrak{m}_X \times \mathfrak{m}_Y)$.

In the proof of this result we failed to prove directly tensorization of the EVI_K, so we rely on the above mentioned Sturm's result saying that the product is $\mathrm{CD}(K,\infty)$. Since the nonbranching property tensorizes, and since it is not difficult to show that

$$\mathrm{CD}(K,\infty) + \text{nonbranching} \implies \text{strong } \mathrm{CD}(K,\infty)$$

the product is also strong $\mathrm{CD}(K,\infty)$.

Finally, we have to prove that linearity of the heat flow tensorizes as well. This is equivalent to saying that

$$\mathrm{Ch}^{X\times Y}(f) = \int \mathrm{Ch}^X(f^y)\,\mathrm{d}\mathfrak{m}_Y(y) + \int \mathrm{Ch}^Y(f^x)\,\mathrm{d}\mathfrak{m}_X(x)$$

or equivalently that (with $f^x(y) = f(x,y) = f^y(x)$)

$$|\nabla f|_*^2(x,y) = |\nabla f^x|_*^2(y) + |\nabla f^y|_*^2(x) \quad \mathfrak{m}_X \times \mathfrak{m}_Y\text{-a.e. in } X \times Y.$$

The proof of the inequality $|\nabla f|_*^2(x,y) \le |\nabla f^x|_*^2(y) + |\nabla f^y|_*^2(x)$ is based on the following calculus lemma and on a smoothing argument with the product semigroup $\mathbf{h}_t^X \times \mathbf{h}_t^Y$.

Lemma 9.5. *If* $f\colon X \times Y \to \mathbb{R}$ *is Lipschitz and* $\gamma = (\gamma^X, \gamma^Y) : [0,1] \to X \times Y$ *is absolutely continuous, then, for a.e.* t, $|(\mathrm{d}/\mathrm{d}t)(f \circ \gamma)(t)|$ *is bounded from above by*

$$\limsup_{h\downarrow 0} \frac{|f(\gamma^X_{t-h}, \gamma^Y_t) - f(\gamma^X_t, \gamma^Y_t)|}{h} + \limsup_{h\downarrow 0} \frac{|f(\gamma^X_t, \gamma^Y_{t+h}) - f(\gamma^X_t, \gamma^Y_t)|}{h}.$$

The proof of the converse inequality $|\nabla f|_*^2(x,y) \ge |\nabla f^x|_*^2(y) + |\nabla f^y|_*^2(x)$ is much more delicate. It relies indeed in a version of Kuwada's lemma in product spaces, where we consider the product semigroup $\mathbf{h}_t^X \times \mathbf{h}_t^Y$, correspondent to the gradient flow of $\frac{1}{2}\int \big(|\nabla f^x|_*^2(y) + |\nabla f^y|_*^2(x)\big)\,\mathrm{d}\mathfrak{m}_X \times \mathfrak{m}_Y(x,y)$:

Proposition 9.6 (Kuwada's lemma in product spaces). *Let* $f \in L^2(X \times Y, \mathfrak{m}_X \times \mathfrak{m}_Y)$ *be a probability density and let* f_t *be the evolution of* f *under the product semigroup,* $\mu_t = f_t \mathfrak{m}_X \times \mathfrak{m}_Y$. *Then for a.e.* $t > 0$ *it holds*

$$|\dot\mu_t|^2 \le \int_{\{f_t>0\}} \frac{|\nabla f_t^x|_*^2(y) + |\nabla f_t^y|_*^2(x)}{f_t(x,y)}\,\mathrm{d}\mathfrak{m}_X \times \mathfrak{m}_Y(x,y).$$

Once we know this, the machinery of identification theorems provides identification of the two gradients.

9.3. The heat flow in RCD(K, ∞) spaces. The identification between the L^2 gradient flow $\mathbf{h}_t$ and the W_2 gradient flow $\mathbf{H}_t$ allows one to pick the best properties from each of them: for instance, the symmetry of the transition probabilities $\theta_t \colon X \times X \to [0,\infty)$, defined by $\mathbf{H}_t\delta_x := \theta_t(x,\cdot)\mathfrak{m}$, comes from the fact that $\mathbf{h}_t$ is L^2-self-adjoint, while the contractivity properties of $\mathbf{h}_t$ in spaces different from $L^2(X,\mathfrak{m})$ follow from those of $\mathbf{H}_t$.

Proposition 9.7 (Properties of the heat flow). *It holds*:

(1) *The pointwise formula* $\tilde{\mathbf{h}}_t f(x) := \int f \,\mathrm{d}\mathbf{H}_t\delta_x$ *provides a version of* $\mathbf{h}_t f$ *and an extension of* $\mathbf{h}_t$ *to a contraction semigroup in all* $L^p(X,\mathfrak{m})$ *spaces.*

(2) $\tilde{\mathbf{h}}_t$ *leaves* $\mathrm{Lip}(\operatorname{supp}\mathfrak{m})$ *invariant and, by the contractivity estimate*

$$W_2(\mathbf{H}_t\delta x, \mathbf{H}_t\delta_y) \le \mathrm{e}^{-Kt}\, d(x,y),$$

$\mathrm{Lip}(\tilde{\mathbf{h}}_t f) \le \mathrm{e}^{-Kt}\,\mathrm{Lip}(f)$. *Furthermore,* $\tilde{\mathbf{h}}_t$ *maps* $L^\infty(X,\mathfrak{m})$ *into* $C_b(\operatorname{supp}\mathfrak{m})$, *the space of continuous functions on* $\operatorname{supp}\mathfrak{m}$.

(3) *The Bakry–Emery estimate holds:*

$$|\nabla(\mathbf{h}_t f)|_*^2 \le \mathrm{e}^{-2Kt}\,\mathbf{h}_t|\nabla f|_*^2 \qquad \mathfrak{m}\text{-a.e. in } X. \tag{9.1}$$

9.4. RCD(K, ∞) spaces and Dirichlet forms. Since Ch is a quadratic form in RCD(K,∞) spaces, the analysis of the connection with Fukushima's theory of Dirichlet forms is useful and mandatory. Let

$$\mathcal{E}(u,v) := \tfrac{1}{4}\big(\mathrm{Ch}(u+v) - \mathrm{Ch}(u-v)\big)$$

be the symmetric bilinear form associated to Ch. It is a Dirichlet form (i.e., closable and Markovian) because Ch is $L^2(X,\mathfrak{m})$-lower semicontinuous and decreases, by the chain rule, under left composition with 1-Lipschitz maps.

In the theory of Dirichlet forms, two objects are naturally defined, namely the *local energy measure*

$$[u](\varphi) := \mathcal{E}(u, u\varphi) - \mathcal{E}\Big(\frac{u^2}{2}, \varphi\Big)$$

and the induced distance (here the inequality $[\psi] \le \mathfrak{m}$ is understood as an inequality of measures)

$$d_{\mathcal{E}}(x,y) := \sup\{|\psi(x) - \psi(y)| : [\psi] \le \mathfrak{m}\}.$$

We proved that in the class of RCD(K,∞) spaces these objects coincide with the natural ones.

Theorem 9.8. *In a* RCD(K,∞) *space* $(X,d,\mathfrak{m})$, *the local energy measure* $[u]$ *coincides with* $|\nabla u|_*^2\mathfrak{m}$ *and the induced distance* $d_{\mathcal{E}}$ *coincides with* d.

The proof involves the construction of a symmetric bilinear form

$$(u,v) \in \big[D(\mathrm{Ch})\big]^2 \mapsto \nabla u \cdot \nabla v \in L^1(X,\mathfrak{m})$$

satisfying the Leibniz rule and providing integral representation to $\mathcal{E}$, namely $\mathcal{E}(u,v) = \int \nabla u \cdot \nabla v \,\mathrm{d}\mathfrak{m}$.

In addition, since $\mathcal{E}$ is also strongly local, the theory of Dirichlet forms can be applied as a black box to obtain a unique (in law) *Brownian motion* in $(\operatorname{supp}\mathfrak{m}, d, \mathfrak{m})$, i.e., a Markov process $\boldsymbol{X}_t$ with continuous sample paths satisfying

$$\mathbf{P}(\boldsymbol{X}_t \mid \boldsymbol{X}_0 = x) = \mathbf{H}_t\delta_x \quad \forall x \in \operatorname{supp}\mathfrak{m}, t \ge 0.$$

9.5. Strong CD(K,∞) condition and EVI$_K$. In a geodesic metric space, convexity of an energy F can be asked along *some* geodesic connecting a given pair of points, or along *all* geodesics connecting a given pair of points. It is well known and easy to check that the first choice (call it *convexity*) leads to a stable condition, under Gromov – Hausdorff limits, while the second choice (call it *strong convexity*) leads in general to an unstable condition.

The definition given below represents a sort of compromise: we ask for a distinguished geodesic in $\mathscr{P}(X)$ (i.e., an optimal geodesic plan $\boldsymbol{\pi}$), but we ask that the same property persists for all weighted geodesic plans $h\boldsymbol{\pi}$, $h \in C_b\bigl(\mathrm{Geo}(X)\bigr)$.

Definition 9.9 (Strong CD(K,∞)). We say that $(X,d,\mathfrak{m})$ is a *strong* CD$(0,\infty)$ space if for all μ_0, $\mu_1 \in \mathscr{P}(X)$ with finite entropy there exists an optimal geodesic plan $\boldsymbol{\pi}$ between them satisfying

$$\begin{aligned}\mathrm{Ent}_{\mathfrak{m}}\bigl((e_t)_\sharp(h\boldsymbol{\pi})\bigr) \le (1-t)\,\mathrm{Ent}_{\mathfrak{m}}\bigl((e_0)_\sharp(h\boldsymbol{\pi})\bigr) + t\,\mathrm{Ent}_{\mathfrak{m}}\bigl((e_1)_\sharp(h\boldsymbol{\pi})\bigr)\\ - \frac{K}{2}t(1-t)W_2^2\bigl((e_0)_\sharp(h\boldsymbol{\pi}),(e_1)_\sharp(h\boldsymbol{\pi})\bigr)\end{aligned}$$

for all $t \in [0,1]$, $h \in C_b\bigl(\mathrm{Geo}(X)\bigr)$ nonnegative, $\int h\,\mathrm{d}\boldsymbol{\pi} = 1$.

Question. Is there a condition stronger than strong convexity and stable? The answer is yes, it is the existence of EVI$_K$ gradient flows.

If H is Hilbert and $F\colon H \to \mathbb{R}\cup\{+\infty\}$ is K-convex and l.s.c., we can write the differential inclusion $-x'(t) \in \partial F\bigl(x(t)\bigr)$ for a.e. $t>0$ as follows:

$$\forall y \in D(F),\quad \langle -x'(t), y - x(t)\rangle + F\bigl(x(t)\bigr) + \frac{K}{2}|x(t)-y|^2 \le F(y) \qquad \text{for a.e. } t>0.$$

Equivalently

$$\forall y \in D(F),\quad \frac{\mathrm{d}}{\mathrm{d}t}\frac{1}{2}|x(t)-y|^2 + F\bigl(x(t)\bigr) + \frac{K}{2}|x(t)-y|^2 \le F(y) \qquad \text{for a.e. } t>0.$$

Definition 9.10. In a metric space (E,d), a locally absolutely continuous curve $u\colon (0,\infty) \to E$ is an EVI$_K$ solution to the gradient flow of $F\colon X \to \mathbb{R}\cup\{+\infty\}$ if for all $v \in D(F)$ it holds

$$\frac{\mathrm{d}}{\mathrm{d}t}\frac{1}{2}d^2(u(t),v) + F(u(t)) + \frac{K}{2}d^2(u(t),v) \le F(v) \quad \text{for a.e. } t>0.$$

This formulation of gradient flows is equivalent in Hilbert spaces, but in general *stronger* than the one based on energy dissipation, see [2]. A remarkable result, proved by Daneri and Savaré in [10], is that existence of a sufficiently rich family of EVI$_K$ gradient flows implies K-convexity.

Theorem 9.11. *In a geodesic space (X,d), assume that* EVI$_K$ *gradient flows exist starting from any $\bar{x} \in \overline{D(F)}$. Then, F is K-convex along any geodesic contained in $\overline{D(F)}$.*

Sketch of proof. We assume $K=0$ and give a formal proof. Let $\gamma\colon [0,1] \to X$ be a constant speed geodesic contained in $\overline{D(F)}$ and $t \in (0,1)$, let x_s be the EVI solution starting from γ_t. With no loss of generality, $\gamma_0,\gamma_1 \in D(F)$. The EVI property gives (we assume that the derivative exists at $s=0$)

$$\frac{\mathrm{d}^+}{\mathrm{d}s}\frac{1}{2}d^2(x_s,\gamma_0)\bigg|_{s=0} \le F(\gamma_0) - F(x_0), \quad \frac{\mathrm{d}^+}{\mathrm{d}s}\frac{1}{2}d^2(x_s,\gamma_1)\bigg|_{s=0} \le F(\gamma_1) - F(x_0).$$

Multiply the first one by $(1-t)$, the second one by t, and use the fact that $x_0 = \gamma_t$, to get

$$(1-t)F(\gamma_0) + tF(\gamma_1) - F(\gamma_t) \geq \frac{1}{2}\frac{\mathrm{d}^+}{\mathrm{d}s}(1-t)d^2(\gamma_0, x_s) + td^2(x_s, \gamma_1)\Big|_{s=0}.$$

On the other hand, the Young and triangle inequalities easily imply that

$$(1-t)d^2(x,z) + td^2(z,y) \geq t(1-t)d^2(x,y) \quad \forall x, y, z \in X,$$

with equality (only) if $d(x,z) = td(x,y)$, $d(z,y) = (1-t)d(x,y)$.

Hence, choosing $x = \gamma_0$, $y = \gamma_1$, the quantity

$$s \mapsto (1-t)d^2(\gamma_0, x_s) + td^2(x_s, \gamma_1)$$

is minimal at $s = 0$ (recall that $x_0 = \gamma_t$), so that its right derivative is nonnegative. □

Having now defined EVI_K, we can give a sketchy proof of the stability; as in the proof of the analogous result for $\mathrm{CD}(K,\infty)$ spaces, the crucial property is the joint lower semicontinuity property of $(\mu, \mathfrak{m}) \mapsto \mathrm{Ent}_{\mathfrak{m}}(\mu)$. The latter is a direct consequence of the duality formula

$$\mathrm{Ent}_{\mathfrak{m}}(\sigma) = \sup_{f \in C_b} \int f \,\mathrm{d}\sigma - \int F^*(f)\,\mathrm{d}\mathfrak{m}$$

where $F(z) = \mathrm{e}^{z-1}$ is the transform of $z \log z$ (set to $+\infty$ for $z < 0$).

Sketch of proof of stability of $\mathrm{RCD}(K,\infty)$. Let $(X_n, d_n, \mathfrak{m}_n) \to (X, d, \mathfrak{m})$ w.r.t. the distance $\mathbb{D}$, $(X_n, d_n, \mathfrak{m}_n)$ $\mathrm{RCD}(K,\infty)$. Thanks to an embedding theorem and the invariance in the isomorphism class we can (possibly extracting a subsequence) assume that both $X_n = \mathrm{supp}\,\mathfrak{m}_n$ and $X = \mathrm{supp}\,\mathfrak{m}$ are contained in a fixed metric space (Y, d_Y) and that $d_n = d_Y|_{X_n \times X_n}$, $d = d_Y|_{X\times X}$.

Given an initial condition $\bar\mu = \rho\mathfrak{m}$ with $\rho \in L^\infty(X, \mathfrak{m})$, we want to build an EVI solution starting from $\bar\mu$ in the space $(X, d, \mathfrak{m})$. We find $\bar\mu_n := \rho_n \mathfrak{m}_n$, with $\|\rho_n\|_\infty \leq \|\rho\|_\infty$, weakly convergent to $\bar\mu$, and the EVI solutions μ_t^n starting from $\bar\mu_n$:

$$\forall \nu, \quad \frac{\mathrm{d}}{\mathrm{d}t}\frac{1}{2}W_2^2(\mu_t^n, \nu) + \mathrm{Ent}_{\mathfrak{m}_n}(\mu_t^n) \leq \mathrm{Ent}_{\mathfrak{m}_n}(\nu) \qquad \text{for a.e. } t > 0.$$

By standard tightness estimates in space and equicontinuity estimates in time we can assume that $\mu_t^n \to \mu_t$ in $\mathscr{P}(X)$ for all $t \geq 0$, so that μ_t is our candidate EVI solution.

In order to check this, for any $\nu \in D(\mathrm{Ent}_{\mathfrak{m}})$ we can find $\nu_n \in D(\mathrm{Ent}_{\mathfrak{m}_n})$ convergent to ν in $\mathscr{P}(X)$ and satisfying $\mathrm{Ent}_{\mathfrak{m}_n}(\nu_n) \to \mathrm{Ent}_{\mathfrak{m}}(\nu)$.

In this way the right hand sides in the EVI converge and also the time derivatives (in the sense of distributions). It remains to prove that

$$\liminf_{n\to\infty} \mathrm{Ent}_{\mathfrak{m}_n}(\mu_t^n) \geq \mathrm{Ent}_{\mathfrak{m}}(\mu_t) \quad \forall t \geq 0$$

and this follows, as we said, by the joint lower semicontinuity of the entropy. □

9.6. Some auxiliary results. In this section I just state some auxiliary results needed to prove the equivalence of conditions (i), (ii), (iii) in Definition 9.1.

Lemma 9.12 (Derivative of squared Wasserstein distance). *Let $(X, d, \mathfrak{m})$ be a* $\mathrm{CD}(K, \infty)$ *space,* $\mu = \rho\mathfrak{m} \in \mathscr{P}(X)$ *such that* $0 < c \leq \rho \leq C < \infty$, *and set* $\mu_t := \mathbf{H}_t(\mu) = \rho_t\mathfrak{m}$. *Let* $\nu = \sigma\mathfrak{m} \in \mathscr{P}(X)$ *and let* φ_t *be a Kantorovich potential relative to* (μ_t, ν). *Then for a.e.* $t > 0$ *it holds*

$$\frac{\mathrm{d}}{\mathrm{d}t}\frac{1}{2}W_2^2(\rho_t\mathfrak{m}, \sigma\mathfrak{m}) \leq \frac{\mathrm{Ch}(\rho_t - \epsilon\varphi_t) - \mathrm{Ch}(\rho_t)}{\epsilon} \qquad \forall \epsilon > 0.$$

Lemma 9.13 (Derivative of the entropy along a geodesic). *Let* $(X, d, \mathfrak{m}) \in \mathbb{X}$ *be a strong* $\mathrm{CD}(K, \infty)$ *space and let* ρ, σ *be bounded probability densities. Assume that* σ *has bounded support and that* $\rho \geq c > 0$. *Then there exists an optimal geodesic plan* $\boldsymbol{\pi}$ *from* $\rho\mathfrak{m}$ *to* $\nu := \sigma\mathfrak{m}$ *satisfying*

$$\frac{\mathrm{Ch}(\varphi) - \mathrm{Ch}(\varphi + \epsilon\rho)}{\epsilon} \leq \lim_{s \downarrow 0} \frac{\mathrm{Ent}_{\mathfrak{m}}(\sigma_s\mathfrak{m}) - \mathrm{Ent}_{\mathfrak{m}}(\rho\mathfrak{m})}{s} \qquad \forall \epsilon > 0,$$

where $(e_s)_\sharp\boldsymbol{\pi} = \sigma_s\mathfrak{m}$ *and* φ *is any Kantorovich potential relative to* $(\rho\mathfrak{m}, \nu)$.

Theorem 9.14 (Metric Brenier theorem for strong $\mathrm{CD}(K, \infty)$ spaces). *Let* $(X, d, \mathfrak{m}) \in \mathbb{X}$ *be a strong* $\mathrm{CD}(K, \infty)$ *space,* $\mu = \rho\mathfrak{m} \in \mathscr{P}(X)$ *with* $c^{-1} \geq \rho \geq c > 0$ *and* $\nu \in \mathscr{P}(X)$ *with bounded support and density. Then there exist an optimal geodesic plan* $\boldsymbol{\pi}$ *and* $L \colon X \to [0, \infty)$ *satisfying*

$$L(\gamma_0) = d(\gamma_0, \gamma_1) \quad \text{for } \boldsymbol{\pi}\text{-a.e. } \gamma \in \mathrm{Geo}(X).$$

Furthermore,

$$L(x) = |\nabla\varphi|_*(x) = |\nabla^+\varphi|(x) \quad \text{for } \mu\text{-a.e. } x \in X,$$

where φ *is any Kantorovich potential relative to* (μ, ν).

10. Open problems and perspectives

(1) It would be interesting to examine the impact of the additional axiom, i.e., linearity of the heat flow, at the level of the finite-dimensional theory, i.e., $\mathrm{CD}(K, N)$ with $N < \infty$. Or to understand, at least in the case $K = 0$, where the Reny entropy is available, the role of the EVI formulation

(2) At least formally, one can use this calculus to write down a differential $\mathrm{CD}(K, N)$ inequality

$$\Delta\frac{|\nabla f|_*^2}{2} \geq \langle\nabla\Delta f, \nabla f\rangle + \frac{(\Delta f)^2}{N} + K|\nabla f|_*^2,$$

and try to investigate its relation with the existing theories. Also, in this differential perspective, one might try to reverse the implication from the Bakry-Emery condition (9.1) to $(\mathrm{R})\mathrm{CD}(K, \infty)$.

(3) In presence of doubling & Poincaré, Cheeger's theory applies and provides, in a suitable and very weak sense, local coordinates and a tangent bundle. The relations with the calculus described in these lectures are still not completely understood.

(4) What about the behavior on small scales of $\mathrm{RCD}(K, \infty)$ spaces? The question makes sense, if one adds a doubling condition. The natural conjecture is that tangent metric spaces, in the measured Gromov – Hausdorff sense, are Euclidean. This has been proved by Cheeger – Colding, but for limits of Riemannian manifolds.

References

1. L. Ambrosio, N. Fusco, and D. Pallara, *Functions of bounded variation and free discontinuity problems*, Oxford Math. Monogr., Clarendon Press, New York, 2000.
2. L. Ambrosio and N. Gigli, *A user's guide to optimal transport*, Lecture Notes in Math., Springer, Berlin, to appear, available at `http://cvgmt.sns.it/paper/195/`.
3. L. Ambrosio, N. Gigli, and G. Savaré, *Gradient flows in metric spaces and in the space of probability measures*, 2nd ed., Lectures Math. ETH Zurich, Birkhäuser, Basel, 2008.
4. ———, *Calculus and heat flows in metric measure spaces with Ricci curvature bounded from below*, available at `arXiv:1106.2090`.
5. ———, *Metric measure spaces with Riemannian Ricci curvature bounded from below*, available at `arXiv:1109.0222`.
6. ———, *Density of Lipschitz functions and equivalence of weak gradients in metric measure spaces*, available at `arXiv:1111.3730`.
7. P. Bernard and B. Buffoni, *Optimal mass transportation and Mather theory*, J. Eur. Math. Soc. (JEMS) **9** (2007), no. 1, 85–121.
8. J. Cheeger, *Differentiability of Lipschitz functions on metric measure spaces*, Geom. Funct. Anal. **9** (1999), no. 3, 428–517.
9. J. Cheeger and T. H. Colding, *On the structure of spaces with Ricci curvature bounded below. I*, J. Differential Geom. **46** (1997), no. 3, 406–480.
10. S. Daneri and G. Savaré, *Eulerian calculus for the displacement convexity in the Wasserstein distance*, SIAM J. Math. Anal. **40** (2008), no. 3, 1104–1122.
11. M. Erbar, *The heat equation on manifolds as a gradient flow in the Wasserstein space*, Ann. Inst. Henri Poincaré Probab. Stat. **46** (2010), no. 1, 1–23.
12. S. Fang, J. Shao, and K.-T. Sturm, *Wasserstein space over the Wiener space*, Probab. Theory Related Fields **146** (2010), no. 3-4, 535–565.
13. N. Gigli, *On the heat flow on metric measure spaces: existence, uniqueness and stability*, Calc. Var. Partial Differential Equations **39** (2010), no. 1-2, 101–120.
14. N. Gigli, K. Kuwada, and S.-i. Ohta, *Heat flow on Alexandrov spaces*, Comm. Pure Appl. Math., to appear, available at `arXiv:1008.1319`.
15. J. Heinonen, *Nonsmooth calculus*, Bull. Amer. Math. Soc. (N.S.) **44** (2007), no. 2, 163–232.
16. R. Jordan, D. Kinderlehrer, and F. Otto, *The variational formulation of the Fokker–Planck equation*, SIAM J. Math. Anal. **29** (1998), no. 1, 1–17.
17. B. Levi, *Sul principio di Dirichlet*, Rend. Circ. Mat. Palermo **22** (1906), no. 1, 293–359.
18. S. Lisini, *Characterization of absolutely continuous curves in Wasserstein spaces*, Calc. Var. Partial Differential Equations **28** (2007), no. 1, 85–120.
19. J. Lott and C. Villani, *Ricci curvature for metric-measure spaces via optimal transport*, Ann. of Math. (2) **169** (2009), no. 3, 903–991.
20. S.-i. Ohta and K.-T. Sturm, *Heat flow on Finsler manifolds*, Comm. Pure Appl. Math. **62** (2009), no. 10, 1386–1433.
21. ———, *Non-contraction of heat flow on Minkowski spaces*, Arch. Ration. Mech. Anal., to appear, available at `arXiv:1009.2312`.
22. F. Otto, *The geometry of dissipative evolution equations: the porous medium equation*, Comm. Partial Differential Equations **26** (2001), no. 1-2, 101–174.
23. N. Shanmugalingam, *Newtonian spaces: an extension of Sobolev spaces to metric measure spaces*, Rev. Mat. Iberoamericana **16** (2000), no. 2, 243–279.
24. K.-T. Sturm, *On the geometry of metric measure spaces. I*, Acta Math. **196** (2006), no. 1, 65–131.
25. ———, *On the geometry of metric measure spaces. II*, Acta Math. **196** (2006), no. 1, 133–177.
26. C. Villani, *Optimal transport: Old and new*, Grundlehren Math. Wiss., vol. 338, Springer, Berlin, 2009.

SCUOLA NORMALE SUPERIORE, PIAZZA CAVALIERI 7, 56126 PISA, ITALY
E-mail address: `l.ambrosio@sns.it`

Centre de Recherches Mathématiques
CRM Proceedings and Lecture Notes
Volume **56**, 2013

Analysis on the Sierpinski Carpet

M. T. Barlow

ABSTRACT. The 'analysis on fractals' and 'analysis on metric spaces' communities have tended to work independently. Metric spaces such as the Sierpinski carpet fail to satisfy some of the properties which are generally assumed for metric spaces. This survey discusses analysis on the Sierpinski carpet, with particular emphasis on the properties of the heat kernel.

1. Background and history

Percolation was introduced by Broadbent and Hammersley [23] in 1957. The easiest version to describe is *bond percolation*. Take any graph $G = (V, E)$ and a probability $p \in (0, 1)$. For each edge e keep it with probability p and delete it with probability $1 - p$, independently of all other edges. This is one of the core models of statistical physics, and has many applications in other areas — one is contact networks for infectious diseases. For books on percolation see [20, 40].

Write $G_p = (V, E_p)$ for the (random) graph obtained by the percolation process. The connected component of x is called the *cluster* containing x and is denoted $\mathcal{C}(x)$. Percolation on $\mathbb{Z}^d$ with $d \geq 2$ has a *phase transition.*

Set

$$\theta(p) = \mathbb{P}_p(|\mathcal{C}(0)| = \infty), \quad p_c = p_c(d) = \inf\{p : \theta(p) > 0\}.$$

Theorem 1.1 ([23]). *For the lattice* $\mathbb{Z}^d$, $p_c \in (0, 1)$.

When p is small G_p consists of a number of (mainly small) finite clusters. For large p the cluster $\mathcal{C}(0)$ may or may not be infinite, but a zero-one law combined with the stationary ergodic nature of the percolation process gives that infinite clusters exist with probability 1. A less elementary argument gives that if $p > p_c$ then there is exactly one infinite cluster. (As always in such contexts, this statement has to be qualified by 'with probability 1.')

In spite of its 50 year history, many open problems remain. The most fundamental of these is what happens at p_c in low dimensions:

Open Problem 1. Is $\theta(p_c) = 0$ if $3 \leq d \leq 18$?

It is known that $\theta(p_c) = 0$ if $d = 2$ or $d \geq 19$ — see [43, 55].

Physicists are interested in 'transport' problems of percolation clusters, that is, in the behavior of solutions to the wave or heat equation. A percolation cluster is

---

2010 *Mathematics Subject Classification.* Primary 60J60; Secondary 58G65.

This is the final form of the paper.

a graph, so one can define the graph (discrete) Laplacian

$$\Delta_{G_p} f(x) = \frac{1}{N_p(x)} \sum_{y \sim_p x} \big(f(y) - f(x)\big).$$

Here $y \sim_p x$ means that $\{x, y\}$ is an edge in G_p, and $N_p(x)$ is the number of neighbors of x. One can then look at the heat equation on G_p (discrete space, continuous time):

$$\frac{\partial u}{\partial t} = \Delta_{G_p} u.$$

(The wave equation would also be of great interest, but I do not know of any work on this for percolation clusters.) For the heat equation the broad situation for the three phases of percolation is as follows:

$\boldsymbol{p < p_c}$ **(subcritical):** No large scale structure.
$\boldsymbol{p > p_c}$ **(supercritical):** For HE get Gaussian limits, homogenization.
$\boldsymbol{p = p_c}$ **(critical):** Hard and interesting, only a few results.

The supercritical case is now well understood — see [4, 19, 70, 79].

In the critical case (or as $p \to p_c$) the main open problem is the existence and values of critical exponents. For example it is conjectured that there exist exponents γ, β such that

$$\mathbb{E}_p|\mathcal{C}(0)| \approx (p_c - p)^{-\gamma} \quad \text{as } p \uparrow p_c,$$
$$\theta(p) \approx (p - p_c)^{\beta} \quad \text{as } p \downarrow p_c.$$

The existence of these exponents has not been established, with two exceptions. The first is in high dimensions, where they take the 'mean field' values — see [42]. The other is for one particular percolation model in $d = 2$ (site percolation on the triangular lattice) where connections with SLE give some of these exponents — see [80].

Physicists believe that these exponents are 'universal.' This means that is it conjectured that they should depend on the dimension d, but not on the particular lattice or specific local features of the model. At first sight this is a surprise to a mathematician, since coarser features of the model, such as the value of p_c, do depend on the lattice. 'Universality' is crucial for the physical interest of percolation and other models in statistical physics, since the main point of these models (for a physicist) is their capacity to account for the behavior of real systems.

In 1976 P. de Gennes in a survey [26] suggested the study of random walks on critical percolation clusters as a tool to study their electrical resistance properties. It was believed then (and has now been proved in some cases — see [84]) that critical percolation clusters are in some senses 'fractal.' To help understand how random walks would behave on such 'fractal' graphs, in the early 1980s mathematical physicists began to study random walks on regular exact fractals, such as the Sierpinski gasket.

While this subject was called by mathematical physicists 'diffusion on fractals,' really they looked at random walks on various fractal graphs, mainly those with considerable regularity. In the late 1980s mathematicians began to study analysis on true fractals, starting with the easiest case, the Sierpinski gasket — see [17, 32, 57, 63].

In the period 1990 – 2005 mathematical work diverged from physics. Mathematicians obtained quite detailed information about solutions of the heat equation

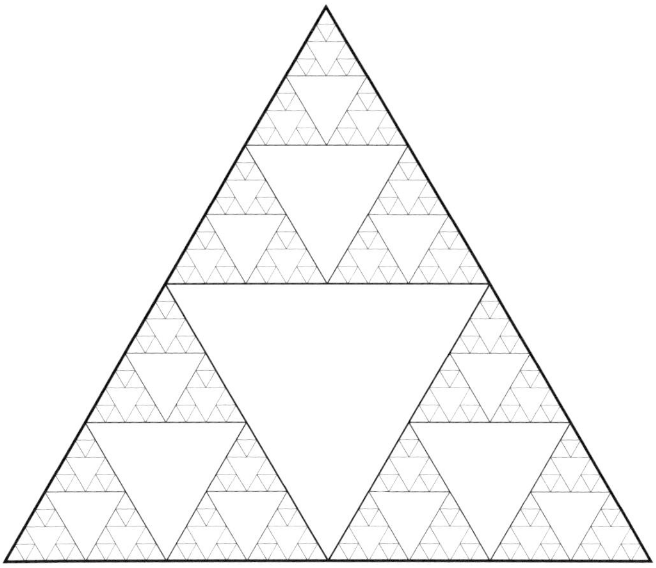

FIGURE 1. The graphical Sierpinski gasket

on regular exact fractals such as the Sierpinski gasket and Sierpinski carpet. This work gave quite accurate results, but needed very strong regularity for the space. Of more interest for physics applications such as percolation would have been cruder results obtained under weaker hypotheses.

From about 2005 the mathematical theory developed tools which can handle some problems for random walks on critical percolation clusters — see for example [16,60]. (Note also the early paper [56], which rather remarkably predates work on the ultimately much easier case of supercritical percolation.)

In these notes, I will discuss analysis on true (regular) fractals, which of course are also metric spaces. For other surveys or books on this topic see [3,58,81].

The Sierpinski gasket (SG) shown above is rather special because of existence of 'cut points.' This feature of the SG enables one to separate exactly the different levels of the set, and so allows various exact calculations to be performed. For example it is not hard to show that the mean time for a random walk on the Sierpinski gasket graph to cross a triangle of side 2^n is exactly 5^n. The resulting space-time scaling index, denoted by physicists $d_w = d_w(SG)$ is then $\log 5/\log 2$, and this number arises in the subsequent analysis of the heat equation on the fractal SG.

While useful, the existence of these cuts points may in some sense have proved to be a snare, since it allowed questions to be solved by complicated calculations, rather than by seeking a deeper insight into the underlying issues. The Sierpinski carpet does not have such cut points, and so is rather harder to study, but has proved its value by forcing the development of more robust tools.

2. Sierpinski carpet

The basic Sierpinski carpet (SC) is a fractal subset of $[0,1]^2$ defined in a similar way to the classical Cantor set, except that one removes the middle square out of a

FIGURE 2. The Sierpinski gasket and Sierpinski carpet

3×3 block. It can be regarded as a model space for studying diffusion in irregular media, with irregularities at many different length scales.

One can define the basic SC in d dimensions, by starting with $[0,1]^d$, dividing each cube into 3^d subcubes and removing the middle cube. One can also define *Generalized Sierpinski carpets* in $d \ge 2$ by removing other (symmetric) patterns of cubes. One example is the Menger Sponge — good pictures of this are easily found on the internet. For simplicity, in these notes I will mainly focus on the basic SC in d dimensions.

It is useful to have a more systematic description of these sets. A map $\psi\colon \mathbb{R}^d \to \mathbb{R}^d$ is a *similitude* if there exists $L > 1$ such that $|\psi(x) - \psi(y)| = L^{-1}|x-y|$ for all $x, y \in \mathbb{R}^d$. We call L the *contraction factor* of ψ. Let $M \ge 1$, and let $\psi_1, \dots, \psi_M$ be similitudes with contraction factors L_i. For $A \subset \mathbb{R}^d$ set

$$\Psi(A) = \bigcup_{i=1}^{M} \psi_i(A),$$

and let $\Psi^{(n)}$ denote the n-fold composition of Ψ.

Let $\mathcal{K}$ be the set of nonempty compact subsets of $\mathbb{R}^d$. For $A \subset \mathbb{R}^d$ set $A^{(\varepsilon)} = \{x : |x-a| \le \varepsilon \text{ for some } a \in A\}$. The *Hausdorff metric* d on $\mathcal{K}$ is defined by

$$d_{\mathcal{K}}(A,B) = \inf\{\varepsilon > 0 : A \subset B^{(\varepsilon)} \text{ and } B \subset A^{(\varepsilon)}\};$$

$d_{\mathcal{K}}$ is a metric on $\mathcal{K}$, and $(\mathcal{K}, d_{\mathcal{K}})$ is complete.

Theorem 2.1 (See [29, 51]). *Let $(\psi_1, \dots, \psi_M)$ be as above, with $L_i > 1$ for each $1 \le i \le M$. Then there exists a unique $F \in \mathcal{K}$ such that $F = \Psi(F)$. Further, if $G \in \mathcal{K}$ then $\Psi^n(G) \to F$ in $d_{\mathcal{K}}$. If $G \in \mathcal{K}$ satisfies $\Psi(G) \subset G$ then $F = \bigcap_{n=0}^{\infty} \Psi^{(n)}(G)$.*

To obtain the Hausdorff dimension $d_f(F)$ of the limiting set F from the contraction factors L_i a "nonoverlap" condition is necessary. We say that $(\psi_1, \dots, \psi_M)$ satisfies the *open set condition* if there exists an open set U such that $\psi_i(U)$, $1 \le i \le M$, are disjoint, and $\Psi(U) \subset U$. Note that, since $\Psi(\overline{U}) \subset \overline{U}$, then the fixed point F of Ψ is given by $F = \bigcap \Psi^{(n)}(\overline{U})$.

Theorem 2.2 (See [29, p. 119]). *Let $(\psi_1, \dots, \psi_M)$ satisfy the open set condition, and let F be the fixed point of Ψ. Let α be the unique real such that*

$$\sum_{i=1}^{M} L_i^{-\alpha} = 1.$$

Then the Hausdorff dimension of F is given by $d_f(F) = \alpha$. Further the x^α-Hausdorff measure of F is strictly positive and finite.

In these notes I will only be interested in the case when L_i all take the same value L. Then it is easy to see that

$$\alpha = d_f(F) = \frac{\log M}{\log L}. \tag{2.1}$$

We call L the *length scale factor*, and M the *mass scale factor* of F.

Let $L = 3$, $M = 3^d - 1$, and let ψ_i, $i = 1, \dots M$ be the (rotation free) similitudes which map $F_0 = [0,1]^d$ onto each of the M sub-cubes of side L^{-1}. We assume that $\psi_1(x) = L^{-1}x$ is the similitude which maps $[0,1]^d$ onto $[0, L^{-1}]^d$. The open set condition clearly holds – one can take $U = (0,1)^d$. If $F_n = \Psi^n(F_0)$ then F_n is the nth approximation to the SC, which is given by

$$F = \bigcap_{n=0}^{\infty} F_n.$$

Note that F_n is the union of M^n cubes each of side L^{-n}. Let μ_n be Lebesgue measure on F_n renormalized to have mass 1 and μ be Hausdorff $x^{d_f(F)}$-measure on F, normalized so that $\mu(F) = 1$. We have that $\mu = \lim_n \mu_n$.

We will wish to consider two other sets, based on the same construction, but which give unbounded subsets of $\mathbb{R}^d$. Let

$$\widetilde{F}_n = L^n F_n = \{L^n x, x \in F_n\};$$

thus $\widetilde{F}_n$ is a subset of $[0, L^n]^d \subset \mathbb{R}^d_+$ and is the union of M^n unit cubes. We have

$$\widetilde{F}_n \cap [0, L^{n-1}]^d = \widetilde{F}_{n-1}.$$

So set

$$\widetilde{F} = \bigcup_{n=0}^{\infty} \widetilde{F}_n.$$

This is called the *precarpet* — see [77]. It is the closure of an open domain in $\mathbb{R}^d$ with piecewise smooth boundary, so is locally regular. It has fractal structure 'at infinity,' which mimics the local structure of the compact Sierpinski carpet F. Define also:

$$\widehat{F} = \bigcap_{n=0}^{\infty} L^{-n} \widetilde{F}.$$

This is the *unbounded fractal Sierpinski carpet*; it is the union of a countable number of copies of F. Note that $\psi_1(\widehat{F}) = \widehat{F}$.

We next look at the metric space structure of F. Using the fact that the boundary of a cube is never removed, it is not hard to prove the following

Lemma 2.3. *Let $x, y \in F$. Then there exists a path γ connecting x, y consisting of countably many line segments, with length*

$$L(\gamma) \le C|x-y|.$$

Thus if $d(x,y)$ is the shortest path (geodesic) distance on F then

$$|x-y| \le d(x,y) \le C|x-y|.$$

In these notes I will always use the geodesic or shortest path metric d when working in subsets of $\mathbb{R}^d$, and will write $B(x,r)$ for balls in this metric.

Lemma 2.4. *(F, d, μ) is a metric measure space, and if $\alpha = d_f(F)$ one has*

$$\mu\big(B_d(x,r)\big) \asymp r^\alpha, \quad x \in F, 0 < r < 1. \tag{2.2}$$

Thus the Sierpinski carpet is *Ahlfors regular*—see [47].

For analysis on the Sierpinski carpet F, a natural starting point is to consider upper gradients.

Definition 2.5. Let $f \in C(F)$. A function $g\colon F \to \mathbb{R}$ is an *upper gradient for* f if for any $x, y \in F$ and (rectifiable) path $(\gamma(t), t \in [0,1])$ connecting x and y then

$$|f(y) - f(x)| \le \int_\gamma g = \int_0^1 g\big(\gamma(t)\big)|\gamma'(t)|\,\mathrm{d}t.$$

We write $|\nabla f|$ for an upper gradient of f.

In many situations a (weak) (q,p)-Poincaré inequality of the type

$$\inf_a \int_{B(x,r)} |f(x) - a|^q \,\mu(\mathrm{d}x) \le C(r) \int_{B(x,\lambda r)} |\nabla f|^p \,\mu(\mathrm{d}x), \tag{2.3}$$

where $|\nabla f|$ is any upper gradient of f, gives useful information about the space—see for example [46, 47].

Theorem 2.6. *Let H_i, $i = 0, 1$ be the left and right edges of the SC in $d = 2$: $H_i = \{x = (x_1, x_2) \in F : x_1 = i\}$. Let*

$$\mathcal{A} = \{f \in C(F) : f|_{H_i} = i, i = 0, 1\}.$$

Let $p > 0$. Then there exists a sequence of functions $f_n \in \mathcal{A}$ and upper gradients $|\nabla f_n|$ such that

$$\int_F |\nabla f_n|^p \,\mu(\mathrm{d}x) \to 0.$$

Consequently no weak Poincaré inequality holds on the Sierpinski carpet.

PROOF. Recall that $F_0 = [0,1]^2$ and that F_n is the nth step in the approximation of the SC. Let

$$f_0(x_1, x_2) = x_1; \quad \text{so} \quad \int_F |\nabla f_0|^p \,\mathrm{d}\mu = 1.$$

Choose $h_1\colon [0,1] \to [0,1]$ to be continuous and linear on each interval $\big[0, \frac{1}{3}\big]$, $\big[\frac{1}{3}, \frac{2}{3}\big]$, $\big[\frac{2}{3}, 1\big]$, with $h(0) = 0$, $h(1) = 1$, and $h_1' = b$ on $\big(0, \frac{1}{3}\big) \cup \big(\frac{2}{3}, 1\big)$, and $h_1' = 3 - 2b$ on $\big(\frac{1}{3}, \frac{2}{3}\big)$. Set

$$f_1(x_1, x_2) = h_1(x_1); \quad \text{note that } f \in \mathcal{A}.$$

Then $|\nabla f_1|$ is equal to b on 6 out of the 8 sub-squares, and $(3-2b)$ on 2 sub-squares. Hence

$$\int_F |\nabla f_1|^p \,\mu(\mathrm{d}x) = \tfrac{6}{8}b^p + \tfrac{2}{8}(3-2b)^p = \tfrac{1}{4}(3b^p + (3-2b)^p) = \varphi(b).$$

Then $\varphi(1) = 1$ and it is easy to check that $\varphi'(1) = p/4 > 0$. We could minimize over b, but it is enough to note that there exists $b \in (0,1)$ with $\varphi(b) < 1$.

With this choice of b we have

$$\int_F |\nabla f_1|^p \,\mathrm{d}\mu = \varphi(b) < 1.$$

We now iterate this construction, to obtain a sequence f_k such that on each square S of side 3^{-k} we then have

$$\int_S |\nabla f_k|^p \,\mathrm{d}\mu = \varphi(b)^k \to 0 \quad \text{as } k \to \infty. \tag{2.4}$$

We define $f_k(x_1, x_2) = h_k(x_1)$ where h_k are defined by

$$h_{k+1}(t) = \begin{cases} (b/3)h_k(3t), & 0 \le t \le \frac{1}{3} \\ (b/3) + (1-2b/3)h_k(3t-1), & \frac{1}{3} \le t \le \frac{2}{3} \\ (1-b/3) + (b/3)h_k(3t-2), & \frac{1}{3} \le t \le \frac{2}{3}. \end{cases}$$

On each square S of side 3^{-k} we then have

$$\int_S |\nabla f_{k+1}|^p d\mu = \varphi(b) \int_S |\nabla f_k|^p \,\mathrm{d}\mu.$$

Hence

$$\int_F |\nabla f_k|^p \,\mathrm{d}\mu = \varphi(b)^k \to 0 \quad \text{as } k \to \infty.$$

The functions f_k constructed above are piecewise linear; if we wished we could then smooth them to obtain a sequence $\tilde{f}_k \in C^1(F) \cap \mathcal{A}$ with the same properties.

To see that no weak Poincaré inequality can hold, we can use the functions f_k to construct functions g_k such that $g_k(x,y) = 1$ if $x \in [0, \frac{1}{3}]$, $g_k(x,y) = -1$ if $x \in [\frac{2}{3}, 1]$, $|g_k| \le 1$ on F, and

$$\int_F |\nabla g_k|^p \,\mathrm{d}\mu \to 0.$$

Since for each k, $a \in \mathbb{R}$,

$$\int_F |g_k(x) - a|^q \,\mu(\mathrm{d}x) \ge \frac{3}{8}(|1-a|^q + |1+a|^q) \ge \frac{3}{8},$$

the weak Poincaré inequality fails. □

We can define a quadratic form Q by

$$Q(f,f) = \int_F |\nabla f|^2 \,\mathrm{d}\mu,$$

and (if one had not seen the result above) one might hope that this would yield a suitable Dirichlet form on some subspace of $C(F)$.

Corollary 2.7. *Q is not closeable; there exists $f_n \to 0$ uniformly such that $Q(f_n - f_m, f_n - f_m) \to 0$ but $Q(f_n, f_n) \ge 1$ for all n.*

This result states that the pre-Hilbert space $H = C^1(F)$ with inner product $Q(f,f)+\|f\|_2^2$ is not closeable in $L^2(F)$. The functions (f_n) given above are Cauchy in H, and since they converge to 0 in L^2, if they had a limit in H it would have to be 0. However, f_n do not converge to 0 in H.

PROOF. Let f be continuous on F. Using Theorem 2.6 we can find for each n a function h_n such that

$$\|f-h_n\|_\infty \le \tfrac{1}{2}\|f|_\infty, \quad Q(h_n,h_n) \le 2^{-n}.$$

(So continuous functions can be closely approximated by functions of low energy.)

Let f_0 satisfy $Q(f_0,f_0) = 2$, $\|f_0\|_\infty = C < \infty$. Choose (g_n) such that $Q(g_n,g_n) \le \delta^2 4^{-n}$ and if $f_{n+1} = f_n - g_n$ then $\|f_{n+1}\|_\infty \le \frac{1}{2}\|f_n\|_\infty$. Then $f_n \to 0$ and

$$\begin{aligned} Q(f_{n+1},f_{n+1}) &= Q(f_n,g_n) - 2Q(f_n,g_n) + Q(g_n,g_n) \\ &\ge Q(f_n,f_n) - 2\delta 2^{-n} Q(f_n,f_n)^{1/2}. \end{aligned}$$

Choosing δ small enough we have $1 \le Q(f_n,f_n) \le 3$ for all n. □

This calculation suggests that the Dirichlet forms

$$Q_n(f,f) = \int_{F_n} |\nabla f|^2 \, d\mu_n$$

are 'too small.' If we are to obtain a useful limit some renormalization is necessary; so we seek constants $a_n \uparrow \infty$ such that if

$$\mathcal{E}_n(f,f) = a_n Q_n(f,f),$$

then $\mathcal{E}_n$ has (at least) subsequential limits. The obvious choice (which works) is to choose a_n so that

$$\inf\{\mathcal{E}_n(f,f) : f \in \mathcal{A}\} = O(1).$$

The renormalization argument involves various constants, and these turn out to have a more intuitive form if we work with the big sets $\widetilde{F}_n = L^n F_n$ rather than the small sets F_n. Let H_0^n, H_1^n be the left and right sides of $\widetilde{F}_n$:

$$H_i^n = \{x = (x_1, \dots x_d) \in \widetilde{F}_n : x_1 = L^n i\}, \quad i = 0,1,$$

and

$$\tilde{\mathcal{A}}_n = \{f \in C(\widetilde{F}_n) : f|_{H_i^n} = i, i = 0,1\}$$

be the set of functions which are zero on the LHS and 1 on the RHS. Let

$$R_n^{-1} = \inf\Big\{ \int_{\widetilde{F}_n} |\nabla f|^2 \, dx : f \in \tilde{\mathcal{A}}_n \Big\}. \tag{2.5}$$

This is the minimal energy of functions in $\tilde{\mathcal{A}}_n$; and so R_n can be interpreted as the 'effective resistance' in $\widetilde{F}_n$ between H_0^n and H_1^n. So it is natural to try

$$\mathcal{E}_n(f,f) = L^{(d-2)n} R_n \int_{F_n} |\nabla f|^2 \, dx.$$

(The term $L^{n(d-2)}$ arises from rescaling the integral of the gradient.) We then have that $\mathcal{E}_n(f,f) \ge 1$ for all $f \in C(F_n)$ with $f = 0$ on the left side of F_n and $f = 1$ on the right side.

We now wish to use the $(\mathcal{E}_n)$ to construct a limiting Dirichlet form $\mathcal{E}$ on F or $\widehat{F}$. Unfortunately, the argument does not take what I would now regard as conceptually the clearest course. It would be nice to be able to proceed as follows:

(1) Using some (as minimal as possible) properties of F, prove that the sequence $\mathcal{E}_n$ has subsequential limits, in the sense of Mosco convergence.
(2) Using some stronger regularity results on F, obtain 'good properties' of any subsequential limit $\mathcal{E}$.
(3) Prove the uniqueness of the limit, and that $\mathcal{E}_n \to \mathcal{E}$.

It is quite reasonable to expect that, at least for (1) and (2), such a program would be possible. See Remark 2.13 below for comments on the situation as far as (3) is concerned. The first construction of limiting processes on Sierpinski carpets was done in [7], and used probabilistic methods (tightness and weak convergence) rather than Dirichlet form methods. An alternative approach using Dirichlet forms can be found in [41, 65]. However, all these papers use quite strong regularity properties of the approximating forms $\mathcal{E}_n$ in order to show the existence of subsequential limits. Thus these arguments combine the 'existence' and 'regularity' parts of the proof in a fashion which makes it unclear exactly which properties of F are needed at each step of the argument.

Since (so far) the ideal argument suggested in (1) and (2) above does not exist, I will sketch what has been done. The proof has two main inputs:

(1) Control of the constants R_n involved in the renormalization.
(2) A Harnack inequality which gives enough regularity of $\mathcal{E}_n$ to give good properties of the limit.

Theorem 2.8 ([8, 71]). *There exists C such that*

$$C^{-1} R_n R_m \le R_{n+m} \le C R_n R_m, \quad n, m \ge 0.$$

Hence there exists $\rho = \lim_n R_n^{1/n}$ such that

$$C^{-1}\rho^n \le R_n \le C\rho^n, \quad n \ge 0. \tag{2.6}$$

Proof. R_n^{-1} is given as the energy of a function h_n in $\tilde{\mathcal{A}}_n$. Using copying and pasting of the functions h_n and h_m one can construct a function $f^{n,m} \in \tilde{\mathcal{A}}_{n+m}$ with energy less than $cR_n^{-1}R_m^{-1}$. (See the papers cited above for the full argument, which requires a comparison between R_n and discrete network approximations.) Thus

$$R_{n+m}^{-1} \le \mathcal{E}_{n+m}(f^{n,m}, f^{n,m}) \le cR_n^{-1}R_m^{-1},$$

which gives the first inequality. A dual characterization of resistance as the minimal energy of a unit flow (see [27] for the classical case of electrical networks) gives a similar upper bound on R_{n+m}.

A sequence x_n is subadditive if $x_{n+m} \le x_n + x_m$, and Fekete's theorem states that for a subadditive sequence,

$$\lim_n \frac{x_n}{n} = \inf_n \frac{x_n}{n} \in [-\infty, x_1].$$

Since $\log(CR_n)$ is subadditive, and $\log(C^{-1}R_n)$ is superadditive, (2.6) follows. □

The subadditivity used to prove the existence of the limit ρ gives no information on its value. For the basic SC with $d = 2$ shorting and cutting arguments give $\frac{7}{6} \le \rho \le \frac{3}{2}$, and numerical calculations suggest that $\rho \simeq 1.251$. For the shorting argument, recall that 'shorts decrease resistance' (see [27]), and starting with the set $\widetilde{F}_n$ impose shorts on each of the lines $x_1 = \frac{1}{3}L^n$, $x_1 = \frac{2}{3}L^n$. (In terms of the definition (2.5) this corresponds to restricting the minimization to functions which are constant on theses lines.) This replaces the resistance problem for the set $\widetilde{F}_n$ with that for 8 copies of $\widetilde{F}_{n-1}$, and one obtains

$$R_n \geq \tfrac{1}{3}R_{n-1} + \tfrac{1}{2}R_{n-1} + \tfrac{1}{3}R_{n-1} = \tfrac{7}{6}R_{n-1},$$

which implies that $\rho \geq 7/6$. A similar argument with cuts gives the upper bound.

In general for $d \geq 2$ one has

$$\frac{2}{3^{d-1}} + \frac{1}{3^{d-1}-1} \leq \rho \leq \frac{3}{3^{d-1}-1}, \tag{2.7}$$

— see [10, Remark 5.4].

The second input is an elliptic Harnack inequality (EHI) for the precarpet $\widetilde{F}$. Note that as the precarpet is the closure of a domain in $\mathbb{R}^d$, one can define the Laplacian and harmonic functions on $\widetilde{F}$ in the usual way. We can also define a reflecting Brownian motion W on $\widetilde{F}$. This is the continuous strong Markov process which behaves locally like a Brownian motion in $\widetilde{F}^o$, and has normal reflection on $\partial\widetilde{F}$ — see [18, 30].

Recall that we are using $B(x,r)$ to denote balls in the pre-SC with respect to the geodesic metric – if we used the Euclidean metric we might have to be careful about connectivity.

Theorem 2.9 ([9]). *Let $x \in \widetilde{F}$, $r > 0$, $B = B(x,r)$ and h be nonnegative on $\overline{B}$ and harmonic on B. There exists a constant C_H (depending only on the Sierpinski carpet) such that if $B' = B(x,r/2)$ then*

$$\sup_{B'} h \leq C_H \inf_{B'} h.$$

The proof uses a probabilistic coupling argument which relies on the symmetry of F very heavily. I will need some notation for hitting and exit times for a stochastic process X.

Definition 2.10. Let $X = (X_t, t \in \mathbb{R}_+)$ be a stochastic process on a metric space $\mathcal{X}$. For $A \subset \mathcal{X}$ set

$$T_A = T_A(X) = \inf\{t \geq 0 : X_t \in A\},$$
$$\tau_{x,r} = \tau_{x,r}(X) = T_{B(x,r)^c}(X).$$

Given two processes X, Y define their collision time by

$$T_C = T_C(X,Y) = \inf\{t > 0 : Y_t = X_t\},$$

with the usual convention that $\inf \varnothing = +\infty$.

The coupling gives that there exists a constant $p_F > 0$, depending on the carpet F, such that the following holds. Given $x_0 \in \widetilde{F}$, $R > 0$, and points $x, y \in B(x_0, R/2dL)$, there exist reflecting Brownian motions W^x, W^y on $\widetilde{F}$ with $W_0^x = x$, $W_0^y = y$ such that $W_t^x = W_t^y$ for $t \geq T_C$ and

$$\mathbb{P}\big(T_C(W^x, W^y) \leq \tau_{x_0,R}(W^x) \wedge \tau_{x_0,R}(W^y)\big) > p_F. \tag{2.8}$$

The processes W^x and W^y are of course not independent — the coupling uses reflection methods as in [69].

A simpler argument, also using the (local) reflection symmetry of $\widetilde{F}$, gives a weak lower bound on the probability of hitting small balls: if $y \in B(x_0, R/2dL)$, $\lambda \in (0, 1/10dL)$, then for the process $W = W^{x_0}$,

$$\mathbb{P}^x(T_{B(y,\lambda R)} < \tau_{x_0,R}) > p_0\lambda^\gamma. \tag{2.9}$$

Here p_0 and $\gamma > 0$ depend only on F.

If h is harmonic on $B = B(x_0, R)$ then $h(W^x)$ is a martingale. Hence if $T = \tau_B(W^x) \wedge \tau_B(W^y)$,

$$\begin{aligned} h(x) - h(y) &= \mathbb{E}\big(h(W_T^x) - h(W_T^y)\big) \\ &= \mathbb{E}(h(W_T^x) - h(W_T^y); T < T_C) \\ &\leq \mathbb{P}(T < T_C) \sup_{x,y \in B} \big(h(x) - h(y)\big) \\ &\leq (1 - p_F) \sup_{x,y \in B} \big(h(x) - h(y)\big). \end{aligned}$$

Hence, writing

$$\mathrm{Osc}_A f = \sup_A f - \inf_A f,$$

and $B' = B(x_0, R/2dL)$, we have

$$\mathrm{Osc}_{B'} h \leq (1 - p_F) \, \mathrm{Osc}_B h. \tag{2.10}$$

This oscillation inequality is not quite enough on its own to give the Harnack inequality — see the example in [5]. However, combined with (2.9) a standard argument (see for example [28]) gives the elliptic Harnack inequality Theorem 2.9.

We have three 'scale factors' for the SC:

(1) $L = L_F$, the length scaling factor.
(2) $M = M_F$, the volume scaling factor.
(3) ρ, the 'resistance scaling factor.'

As stated above, for the basic SC in d dimensions, $L = 3$, and $M = 3^d - 1$. For $[0,1]^d$ (which can be regarded as a trivial SC) if $L = 3$ then $M = 3^d$ and $\rho = 3^{2-d}$. Recall the definition of d_f from (2.1), and set

$$d_w(F) = d_w = \frac{\log M\rho}{\log L}.$$

d_w was called by physicists the *walk dimension* and is related to space time scaling of the heat equation. It turns out that one always has $d_w \geq 2$: for the SC in d dimensions this follows from the lower bound in (2.7).

Given a regular fractal F, since L and M are given by the construction, one can calculate d_f easily. The constant ρ which gives d_w is somehow deeper, and seems to require some analysis on the set or its approximations. Loosely one can say that d_f is a 'geometric' constant, while d_w is an 'analytic' constant. One may guess that in some sense ρ or β are in general inaccessible by any purely geometric argument. (An exception is for trees, where one has $d_w = 1 + d_f$.)

The two inputs (Theorem 2.8 and Theorem 2.9) lead to good control of the heat equation in $\widetilde{F}$.

Theorem 2.11 ([10]). *Let $p_t(x,y)$ be the heat kernel on the pre-SC $\widetilde{F}$. Then writing $\beta = d_w$*

$$p_t(x,y) \overset{(c)}{\asymp} c\mu\big(B(x, t^{1/\beta})\big)^{-1} \exp\big(-c(d(x,y)^\beta/t)^{1/(\beta-1)}\big), \tag{2.11}$$

for (t,x,y) such that $t \geq 1 \vee d(x,y)$.

Remark 2.12. (1) Here $\overset{(c)}{\asymp}$ means that upper bound holds with constants c_1, c_2 and lower bounds hold with constants c_3, c_4.
(2) One gets the usual Gaussian type bounds if $t \leq 1 \vee d(x,y)$.

(3) Since $\widetilde{F}$ looks locally like $\mathbb{R}^d$ at length scales less than 1, one expects different bounds when t, $d(x,y)$ are small.
(4) Integrating these bounds, if W_t is the associated diffusion process one gets

$$\mathbb{E}^x d(x,W_t)^2 \asymp t^{2/\beta}, \quad t \geq 1.$$

When $\beta \neq 2$ this is called 'anomalous diffusion.'

I will defer discussing the proof of this theorem until the next section, where it will follow from Theorem 2.8, Theorem 2.9, and a general result on heat kernel bounds in metric measures spaces, Theorem 3.14.

The theorem above shows that the space-time scaling involves the new parameter $\beta = d_w$. There are several different ways of seeing why this anomalous scaling arises.

One is to look at the Poincaré inequality in a cube in $\widetilde{F}$. Let S be a cube of side $R = L^n$ in $\widetilde{F}$; by symmetry we can assume that $S = [0, L^n]^d \cap \widetilde{F} = \widetilde{F}_n$. Let P_n be the best constant in the PI

$$\int_S (f - \bar{f})^2 \,\mathrm{d}x \leq P_n \int_S |\nabla f|^2 \,\mathrm{d}x.$$

Let g_n be the function in $\tilde{\mathcal{A}}_n$ which attains the minimum for the resistance across S. By symmetry the average value of g_n on $\widetilde{F}_n$ is $\frac{1}{2}$. Then we can use g_{n-1} to build a function f_n on $\widetilde{F}_n$ which satisfies

$$\begin{aligned} f_n(x_1,x') &= -1, & 0 \leq x_1 \leq \tfrac{1}{3}L^n, \\ &\in [-1,1], & \tfrac{1}{3}L^n \leq x_1 \leq \tfrac{2}{3}L^n, \\ &= 1, & \tfrac{2}{3}L^n \leq x_1 \leq 1. \end{aligned}$$

Here I have used the notation $x = (x_1, x')$ for points in $\mathbb{R}^d = \mathbb{R} \times \mathbb{R}^{d-1}$. Then

$$\int_S (f_n - \bar{f}_n)^2 \simeq |S| = M^n, \quad \int_S |\nabla f_n|^2 \leq c R_{n-1}^{-1},$$

and so

$$P_n \geq c M^n R_{n-1} \geq (M\rho)^n = c R^{\log(M\rho)/\log L} = c R^\beta.$$

The regularity given by EHI gives that in fact $P_n \asymp (M\rho)^n$ — see [65]. Since P_n gives the time scale over which heat converges to equilibrium, this suggests that heat (or the diffusion process X) takes time order R^β to move a distance R, rather than the classical time $O(R^2)$.

The heat kernel bounds on the pre-SC then enables one to prove that the rescaled Dirichlet forms $\mathcal{E}_n$ on F_n have subsequential limits. Taking any one of these limits, one obtains a Dirichlet form $(\mathcal{E}, \mathcal{F})$ on $L^2(F,\mu)$. Different subsequences might give different limits, but any limit satisfies symmetry conditions and has a heat kernel which satisfies (2.11).

Remark 2.13. A recent result [12] proves that any Dirichlet form which is symmetric with respect to suitable local symmetries of the Sierpinski carpet is unique, up to constants. This shows that if $\mathcal{E}'$ and $\mathcal{E}''$ are both subsequential limits of the $(\mathcal{E}_n)$ then $\mathcal{E}'' = \lambda \mathcal{E}'$ for some $\lambda \in (C^{-1}, C)$, where C is the constant in Theorem 2.8.

This does not seem enough to prove that $\lim_n \mathcal{E}_n$ exists. One difficulty is that we do not know if $\lim_n \rho^{-n} R_n$ exists.

3. Heat kernels on metric measure spaces

Examples of these include manifolds, $\mathbb{R}^d$, domains in $\mathbb{R}^d$, 'cable systems' for graphs, and true fractals such as the Sierpinski carpet described above.

Metric measure spaces. Let $(\mathcal{X}, d, \mu)$ be a metric measure space. We assume that $(\mathcal{X}, d)$ is complete and connected, that d is a length metric, μ is Radon, $\mathcal{X}$ has infinite radius, and the balls

$$B(x,r) = \{y : d(x,y) < r\}, x \in \mathcal{X}, r > 0, \text{ are precompact.}$$

Set $V(x,r) = \mu\big(B(x,r)\big)$.

Definition 3.1. $\mathcal{X}$ satisfies V_α if

$$c_1 r^\alpha \le V(x,r) \le c_2 r^\alpha, \quad x \in \mathcal{X}, r > 0.$$

$\mathcal{X}$ satisfies *volume doubling* VD if

$$V(x,2r) \le c_D V(x,r), \quad x \in \mathcal{X}, r > 0.$$

It is easy to show that V_α implies VD, and that VD implies polynomial volume growth: there exists $\alpha_1 < \infty$ such that for $x \in \mathcal{X}$, $0 < r < R$ one has

$$\frac{V(x,R)}{V(x,r)} \le C(R/r)^{\alpha_1}.$$

An example of a set which satisfies VD but not V_α for any α is the precarpet $\widetilde{F}$, where one has

$$V(x,r) \asymp r^d, \qquad r \in (0,1), \tag{3.1}$$

$$V(x,r) \asymp r^{d_f(F)}, \quad r \in (1,\infty). \tag{3.2}$$

Dirichlet forms. Let $(\mathcal{E}, \mathcal{F})$ be a regular strongly local Dirichlet form on $L^2(\mathcal{X}, \mu)$ — see [31]. This means:

(1) $\mathcal{E}(f,g)$ is a symmetric bilinear form defined on a subspace $\mathcal{F} \subset L^2(\mathcal{X})$.
(2) $\mathcal{E}$ is 'Markov': $\mathcal{E}(f_+ \wedge 1, f_+ \wedge 1) \le \mathcal{E}(f,f)$.
(3) $\mathcal{E}$ is closed: if $\|f\|_H = \mathcal{E}(f,f) + \|f\|_2^2$ then $\mathcal{F}$ is a complete Hilbert space in $\|\cdot\|_H$.
(4) $\mathcal{E}$ is regular: $\mathcal{F} \cap C_0(\mathcal{X})$ is dense in $\mathcal{F}$ w.r.t. $\|\cdot\|_H$ and dense in $C_0(\mathcal{X})$ w.r.t. $\|\cdot\|_\infty$.
(5) $\mathcal{E}$ is strongly local: if f_1, f_2 have compact support and f_1 is constant on an open set U_1 containing $\operatorname{supp}(f_2)$, then $\mathcal{E}(f_1, f_2) = 0$.

If D is a domain in $\mathcal{X}$ we define

$$\mathcal{F}_D = \{f \in \mathcal{F} : f = 0 \text{ on } D^c\}.$$

For $f \in \mathcal{F}$ there exists a measure $\Gamma(f,f)$ (called the 'energy measure') which gives $\mathcal{E}$ by integration:

$$\mathcal{E}(f,f) = \int_{\mathcal{X}} \mathrm{d}\Gamma(f,f).$$

For bounded f the measure $\Gamma(f,f)$ is defined by requiring that

$$\int g \,\mathrm{d}\Gamma(f,f) = 2\mathcal{E}(gf,f) - \mathcal{E}(g,f^2), \quad g \in C(\mathcal{X}) \cap \mathcal{F}.$$

The measure $\Gamma(f,g)$ is then defined by polarization. (See [31, p. 113], where the alternative notation $\mu^c_{\langle f \rangle}$ is used.) For a manifold $d\Gamma(f,g) = (\nabla f . \nabla g) d\mu$, but in general Γ need not be absolutely continuous with respect to μ. $d\Gamma$ satisfies Leibniz-type rules — e.g.,

$$\mathrm{d}\Gamma(fg,h) = f\,\mathrm{d}\Gamma(g,h) + g\,\mathrm{d}\Gamma(f,h). \tag{3.3}$$

(Strictly speaking we may have to take a 'quasi-continuous version' of the functions f, g, but in these notes I will ignore regularity issues of this kind.)

Definition 3.2. We call a metric measure space with such a Dirichlet form a MMD space.

Example 3.3. (1) Manifolds.

(2) Divergence form operators on a domain $D \subset \mathbb{R}^d$. Let $a = \big(a_{ij}(x)\big)$ be bounded and measurable, and set

$$\mathcal{E}(f,f) = \int \nabla f \cdot a \nabla f.$$

(3) The 'cable system' for a graph — see [85]. If $G = (V, E)$ is a graph, let $\mathcal{X}$ be the metric space obtained by replacing each edge $e \in E$ by a unit line segment I_e, joined at the vertices. Then

$$\mathcal{E}(f,f) = \sum_e \int_{I_e} (f')^2 \,\mathrm{d}x.$$

(These are now often, rather unfortunately, called 'quantum graphs').

(4) Sierpinski carpets, where $\mathcal{E}$ is a subsequential limit of the approximations $\mathcal{E}_n$.

Semigroup and heat kernel. On a MMD space we have a Laplacian type operator $\big(\mathcal{L}, \mathcal{D}(\mathcal{L})\big)$ which satisfies

$$\mathcal{E}(f,g) = \int (-\mathcal{L}f) g \,\mathrm{d}\mu \quad \text{for } f \in \mathcal{D}(\mathcal{L}),\ g \in \mathcal{F}.$$

Associated with $\mathcal{L}$ is a semigroup, formally given by $P_t = \exp(t\mathcal{L})$. This semigroup defines a Markov process $(X_t, \mathbb{P}^x)$ on $\mathcal{X}$, and

$$\mathbb{E}^x f(X_t) = \mathbb{E}(f(X_t) \mid X_0 = x) = P_t f(x), \quad x \in \mathcal{X}, t \geq 0.$$

If P_t has a density $p_t(x,y)$ with respect to μ then this is the *heat kernel* (or transition density of the process X):

$$\mathbb{P}^x(X_t \in A) = P_t 1_A(x) = \int_A p_t(x,y)\,\mu(\mathrm{d}y). \tag{3.4}$$

The heat kernel is symmetric:

$$p_t(x,y) = p_t(y,x), \quad \mu \times \mu \text{ a.e.}, \tag{3.5}$$

satisfies the Chapman – Kolmogorov equation

$$\int p_s(x,y) p_t(y,z)\,\mu(\mathrm{d}y) = p_{s+t}(x,z), \tag{3.6}$$

and the heat equation

$$\frac{\partial}{\partial t} p_t(x,y) = \mathcal{L} p_t(x,y).$$

The definition (3.4) only defines $p_t(x,y)$ for μ-a.a. y, but using the Chapman – Kolmogorov equation one can regularize $p_t(x,y)$ so that (3.4) – (3.6) hold except on a set of capacity zero — see [39, 86].

If $D \subset \mathcal{X}$ is open, we can also define the *killed semigroup*

$$P_t^D f(x) = \mathbb{E}^x(f(X_t); t < \tau_D);$$

we write $p_t^D(x,y)$ for the corresponding killed heat kernel.

Heat kernel bounds and Hölder continuity. This problem for divergence form operators in $\mathbb{R}^d$ was solved by De Giorgi, Moser and Nash in the late 1950s — see in particular [73, 76]. Moser's methods were extended to manifolds by Bombieri and Giusti in [21]. These papers were followed by work of Aronson [2], Li and Yau [68], Fabes and Stroock [28], Grigor′yan [33], Saloff-Coste [78], Sturm [83], and others. The following theorem summarizes much of this work.

Theorem 3.4 **([33, 78, 83]).** *Let* $(\mathcal{X}, \mathcal{E})$ *be a MMD space. The following are equivalent*:

(1) $p_t(x,y)$ *satisfies* (GB) — *that is Gaussian bounds*;
(2) *A parabolic Harnack inequality* (PHI) *holds on* $\mathcal{X}$;
(3) $\mathcal{X}$ *satisfies* VD *plus* PI (*Poincaré inequality*).

This theorem gives the equivalence between three possible conditions relating to the heat equation on the MMD space $(\mathcal{X}, \mathcal{E})$. Before I discuss further this result, and its usefulness, I need to define the various terms in the theorem.

Definition 3.5. (p_t) satisfies Gaussian bounds (GB) if

$$p_t(x,y) \overset{(c)}{\asymp} V(x, ct^{1/2})^{-1} \exp(-cd(x,y)^2/t).$$

Definition 3.6. $(\mathcal{X}, \mathcal{E})$ satisfies PI if for all balls $B = B(x,r)$, $f\colon B \to \mathbb{R}$,

$$\min_a \int_B (f-a)^2 \, \mathrm{d}\mu = \int_B (f - \bar{f})^2 \, \mathrm{d}\mu \le C_P \, r^2 \int_B \mathrm{d}\Gamma(f,f).$$

For the third definition, of a Harnack inequality, we need to define harmonic functions (solutions of $\mathcal{L}h = 0$) and *caloric* functions (solutions of $\partial_t u(x,t) = \mathcal{L}u$) in the MMD context. If $D \subset \mathcal{X}$ is open, we say $h\colon D \to \mathbb{R}$ is *harmonic* in D if

$$\mathcal{E}(f,h) = 0 \quad \text{for all } f \in C_0(X) \cap \mathcal{F}, \ \operatorname{supp} f \subset D.$$

There are various definitions of caloric, all of which essentially say that

$$\frac{\partial}{\partial t} u(x,t) = \mathcal{L}u(x,t), \quad (x,t) \in Q \subset \mathcal{X} \times \mathbb{R}.$$

Lack of regularity in general means they may not be exactly equivalent, but in all cases the heat kernel $p_t(x,y)$ and $p_t^D(x,y)$ are caloric.

Harnack inequalities.

Definition 3.7. $(\mathcal{X}, \mathcal{E})$ satisfies the elliptic Harnack inequality (EHI) if whenever $B = B(x,R)$ and $h\colon \overline{B} \to \mathbb{R}_+$ is harmonic in B then if $B' = B(x, R/2)$

$$\sup_{B'} h \le C_E \inf_{B'} h.$$

(Strictly speaking, on a MMD space we may have a small exceptional set, so should write ess sup and ess inf above.)

Easy iteration arguments show that if EHI holds then harmonic functions are Hölder continuous, with index $\delta = \delta(C_E)$.

The parabolic Harnack inequality (PHI) is more complicated to state.

Definition 3.8. Let $T = R^2$, $B = B(x,R)$, $Q = B \times (0,T)$, and

$$Q_- = B' \times \left[\tfrac{1}{4}T, \tfrac{1}{2}T\right], \quad Q_+ = B' \times \left[\tfrac{3}{4}T, T\right).$$

Then $(\mathcal{X}, \mathcal{E})$ satisfies PHI if whenever $u = u(x,t)$ is nonnegative and caloric in Q then

$$\sup_{Q_-} u \le C_H \inf_{Q_+} u.$$

Remark 3.9. (1) Note that PHI implies EHI since if h is harmonic then $u(x,t) = h(x)$ is caloric.

(2) Iteration arguments show that PHI implies Hölder continuity of caloric functions (except on an exceptional set).
(3) Theorem 3.4 gives necessary and sufficient conditions for PHI. It also proves that GB and PHI are *stable*: if $\mathcal{E}'$ is another Dirichlet form and $\mathcal{E}' \asymp \mathcal{E}$ then PHI holds for $(\mathcal{X}, \mathcal{E}')$ iff it holds for $(\mathcal{X}, \mathcal{E})$. The PI is clearly stable, while the stability of PHI or GB is far from evident.
(4) Necessary and sufficient conditions for EHI are not known. It is also not known whether or not EHI is stable.
(5) The theorem provides an effective tool for proving that GB and PHI hold for a space $(\mathcal{X}, \mathcal{E})$, since VD and PI are often quite straightforward to prove. (PI can often be obtained from an isoperimetric inequality.)

Extensions to (fractal) MMD spaces. We can replace the Gaussian bound on the heat kernel by more general bounds. Since I want to discuss spaces, such as the pre-Sierpinski carpet or quantum graphs, where the local and global structures are different, introduce the space-time scaling function:

$$\Psi(r) = \begin{cases} r^{\beta_L} & \text{if } 0 \le r \le 1, \\ r^{\beta} & \text{if } r \ge 1. \end{cases} \tag{3.7}$$

(If the space is locally Euclidean we will have $\beta_L = 2$; this is the case for the pre-SC and quantum graphs).

Definition 3.10. We say (p_t) satisfies HK(Ψ) if

$$p_t(x,y) \overset{(c)}{\asymp} V(x, ct^{1/\beta_L})^{-1} \exp\big(-c(d(x,y)^{\beta_L}/t)^{1/(\beta_L-1)}\big) \quad \text{in } I_{\text{loc}}$$
$$p_t(x,y) \overset{(c)}{\asymp} V(x, ct^{1/\beta})^{-1} \exp\big(-c(d(x,y)^{\beta}/t)^{1/(\beta-1)}\big) \quad \text{in } I_{\text{glob}}.$$

Here

$$I_{\text{loc}} = \{(t,x,y) : t \le 1 \vee d(x,y)\}, \quad I_{\text{glob}} = \{(t,x,y) : t \ge 1 \vee d(x,y)\}.$$

If $\Psi(r) = r^\beta$ we write this condition as HK(β).

Example 3.11. (1) The pre-SC satisfies HK(Ψ) with $\beta_L = 2$, $\beta = d_w$.

(2) The true (infinite) SC $\widehat{F}$ satisfies HK(β) with $\beta = d_w$.
(3) GB are just HK(2).
(4) The cable graph associated with the Sierpinski gasket (see Figure 1) satisfies HK(Ψ) with $\beta_L = 2$, $\beta = \log 5/\log 2$.

It is natural to ask what values β can take.

Theorem 3.12. *Suppose an (infinite, connected) MMD space $(\mathcal{X}, \mathcal{E})$ satisfies (V_α) and* HK(β). *Then $\alpha \ge 1$, $2 \le \beta \le 1 + \alpha$, and all these values are possible.*

PROOF. These bounds on α, β were given (without proof) in my Saint Flour notes [3, Section 3]. See [36] for a proof that $2 \leq \beta \leq 1 + \alpha$ in the MMD context. That $\alpha \geq 1$ is obvious from the existence of points x_n with $d(x_n, x_0) = n$. The bound $\beta \leq 1 + \alpha$ comes from the fact that when $\beta > \alpha$ then harmonic functions are Hölder continuous of order $\beta - \alpha$.

Hino [49] proved that $\beta \geq 2$ by showing in general that

$$\liminf_{t \downarrow 0} t \log p_t(x, y) \geq -C(x, y) > -\infty.$$

If HK(β) holds then the LHS is $ct \log t - c' d(x, y)^c t^{(\beta-2)/(\beta-1)}$, so would diverge to $-\infty$ if $\beta < 2$.

An alternative argument would be to use the Davies – Gaffney bound (see [25]), which gives for functions $f_i \in L^2$ with supports $B(x_i, r)$ that if $R \geq 3r$ then

$$\langle P_t f_1, f_2 \rangle \leq c_1 \|f_1\|_2 \|f_2\|_2 \exp(-c_2 R^2/t). \tag{3.8}$$

In [3] I conjectured that if $\beta = 2$ then only $\alpha \in \mathbb{N}$ was possible. Bourdon and Pajot [22], and Laakso [66], in answering a question in [48], gave examples which showed otherwise. In the graph case I gave constructions for all α, β satisfying the inequalities above using Laakso's method — see [6]. The same construction will work in the MMD setting. □

The condition $\beta \geq 2$, combined with polynomial volume growth, means that heat (or the diffusion X_t) can move at most distance $O(t^{1/2})$ in time t. So (as the Davies – Gaffney bound shows) Euclidean space gives essentially the fastest possible speed of heat diffusion. Obstacles such as the cut out cubes in the Sierpinski carpet can slow X down, but not speed it up.

EHI and PHI. The precarpet gives an example where EHI holds but PHI fails. Let $R \gg 1$, $Q = B(x_0, R) \times [0, T]$ where $T = R^2$, and let $d(x_1, x_0) = R/3$. Set $u(x, t) = p_t(x_0, x)$. Then

$$\sup_{Q_-} u \asymp p_{T/4}(x_0, x_0) \asymp cT^{-\alpha/\beta},$$

while

$$\begin{aligned} \inf_{Q_+} u \leq p_T(x_0, x_1) &\leq cT^{-\alpha/\beta} \exp\bigl(-c(R^\beta/T)^{1/(\beta-1)}\bigr) \\ &= cT^{-\alpha/\beta} \exp\bigl(-c' R^{(\beta-2)/(\beta-1)}\bigr) \ll \sup_{Q_-} u. \end{aligned}$$

The PHI fails in this example because heat needs time $O(R^\beta)$ rather than $O(R^2)$ to flow from x_0 to x_1. The fix is clear: define a modified PHI to take account of the space time scaling $t = \Psi(r)$.

Definition 3.13. $(\mathcal{X}, \mathcal{E})$ satisfies PHI(Ψ) if when $R > 0$, $T = \Psi(R)$, $B = B(x, R)$, $Q = B \times (0, T)$, $Q_- = B' \times \left[\frac{1}{4}, \frac{1}{2}T\right]$, $Q_+ = B' \times \left[\frac{3}{4}, T\right)$, and $u = u(x, t)$ is nonnegative and caloric in Q then

$$\sup_{Q_-} u \leq C_H \inf_{Q_+} u.$$

In the same way one defines the rescaled Poincaré inequality PI(Ψ):

$$\min_a \int_{B(x,r)} (f - a)^2 \, d\mu = \int_{B(x,r)} (f - \bar{f})^2 \, d\mu \leq C_P \, \Psi(r) \int_{B(x,r)} d\Gamma(f, f).$$

If $\Psi(r) = r^\beta$ we write PHI(β), PI(β). The original PI and PHI are just PHI(2) and PI(2).

Given Theorem 3.4 a natural first guess would be that

$$\text{HK}(\Psi) \iff \text{PHI}(\Psi) \iff \text{VD} + \text{PI}(\Psi). \tag{3.9}$$

The first double implication in (3.9) is correct.

Introduce two additional conditions. We say $T(\Psi)$ holds for the diffusion X if

$$(T(\Psi)) \qquad \mathbb{E}^x \tau_{x,R} \asymp \Psi(R), \quad x \in \mathcal{X}, R > 0.$$

We can define the effective resistance between two sets in the MMD context by setting

$$R_{\text{eff}}(B_1, B_2)^{-1} = \inf\{\mathcal{E}(f,f) : f = 1 \text{ on } B_1,\ f = 0 \text{ on } B_2\}. \tag{3.10}$$

Then we say RES(Ψ) holds for $(\mathcal{X}, \mathcal{E})$ if

$$R_{\text{eff}}(B(x,R), B(x,2R)^{\text{c}}) \asymp \frac{\Psi(R)}{V(x,R)}, \quad x \in \mathcal{X}, R > 0.$$

Theorem 3.14 ([13, 39, 45]). *The following are equivalent*:
(a) $(\mathcal{X}, \mathcal{E})$ *satisfies* HK(Ψ).
(b) $(\mathcal{X}, \mathcal{E})$ *satisfies* PHI(Ψ).
(c) $(\mathcal{X}, \mathcal{E})$ *satisfies* VD, EHI *and* $T(\Psi)$.
(d) $(\mathcal{X}, \mathcal{E})$ *satisfies* VD, EHI *and* RES(Ψ):

These all imply PI(Ψ).

Note that this theorem does not give stability of PHI(Ψ). The conditions VD and RES(Ψ) are clearly stable, but this is not apparent for any of the other ones.

It is easy to see that one cannot have HK(Ψ) $\Leftrightarrow$ VD + PI(Ψ) in general. If $\Psi_1 \geq \Psi$ then PI(Ψ) implies PI(Ψ_1), while if $\Psi_1(r)/\Psi(r) \to \infty$ then HK(Ψ) and HK(Ψ_1) cannot both hold.

The inequality PI(Ψ) tells us, roughly, that heat homogenizes in a ball size R in time at most $\Psi(R)$, but does not preclude the possibility it might homogenize more quickly. To 'capture' HK(Ψ) one needs an upper bound on the rate of heat diffusion. (This is not necessary in the classical case $\Psi(r) = r^2$ because this is the fastest possible.) This theorem shows that, combined with the regularity coming from EHI, control of either resistance or exit times is sufficient for this.

Sketch proof of Theorem 3.14. The equivalence of (a) – (d) in the graph context is given in [38], and except for technical issues arising from the need for regularity, the same arguments work for metric measure space. Nearly all the implications have been written out in the MMD case: the equivalence of (a) and (c) is proved in [39], of (a) and (b) in [45], and that (d) implies (c) in [13].

(1) The equivalence HK(Ψ) $\Leftrightarrow$ PHI(Ψ) is proved very much as in the classical $\Psi(r) = r^2$ case – see in particular [28].

(2) Given HK(Ψ) + PHI(Ψ) one gets EHI immediately, while the proof that PHI(Ψ) implies VD runs as in the classical case given in Theorem 3.4. $T(\Psi)$ is easy since integration of the heat kernel bounds gives

$$\mathbb{P}^x\big(X_t \notin B(x, \lambda t^{1/\beta})\big) \leq \exp\big(-c\lambda^{\beta/(\beta-1)}\big).$$

(3) A general result using potential theory gives that if $B' = B(x,r/2)$, $B = B(x,r)$ then there exists a probability measure π on $\partial B'$ such that if τ_B is the first exit by X from B and T' the first hit on B' then

$$\int \pi(\mathrm{d}z)\mathbb{E}^z\tau_B = R_{\mathrm{eff}}(B',B^{\mathrm{c}})\int_{B-B'}\mu(\mathrm{d}y)\mathbb{P}^y(T'\le\tau_B).$$

Using the regularity from EHI one can then show that

$$\mathbb{E}^z\tau_B \asymp R_{\mathrm{eff}}(B',B^{\mathrm{c}})V(x,r), \quad z\in\partial B'.$$

So VD + EHI + $T(\Psi)$ $\Leftrightarrow$ VD + EHI + RES(Ψ).

(4) What remains is the hardest (and most useful) implication:

$$\mathrm{VD}+\mathrm{EHI}+T(\Psi) \implies \mathrm{HK}(\Psi).$$

Like most heat kernel bounds, this proceeds in several steps.

Step 1. Obtain the upper bound

$$\text{(DUE)} \qquad \sup_x p_t^B(x,x) \le \frac{c(\kappa)}{V(x,t^{1/\beta})},$$

where $B = B(x,\kappa t^{1/\beta})$. Write $\lambda_1(D)$ for the smallest eigenvalue in a domain $D\subset\mathcal{X}$: $\lambda_1(D)$ is given by the variational formula

$$\lambda_1(D) = \inf\Big\{\frac{\mathcal{E}(f,f)}{\|f\|_2^2} : f\in\mathcal{F}_D, f\ne 0\Big\}. \tag{3.11}$$

Then the main step is to prove that there exists $\nu>0$ such that if $B=B(x_0,R)$ and $D\subset B$ then

$$\lambda_1(D) \ge \frac{c}{\Psi(R)}\Big(\frac{\mu(B)}{\mu(D)}\Big)^{\nu}. \tag{3.12}$$

This called a Faber – Krahn inequality, and this is known to imply (DUE) — see [34] for the manifold case, and [35] for the MMD case.

To prove (3.12) write $g_B(x,y)$ for the Green function in a domain $B\subset\mathcal{X}$. The condition $T(\Psi)$ gives bounds on

$$h(x) = \mathbb{E}^x\tau_{x,r} = \int_{B(x,r)} g_{B(x,r)}(x,y)\,\mu(\mathrm{d}y).$$

The EHI enables one to pass from integrals to pointwise bounds on Green's functions in balls, and using this one can construct a 'candidate function' f which when used in (3.11) gives the Faber – Krahn inequality.

Step 2. Establish the full upper bound in HK(Ψ). In general obtaining the full upper bound from (DUE) can be quite challenging, and usually requires a lengthy proof. However, in this situation the estimate $T(\Psi)$ makes it reasonably straightforward. $T(\Psi)$ states that the time taken to leave a ball $B(x,r)$ is of order $\Psi(r)$. It follows that there exists p_0, $\delta>0$ such that

$$\mathbb{P}^x\big(\tau_{x,r}\le\delta\Psi(r)\big)\le\tfrac{1}{2}.$$

Given a ball $B(x,R)$ divide the 'journey' of X from x to $B(x,R)^{\mathrm{c}}$ into steps of length $r=R/n$; clearly there have to be at least n of these. Call a part of the journey (i.e., across a ball $B(y,r)$) 'quick' if it takes time less than $\delta\Psi(r)$, and 'slow' otherwise. Then the number of quick journeys is stochastically dominated by a Binomial random variable, and standard bounds give that

$$\mathbb{P}(\text{more than } 2n/3 \text{ of the journeys are quick}) \le \mathrm{e}^{-cn}.$$

If the number of quick journeys out of the first n is less than $2n/3$ then the time to exit the ball is at least

$$t(n) = \tfrac{1}{3}\delta\Psi(r)n = \tfrac{1}{3}\delta\Psi(R/n)n. \tag{3.13}$$

Thus

$$\mathbb{P}^x\big(\tau_{x,R} < t(n)\big) \le \exp(-cn).$$

Given R and T, we wish to choose n so that $T = t(n)$. Using (3.13) we have $T \asymp R^\gamma n^{1-\gamma}$, where $\gamma = \beta$ if $R/n \ge 1$ and $\gamma = \beta_L$ if $R/n < 1$. It follows that $n^{\gamma-1} \asymp R^\gamma/T$, and so that

$$\frac{R^{\gamma-1}}{n^{\gamma-1}} \asymp \frac{T}{R}. \tag{3.14}$$

So if $T > R$ then $r > 1$, and $\gamma = \beta$, while if $T < R$ then $\gamma = \beta_L$. (This calculation neglects the effect of the constant $\frac{1}{3}\delta$, but both cases give $n = O(R)$ when $T \asymp R$.) We thus obtain

$$\log \mathbb{P}^x(\tau_{x,R} < T) \le \begin{cases} -c(R^\beta/T)^{1/(\beta-1)} & \text{if } R < T, \\ -c(R^{\beta_L}/T)^{1/(\beta_L-1)} & \text{if } R \ge T. \end{cases} \tag{3.15}$$

One can then combine this bound with (DUE) to obtain the full upper bound.

Step 3. Obtain the lower bound. The upper bound leads easily to:

$$p_t(x,x) \ge cV(x, t^{1/\beta})^{-1}.$$

The key is to extend this inequality to a ball around x:

$$p_t(x,y) \ge cV(x,t^{1/\beta})^{-1}, \quad y \in B(x, \delta t^{1/\beta}). \tag{3.16}$$

EHI gives Hölder continuity of harmonic functions. With other estimates, this extends also to give control of oscillations of $p_t(x,y)$, and hence control of

$$|p_t(x,x) - p_t(x,y)|.$$

Once one has the 'near diagonal lower bound' (3.16) a standard chaining argument as in [9, 28] then gives the full lower bound. □

PROOF OF THEOREM 2.11. The heat kernel bounds on the pre-SC $\widetilde{F}$ follow from the implication (d) implies (a) of Theorem 3.14. VD is immediate from (3.1)–(3.2), and EHI is proved in Theorem 2.9. Theorem 2.8 gives that if $r = 3^n$ with $n \ge 0$ then the resistance across a cube of side r is $R_n \simeq \rho^n = r^{d_w - d_f}$. Using this the condition RES(Ψ) follows, where $\beta_L = 2$ and $\beta = d_w$. □

Stable conditions for HK(Ψ). As remarked above, the conditions in Theorem 3.14 are not stable.

Theorem 3.15 ([11])**.** *The following are equivalent*:

(a) $(\mathcal{X}, \mathcal{E})$ *satisfies* HK(Ψ).
(b) $(\mathcal{X}, \mathcal{E})$ *satisfies* VD, PI(Ψ) *and* CS(Ψ).

The condition CS(Ψ) states that there exist 'low energy' cutoff functions.

Definition 3.16. A function $\varphi\colon \mathcal{X} \to \mathbb{R}_+$ is a *cutoff function* for $B(x,R/2) \subset B(x,R)$ if $\varphi(y) = 1$ on $B(x,r/2)$ and is zero on $B(x,r)^c$.

$(\mathcal{X},\mathcal{E})$ satisfies CS(Ψ) if there exists $\theta\in(0,1]$ such that for all x,R there exists a Hölder continuous cutoff function φ for $B(x,R/2)\subset B(x,R)$ such that if $s\in(0,R)$, $y\in B(x,R)$, and $f\colon B=B(y,s)\to\mathbb{R}$ then

$$\int_{B(y,s/2)} f^2 d\Gamma(\varphi,\varphi)\le c(s/R)^{2\theta}\Big(\int_{B(y,s)} \mathrm{d}\Gamma(f,f)+\Psi(s)^{-1}\int_{B(y,s)} f^2\,\mathrm{d}\mu\Big). \tag{3.17}$$

Note that if $\Psi_1\ge\Psi_2$ then CS(Ψ_2) implies CS(Ψ_1): thus increasing Ψ weakens PI($\cdot$) but strengthens CS($\cdot$).

CS(2) always holds on a manifold — just take φ 'linear' between $B(x,R/2)$ and $B(x,R)$, so that $\|\nabla\varphi\|_\infty\le cR^{-1}$. Then

$$\begin{aligned}\int_{B(y,s/2)} f^2\,\mathrm{d}\Gamma(\varphi,\varphi)&=\int_{B(y,s/2)} f^2|\nabla\varphi|^2\,\mathrm{d}\mu\\&\le\int_{B(y,s/2)} f^2\,\mathrm{d}\mu\|\nabla\varphi\|_\infty^2\\&\le cR^{-2}\int_{B(y,s/2)} f^2\,\mathrm{d}\mu\\&\le c(s/R)^2s^{-2}\int_{B(y,s)} f^2\,\mathrm{d}\mu.\end{aligned}$$

Proof of Theorem 3.15. (1) HK(Ψ) implies VD, PI(Ψ) and CS(Ψ). The first two follow by Theorem 3.14. To prove CS(Ψ) one uses properties of Green's functions to construct a suitable φ. Given balls $B'=B(x_0,R/2)\subset B=B(x_0,R)$, let $B_1=B(x_0,2R/3)$. Define the λ-resolvent for the killed process by

$$G_B^\lambda f(x)=\mathbb{E}^x\int_0^{\tau_B} f(X_s)e^{-\lambda s}\,\mathrm{d}s.$$

Set $\lambda=\Psi(R)^{-1}$ and $h(x)=G_B^\lambda 1_{B_1}(x)$. Then h has support $\overline{B}$, $h\le G^\lambda 1=\lambda^{-1}$, and the heat kernel bounds give that $h\ge c\Psi(R)$ on B'. Thus $\varphi=1\wedge c\Psi(R)^{-1}h$ is a cutoff function for B'. The parabolic Harnack inequality gives Hölder continuity of the heat kernel, and hence of φ, so it remains to verify (3.17). A starting point for this is the observation that, writing $\mathcal{E}_\lambda(g,g)=\mathcal{E}(g,g)+\lambda\langle g,g\rangle$, where $\langle\cdot,\cdot\rangle$ denotes the inner product on $L^2(\mathcal{X},\mu)$, then if g has support B, by [31, Theorem 4.4.1]

$$\mathcal{E}_\lambda(g,h)=\mathcal{E}_\lambda(g,G_\lambda^B 1_{B_1})=\langle g,1_{B_1}\rangle.$$

We therefore have, using the inequality $2ab\le\frac12a^2+2b^2$, and the Leibniz rule (3.3),

$$\begin{aligned}\int f^2 d\Gamma(h,h)&=\int d\Gamma(f^2h,h)-2\int fh\,\mathrm{d}\Gamma(f,h)\\&\le\mathcal{E}_\lambda(f^2h,h)+\frac12\int f^2\,\mathrm{d}\Gamma(h,h)+2\int h^2\,\mathrm{d}\Gamma(f,f)\\&=\langle fh^2,1_{B_1}\rangle+\frac12\int f^2\,\mathrm{d}\Gamma(h,h)+2\int h^2\,\mathrm{d}\Gamma(f,f).\end{aligned}$$

This implies (3.17) with $s=R$. Obtaining the inequality on smaller balls requires some more work: see [11, Proposition 3.4] for more details on this argument.

(2) In many arguments one needs to control expressions of the form

$$\int_A f^2|\nabla\varphi|^2\,\mathrm{d}\mu,$$

where $A = B(x, r+h) - B(x,r)$ and φ is a cutoff function between $B(x,r)$ and $B(x,r+h)$. The CS inequalities enable one to do this. The 'extra' term

$$\int_A d\Gamma(f,f)$$

often turns out to be harmless or controllable.

In particular, CS(Ψ) provides a family of cutoff functions which enable one to run the first ('easy') part of Moser's argument. Thus one can use the methods of [73–75] to show that VD, PI(Ψ) and CS(Ψ) imply EHI. Once one has EHI one can use Theorem 3.14, since PI(Ψ) and CS(Ψ) also imply RES(Ψ). □

Remark 3.17. (1) The condition CS(β) is unfortunately rather complicated. However Theorem 3.15 does not preclude the possibility that there might exist a simpler condition ?(Ψ) such that ?(Ψ) + PI(Ψ) + VD $\Leftrightarrow$ HK(Ψ). One candidate for this is RES(Ψ).

(2) If $(\mathcal{E}_i, \mathcal{F})$, $i = 1, 2$ are strongly local Dirichlet forms with

$$C^{-1}\mathcal{E}_1(f,f) \le \mathcal{E}_2(f,f) \le C\mathcal{E}_1(f,f), \quad f \in \mathcal{F},$$

then by [67, Proposition 1.5.5(b)] their energy measures satisfy

$$C^{-1}\Gamma_1(f,f) \le \Gamma_2(f,f) \le C\Gamma_1(f,f), \quad f \in \mathcal{F}.$$

(See also [72, p. 389].) Thus the condition CS(Ψ) is stable, and therefore this theorem does give the stability of HK(Ψ) and PHI(Ψ).

(3) A recent paper [1] provides some simplification of CS(Ψ).

Suppose that V_α holds and $\Psi(r) = r^\beta$, so that HK(β) takes the form

$$p_t(x,y) \overset{(c)}{\asymp} ct^{-\alpha/\beta} \exp\big(-c(d(x,y)^\beta/t)^{1/(\beta-1)}\big).$$

In particular $p_t(x,x) \asymp t^{-\alpha/\beta}$. The behavior of X can be divided into two main cases.

(1) $\alpha < \beta$. In this case the process X_t is recurrent, and in fact 'hits points,' so that the range

$$R_t = \{X_s, 0 \le s \le t\}$$

has positive μ measure. Ultimately the process X will hit every point in the space $\mathcal{X}$ infinitely often.

(2) $\alpha > \beta$. In this case the process X_t is transient, and R_t has zero μ measure. The probability that X hits any specific point after time 0 is zero. The Green function on the whole space $\mathcal{X}$ exists and satisfies

$$g(x,y) = \int_0^\infty p_t(x,y)\,dt \asymp \frac{1}{d(x,y)^{\alpha-\beta}}.$$

In the classical case of $\mathbb{R}^d$ the first case (1) $\alpha < \beta$ only arises when $d = 1$. In the MMD setting we have a richer family of 'low-dimensional' spaces; in fact nearly all the exact fractals studied in the physics literature turn out to have $\alpha < \beta$.

The resistance estimates on R_n are good enough to show that the basic SC is recurrent if $d = 2$, and transient for $d \ge 3$. One can find generalized Sierpinski carpets in $d \ge 3$ for which $d_f > 2$ but $d_w < d_f$, so that X is still recurrent – see [10].

In low dimensions one has a much nicer version of the stability result. Introduce the condition PR(β) (for 'pointwise resistance'):

$$R_{\text{eff}}(x,y) = R_{\text{eff}}(\{x\},\{y\}) \asymp \frac{d(x,y)^\beta}{V\big(x, d(x,y)\big)}.$$

In the recurrent case $R_{\text{eff}}(x,y)$ is a metric — see [59].

Theorem 3.18 ([61]). *Let $\mathcal{X}$ satisfy V_α, and let $\beta > \alpha$. Then the following are equivalent*:

(1) $(\mathcal{X},\mathcal{E})$ *satisfies* HK(β).
(2) $(\mathcal{X},\mathcal{E})$ *satisfies* PR(β).

So in this case Harnack inequalities are no longer difficult, and one just has to calculate resistance between points.

This result was originally proved in the graph context in [14], and has applications to random walks on critical percolation clusters — see [16, 60]. It is interesting to note that while de Gennes initially proposed the use of random walks to study resistance of percolation clusters, the mathematics has been in the opposite direction: using resistances to study random walks.

In the transient case one has $R_{\text{eff}}(x,y) = \infty$, and in spaces such as $\mathbb{R}^d$

$$\limsup_{d(x,y)\to\infty} R_{\text{eff}}\big(B(x,\varepsilon), B(y,\varepsilon)\big) < \infty.$$

Thus resistance between points or small sets gives no useful information.

Rough isometries. Let (X_i, d_i, μ_i) be metric measure spaces. A map $\varphi\colon \mathcal{X}_1 \to \mathcal{X}_2$ is a *rough isometry* if there exist constants C_1, C_2, C_3 such that

$$C_1^{-1}(d_1(x,y) - C_2) \le d_2\big(\varphi(x),\varphi(y)\big) \le C_1(d_1(x,y) + C_2),$$
$$X_2 = \bigcup_{x\in X_1} B_{d_2}(\varphi(x), C_2),$$
$$C_3^{-1}\mu_1(B_{d_1}(x,C_2)) \le \mu_2(B_{d_2}(\varphi(x),C_2)) \le C_3\mu_1\big(B_{d_1}(x,C_2)\big).$$

If there exists a rough isometry between two spaces they are said to be *roughly isometric*. (One can check that this is an equivalence relation.) If two spaces are roughly isometric then they have the same large scale structure, although their local structures may be quite different. For example, $\mathbb{Z}_+^d$ and $\mathbb{R}_+^d$ are roughly isometric, as are the precarpet $\widetilde{F}$ and the infinite SC $\widehat{F}$.

Theorems 3.15 and 3.18 imply that the global part of HK(Ψ), i.e., that relating to the index β, is stable under rough isometries. (One also needs some local regularity of both spaces — see [11, Theorem 2.21] for more details.)

In particular if M is a manifold (with some local regularity) which is roughly isometric to the pre-SC $\widetilde{F}$ then the heat kernel on M also satisfies HK(Ψ') — where Ψ' is given by (3.7) with indices β and 2.

Local structure. In the MMD setting we have a Dirichlet form $(\mathcal{E},\mathcal{F})$ on $L^2(\mathcal{X},\mu)$. Suppose that HK(β) holds: we can then ask about the regularity properties of the functions in $\mathcal{F}$.

Theorem 3.19. *Let $\mathcal{X}$ be the Sierpinski carpet. Suppose that $f \in C^1(\mathbb{R}^2)$ and $g = f|_F \in \mathcal{F}$. Then g is constant.*

However, $\mathcal{F}$ can be described explicitly as a Besov space. This was done for the Sierpinski gasket in [52], but holds much more generally. Set

$$N_\theta(f) = \sup_{0<r\leq 1} r^{-\alpha} \int_F V(x,r)^{-1} \int_{B(x,r)} |f(x) - f(y)|^2 \, \mu(\mathrm{d}x)\, \mu(\mathrm{d}y),$$
$$W(\theta) = \{f \in L^2(F,\mu) : N_\theta(f) < \infty\}.$$

Theorem 3.20 ([36, 62]). *Suppose* HK(β) *holds. Then* $\mathcal{F} = W(\beta)$, *and*

$$\mathcal{E}(f,f) \asymp N_\beta(f).$$

Further,

$$\beta = \sup\{\theta : \dim W(\theta) = \infty\}.$$

Corollary 3.21. *Let* $\mathcal{E}_1, \mathcal{E}_2$ *be Dirichlet forms on a metric measure space* $\mathcal{X}$. *If* $(\mathcal{X}, \mathcal{E}_i)$ *satisfy* HK(β_i), *then* $\beta_1 = \beta_2$.

Open Problem 2. This result gives a characterization of $\beta = d_w$ which appears to be independent of the Dirichlet form $\mathcal{E}$. Can one use it to obtain bounds on the 'walk dimension' d_w of the SC?

Energy measures. These are the measures $\mathrm{d}\Gamma(f,f)$, $f \in \mathcal{F}$. Suppose that $(\mathcal{X}, \mathcal{E})$ satisfies HK(β). If one had $\mathrm{d}\Gamma(f,f) \leq C(f)\,\mathrm{d}\mu$ then we could define a metric associated with $\mathcal{E}$ by setting

$$L = \{f : \mathrm{d}\Gamma(f,f) \leq \mathrm{d}\mu\},$$

and

$$d_{\mathcal{E}}(x,y) = \sup\{f(y) - f(x) : f \in L\}.$$

One would then expect to have Gaussian heat kernel bounds with respect to $d_{\mathcal{E}}$, and on-diagonal heat kernel decay of the form $t^{-d_f(F)/2}$ — see [82].

Theorem 3.22 ([50]). *Let* $\mathcal{X}$ *be a Sierpinski carpet. If* $f \in \mathcal{F}$ *and* $\mathrm{d}\Gamma(f,f) \ll \mathrm{d}\mu$, *then* f *is constant.*

Kusuoka [64] proved a similar result for the Sierpinski gasket (SG). Thus the energy measures and Hausdorff measure are mutually singular.

References

1. S. Andres and M. T. Barlow, *Energy inequalities for cutoff functions and some applications*, preprint.
2. D. G. Aronson, *Bounds for the fundamental solution of a parabolic equation*, Bull. Amer. Math. Soc. **73** (1967), 890–896.
3. M. T. Barlow, *Diffusions on fractals*, Lectures on Probability Theory and Statistics (Saint-Flour, 1995) (P. Bernard, ed.), Lecture Notes in Math., vol. 1690, Springer, Berlin, 1998, pp. 1–121.
4. ______, *Random walks on supercritical percolation clusters*, Ann. Probab. **32** (2004), no. 4, 3024–3084.
5. ______, *Some remarks on the elliptic Harnack inequality*, Bull. London Math. Soc. **37** (2005), no. 2, 200–208.
6. ______, *Which values of the volume growth and escape time exponent are possible for a graph?*, Rev. Mat. Iberoamericana **20** (2004), no. 1, 1–31.
7. M. T. Barlow and R. F. Bass, *The construction of Brownian motion on the Sierpiński carpet*, Ann. Inst. H. Poincaré Probab. Statist. **25** (1989), no. 3, 225–257.
8. ______, *On the resistance of the Sierpiński carpet*, Proc. Roy. Soc. London Ser. A **431** (1990), no. 1882, 345–360.

9. ———, *Transition densities for Brownian motion on the Sierpiński carpet*, Probab. Theory Related Fields **91** (1992), no. 3-4, 307–330.
10. ———, *Brownian motion and harmonic analysis on Sierpinski carpets*, Canad. J. Math. **51** (1999), no. 4, 673–744.
11. M. T. Barlow, R. F. Bass, and T. Kumagai, *Stability of parabolic Harnack inequalities on metric measure spaces*, J. Math. Soc. Japan **58** (2006), no. 2, 485–519.
12. M. T. Barlow, R. F. Bass, T. Kumagai, and A. Teplyaev, *Uniqueness of Brownian motion on Sierpiński carpets*, J. Eur. Math. Soc. (JEMS) **12** (2010), no. 3, 655–701.
13. ———, *Supplementary notes for "Uniqueness of Brownian motion on Sierpiński carpets"*, unpublished, available at `www.math.ubc.ca/~barlow`.
14. M. T. Barlow, T. Coulhon, and T. Kumagai, *Characterization of sub-Gaussian heat kernel estimates on strongly recurrent graphs*, Comm. Pure Appl. Math. **58** (2005), no. 12, 1642–1677.
15. M. T. Barlow, A. Grigor′yan, and T. Kumagai, *On the equivalence of parabolic Harnack inequalities and heat kernel estimates*, J. Math. Soc. Japan, to appear.
16. M. T. Barlow, A. A. Járai, T. Kumagai, and G. Slade, *Random walk on the incipient infinite cluster for oriented percolation in high dimensions*, Comm. Math. Phys. **278** (2008), no. 2, 385–431.
17. M. T. Barlow and E. A. Perkins, *Brownian motion on the Sierpiński gasket*, Probab. Theory Related Fields **79** (1988), no. 4, 543–623.
18. R. F. Bass and P. Hsu, *Some potential theory for reflecting Brownian motion in Hölder and Lipschitz domains*, Ann. Probab. **19** (1991), no. 2, 486–508.
19. N. Berger and M. Biskup, *Quenched invariance principle for simple random walk on percolation clusters*, Probab. Theory Related Fields **137** (2007), no. 1-2, 83–120.
20. B. Bollobás and O. Riordan, *Percolation*, Cambridge Univ. Press, New York, 2006.
21. E. Bombieri and E. Giusti, *Harnack's inequality for elliptic differential equations on minimal surfaces*, Invent. Math. **15** (1972), 24–46.
22. M. Bourdon and H. Pajot, *Poincaré inequalities and quasiconformal structure on the boundary of some hyperbolic buildings*, Proc. Amer. Math. Soc. **127** (1999), no. 8, 2315–2324.
23. S. R. Broadbent and J. M. Hammersley, *Percolation processes.* I: *Crystals and mazes*, Proc. Cambridge Philos. Soc. **53** (1957), 629–641.
24. A. K. Chandra, P. Raghavan, W. L. Ruzzo, R. Smolensky, and P. Tiwari, *The electrical resistance of a graph captures its commute and cover times*, Proceedings of the 21st Annual ACM Symposium on Theory of Computing (Seattle, WA, 1989) (D. S. Johnson, ed.), ACM, New York, 1989, pp. 574–586.
25. E. B. Davies, *Heat kernel bounds, conservation of probability and the Feller property*, J. Anal. Math. **58** (1992), 99–119.
26. P. G. de Gennes, *La percolation: un concept unificateur*, La Recherche **7** (1976), 919–927.
27. P. G. Doyle and J. L. Snell, *Random walks and electric networks*, Carus Math. Monogr., vol. 22, Math. Assoc. America, Washington, DC, 1984.
28. E. B. Fabes and D. W. Stroock, *A new proof of Moser's parabolic Harnack inequality using the old ideas of Nash*, Arch. Rational Mech. Anal. **96** (1986), no. 4, 327–338.
29. K. Falconer, *Fractal geometry: Mathematical foundations and applications*, Wiley, Chichester, 1990.
30. M. Fukushima, *A construction of reflecting barrier Brownian motions for bounded domains*, Osaka J. Math. **4** (1967), 183–215.
31. M. Fukushima, Y. Ōshima, and M. Takeda, *Dirichlet forms and symmetric Markov processes*, de Gruyter Stud. Math., vol. 19, de Gruyter, Berlin, 1994.
32. S. Goldstein, *Random walks and diffusions on fractals*, (Minneapolis, MN, 1984) (H. Kersten, ed.), IMA Vol. Math. Appl., vol. 8, Springer, New York, 1987, pp. 121–129.
33. A. A. Grigor′yan, *The heat equation on non-compact Riemannian manifolds*, Mat. Sb. **182** (1991), no. 1, 55–87 (Russian); English transl., Math. USSR-Sb. **72** (1992), no. 1, 47–77.
34. A. Grigor′yan, *Heat kernel upper bounds on a complete noncompact manifold*, Rev. Mat. Iberoamericana **10** (1994), no. 2, 395–452.
35. A. Grigor′yan and J. Hu, *Upper bounds of heat kernels on doubling spaces*, preprint.
36. A. Grigor′yan, J. Hu, and K.-S. Lau, *Heat kernels on metric measure spaces and an application to semilinear elliptic equations*, Trans. Amer. Math. Soc. **355** (2003), no. 5, 2065–2095 (electronic).

37. A. Grigor′yan and A. Telcs, *Sub-Gaussian estimates of heat kernels on infinite graphs*, Duke Math. J. **109** (2001), no. 3, 451–510.
38. ———, *Harnack inequalities and sub-Gaussian estimates for random walks*, Math. Ann. **324** (2002), no. 3, 521–556.
39. ———, *Two-sided estimates of heat kernels on metric measure spaces*, Ann. Probab. **40** (2012), no. 3, 1212–1284.
40. G. Grimmett, *Percolation*, 2nd ed., Grundlehren Math. Wiss., vol. 321, Springer, Berlin, 1999.
41. B. M. Hambly, T. Kumagai, S. Kusuoka, and X. Y. Zhou, *Transition density estimates for diffusion processes on homogeneous random Sierpinski carpets*, J. Math. Soc. Japan **52** (2000), no. 2, 373–408.
42. T. Hara and G. Slade, *Mean-field critical behaviour for percolation in high dimensions*, Comm. Math. Phys. **128** (1990), no. 2, 333–391.
43. ———, *Mean-field behaviour and the lace expansion*, Probability and Phase Transition (Cambridge, 1993), NATO Adv. Sci. Inst. Ser. C Math. Phys. Sci., vol. 420, Kluwer, Dordrecht, 1994, pp. 87–122.
44. S. Havlin and D. Ben-Avraham, *Diffusion in disordered media*, Adv. in Phys. **36** (1987), 695–798.
45. W. Hebisch and L. Saloff-Coste, *On the relation between elliptic and parabolic Harnack inequalities*, Ann. Inst. Fourier (Grenoble) **51** (2001), no. 5, 1437–1481.
46. J. Heinonen, *Lectures on analysis on metric spaces*, Universitext, Springer, New York, 2001.
47. J. Heinonen and P. Koskela, *Quasiconformal maps in metric spaces with controlled geometry*, Acta Math. **181** (1998), no. 1, 1–61.
48. J. Heinonen and S. Semmes, *Thirty-three yes or no questions about mappings, measures, and metrics*, Conform. Geom. Dyn. **1** (1997), 1–12 (electronic).
49. M. Hino, *On short time asymptotic behavior of some symmetric diffusions on general state spaces*, Potential Anal. **16** (2002), no. 3, 249–264.
50. ———, *On singularity of energy measures on self-similar sets*, Probab. Theory Related Fields **132** (2005), no. 2, 265–290.
51. J. E. Hutchinson, *Fractals and self-similarity*, Indiana Univ. Math. J. **30** (1981), no. 5, 713–747.
52. A. Jonsson, *Brownian motion on fractals and function spaces*, Math. Z. **222** (1996), no. 3, 495–504.
53. M. Kanai, *Rough isometries, and combinatorial approximations of geometries of noncompact Riemannian manifolds*, J. Math. Soc. Japan **37** (1985), no. 3, 391–413.
54. ———, *Analytic inequalities, and rough isometries between noncompact Riemannian manifolds*, Curvature and Topology of Riemannian Manifolds (Katata, 1985) (K. Shiohama, T. Sakai, and T. Sunada, eds.), Lecture Notes in Math., vol. 1201, Springer, Berlin, 1986, pp. 122–137.
55. H. Kesten, *The critical probability of bond percolation on the square lattice equals* $\frac{1}{2}$, Comm. Math. Phys. **74** (1980), no. 1, 41–59.
56. ———, *Subdiffusive behavior of random walk on a random cluster*, Ann. Inst. H. Poincaré Probab. Statist. **22** (1986), no. 4, 425–487.
57. J. Kigami, *A harmonic calculus on the Sierpiński spaces*, Japan J. Appl. Math. **6** (1989), no. 2, 259–290.
58. ———, *Analysis on fractals*, Cambridge Tracts in Math., vol. 143, Cambridge Univ. Press, Cambridge, 2001.
59. ———, *Effective resistances for harmonic structures on p.c.f. self-similar sets*, Math. Proc. Cambridge Philos. Soc. **115** (1994), no. 2, 291–303.
60. G. Kozma and A. Nachmias, *The Alexander–Orbach conjecture holds in high dimensions*, Invent. Math. **178** (2009), no. 3, 635–654.
61. T. Kumagai, *Heat kernel estimates and parabolic Harnack inequalities on graphs and resistance forms*, Publ. Res. Inst. Math. Sci. **40** (2004), no. 3, 793–818.
62. T. Kumagai and K.-T. Sturm, *Construction of diffusion processes on fractals, d-sets, and general metric measure spaces*, J. Math. Kyoto Univ. **45** (2005), no. 2, 307–327.
63. S. Kusuoka, *A diffusion process on a fractal*, Probabilistic Methods in Mathematical Physics (Katata/Kyoto, 1985) (K. Itô and N. Ikeda, eds.), Academic Press, Boston, MA, 1987, pp. 251–274.

64. ———, *Dirichlet forms on fractals and products of random matrices*, Publ. Res. Inst. Math. Sci. **25** (1989), no. 4, 659–680.
65. S. Kusuoka and Z. X. Yin, *Dirichlet forms on fractals: Poincaré constant and resistance*, Probab. Theory Related Fields **93** (1992), no. 2, 169–196.
66. T. J. Laakso, *Ahlfors Q-regular spaces with arbitrary $Q > 1$ admitting weak Poincaré inequality*, Geom. Funct. Anal. **10** (2000), no. 1, 111–123.
67. Y. Le Jan, *Mesures associées à une forme de Dirichlet: Applications*, Bull. Soc. Math. France **106** (1978), no. 1, 61–112.
68. P. Li and S.-T. Yau, *On the parabolic kernel of the Schrödinger operator*, Acta Math. **156** (1986), no. 3-4, 153–201.
69. T. Lindvall and L. C. G. Rogers, *Coupling of multidimensional diffusions by reflection*, Ann. Probab. **14** (1986), no. 3, 860–872.
70. P. Mathieu and A. Piatnitski, *Quenched invariance principles for random walks on percolation clusters*, Proc. R. Soc. Lond. Ser. A Math. Phys. Eng. Sci. **463** (2007), no. 2085, 2287–2307.
71. I. McGillivray, *Resistance in higher-dimensional Sierpiński carpets*, Potential Anal. **16** (2002), no. 3, 289–303.
72. U. Mosco, *Composite media and asymptotic Dirichlet forms*, J. Funct. Anal. **123** (1994), no. 2, 368–421.
73. J. Moser, *On Harnack's theorem for elliptic differential equations*, Comm. Pure Appl. Math. **14** (1961), 577–591.
74. ———, *A Harnack inequality for parabolic differential equations*, Comm. Pure Appl. Math. **17** (1964), 101–134.
75. ———, *On a pointwise estimate for parabolic differential equations*, Comm. Pure Appl. Math. **24** (1971), 727–740.
76. J. Nash, *Continuity of solutions of parabolic and elliptic equations*, Amer. J. Math. **80** (1958), 931–954.
77. H. Osada, *Isoperimetric constants and estimates of heat kernels of pre Sierpiński carpets*, Probab. Theory Related Fields **86** (1990), no. 4, 469–490.
78. L. Saloff-Coste, *A note on Poincaré, Sobolev, and Harnack inequalities*, Internat. Math. Res. Notices **2** (1992), 27–38.
79. V. Sidoravicius and A.-S. Sznitman, *Quenched invariance principles for walks on clusters of percolation or among random conductances*, Probab. Theory Related Fields **129** (2004), no. 2, 219–244.
80. S. Smirnov and W. Werner, *Critical exponents for two-dimensional percolation*, Math. Res. Lett. **8** (2001), no. 5-6, 729–744.
81. R. S. Strichartz, *Analysis on fractals*, Notices Amer. Math. Soc. **46** (1999), no. 10, 1199–1208.
82. K.-T. Sturm, *Analysis on local Dirichlet spaces.* II: *Upper Gaussian estimates for the fundamental solutions of parabolic equations*, Osaka J. Math. **32** (1995), no. 2, 275–312.
83. ———, *Analysis on local Dirichlet spaces.* III: *The parabolic Harnack inequality*, J. Math. Pures Appl. (9) **75** (1996), no. 3, 273–297.
84. R. van der Hofstad and A. A. Járai, *The incipient infinite cluster for high-dimensional unoriented percolation*, J. Statist. Phys. **114** (2004), no. 3-4, 625–663.
85. N. Th. Varopoulos, *Long range estimates for Markov chains*, Bull. Sci. Math. (2) **109** (1985), no. 3, 225–252.
86. J. A. Yan, *A formula for densities of transition functions*, Séminaire de Probabilités, XXII, Lecture Notes in Math., vol. 1321, Springer, Berlin, 1988, pp. 92–100.

Department of Mathematics, University of British Columbia, Vancouver, BC V6T 1Z2, Canada
E-mail address: barlow@math.ubc.ca

Centre de Recherches Mathématiques
CRM Proceedings and Lecture Notes
Volume **56**, 2013

Heat Kernel Estimates, Sobolev-Type Inequalities and Riesz Transform on Noncompact Riemannian Manifolds

Thierry Coulhon

Abstract. Let M be a complete noncompact Riemannian manifold, or more generally a metric measure space endowed with a heat kernel, satisfying the volume doubling property. In this survey, the connection between various kinds of upper and lower heat kernel estimates will be examined. We will explain the connections between heat kernel estimates and Sobolev inequalities, some sufficient conditions in terms of heat kernel gradient estimates for L^p-boundedness of the Riesz transform, and finally, in the polynomial growth setting, the connection between boundedness of Riesz transform and a new version of Sobolev-type inequalities.

Part of the contents of these lecture notes is now classical. The more recent stuff is mainly taken from [6, 10, 11].

1. The setting

1.1. Weighted manifolds. Let (M,g) be a complete connected noncompact Riemannian manifold endowed with a measure μ which has a smooth positive density with respect to the Riemannian measure. A triple (M,g,μ) is called a weighted Riemannian manifold.

For $x \in M$ and $r > 0$, let $B(x,r)$ be the open ball of center x and radius r. Denote by $V(x,r) := \mu(B(x,r))$ the volume of $B(x,r)$.

Assume the volume doubling property: there exists $C > 0$ such that, for all $x \in M$ and $r > 0$,

$$V(x,2r) \le CV(x,r). \tag{D}$$

A simple consequence of (D) is that there exists $\nu > 0$ such that

$$\frac{V(x,r)}{V(x,s)} \le C\left(\frac{r}{s}\right)^{\nu}, \ \forall x \in M,\ r \ge s > 0. \tag{D_ν}$$

It follows easily that, for any $x, y \in M$ and $r > 0$, one has

$$V(x,r) \le C\left(1 + \frac{d(x,y)}{r}\right)^{\nu} V(y,r).$$

2010 *Mathematics Subject Classification.* Primary 58J35; Secondary 46E35.

Thanks are due to my Ph.D. student Chen Li who prepared a first version of the present lecture notes.

This is the final form of the paper.

A less well-known but very useful consequence is that there exists $\nu' > 0$ such that the following lower estimate holds:

$$\frac{V(x,r)}{V(x,s)} \geq C\left(\frac{r}{s}\right)^{\nu'}, \quad \forall x \in M, r \geq s > 0.$$

A particular case of the volume doubling property is when the volume has polynomial growth, i.e., there exists $D > 0$ such that $V(x,r) \simeq r^D$. A basic example is of course the Euclidean space $\mathbb{R}^D$, but one may also think of the connected sum of $\mathbb{R}^D$ with itself: take two copies of $\mathbb{R}^D \setminus B(0,1)$ endowed with the Euclidean metric, and glue them smoothly along the unit circles.

In many cases, polynomial growth is too much to ask. Rather, the behavior of the volume of balls may be governed by two different exponents for small and large radii: $V(x,r) \simeq r^d$, $0 < r \leq 1$, $V(x,r) \simeq r^D$, $r \geq 1$. Again, property (D) holds in that case. Here are a few examples of such volume growth:

(1) the Heisenberg group, endowed with a left invariant Riemannian metric (see [13]).
(2) Co-compact Galois covering manifolds with a nilpotent deck transformation group (see for instance [24]).
(3) Fractal-like manifolds, for instance the Vicsek manifold (see [3]).

On the one hand, if one thinks of general manifolds, one sees that the doubling condition is much more natural than the polynomial growth assumption, even with two exponents, which is quite restrictive. On the other hand, it is often more difficult to deal with the general case. A striking example is the contrast in difficulty between the characterizations of the natural heat kernel upper bounds in both situations (see Section 4 below).

The Laplace – Beltrami operator Δ_μ on (M,g,μ) is defined by

$$(\Delta_\mu f, g) = \int_M \nabla f \cdot \nabla g \, \mathrm{d}\mu, \quad \forall f,g \in \mathcal{C}_0^\infty(M).$$

A nontrivial theorem says that, since M is complete, one can extend Δ_μ to a self-adjoint operator on $L^2(M,d\mu)$ (see [12,16]). In the sequel, we denote Δ_μ by Δ. Since Δ is obviously nonnegative, spectral theory yields the heat semigroup $\mathrm{e}^{-t\Delta}$ as a family of linear contractions of $L^2(M,\mu)$. One proves that $\mathrm{e}^{-t\Delta}$ is an integral operator with kernel $p_t(x,y)$, that is,

$$\mathrm{e}^{-t\Delta} f(x) = \int_M p_t(x,y) f(y) \, \mathrm{d}\mu(y),$$

for a.e. x in M. Basic properties of the heat kernel include (see [16]):

Smoothness: $p_t(x,y)$ is smooth in $t > 0$ and $x,y \in M$.

Sub-Markovian property: For $f \in L^2(M)$, we have $0 \leq f \leq 1 \Rightarrow 0 \leq \mathrm{e}^{-t\Delta} f \leq 1$. In other words, $\int_M p_t(x,y)\, \mathrm{d}\mu(y) \leq 1$. If moreover $\int_M p_t(x,y)\, \mathrm{d}\mu(y) = 1$, (M,g,μ) is said to be stochastic complete. The sub-Markovian property is related to the bilinear form $\mathcal{E}(f,g) = (\Delta f, g)$ being a Dirichlet form. This is the Beurling – Deny theory. For a very nice point of view on this theory, see [22].

One can keep all the relevant features of the Riemannian case in a much more general setting.

1.2. Metric measure spaces. Let M be a connected locally compact separable space and let μ be a positive Radon measure on M. Let $\mathcal{E}$ be a regular and strongly local Dirichlet form on M with domain $\mathcal{D} \in L^2(M, d\mu)$. That is, there is a core $\mathcal{C} \in L^2(M, d\mu) \cap \mathcal{C}_0(M)$ which is dense in $\mathcal{D}$ under the norm $(\|f\|_2^2 + \mathcal{E}(f,f))^{1/2}$. $\mathcal{E}$ admits an "energy measure" Γ:

$$\mathcal{E}(f,g) = \int_M \mathrm{d}\Gamma(f,g)$$

for $f, g \in \mathcal{D}$. On Riemannian manifolds, $\mathrm{d}\Gamma(f,f) = |\nabla f|^2 \, \mathrm{d}\mu$ with μ the Riemannian measure.

For any $x, y \in M$, define

$$d(x,y) = \sup\{f(x) - f(y) : \mathrm{d}\Gamma(f,f) \le \mathrm{d}\mu, f \in \mathcal{C}\}.$$

Assume that d is actually a true metric that generates the original topology on M and that (M, d) is complete.

Thus (M, d, μ) is a metric measure space.

Let A be the self-adjoint operator uniquely associated with the Dirichlet space $\big(M, \mathcal{D}, L^2(M, d\mu)\big)$:

$$(Af, g) = \mathcal{E}(f,g).$$

The generated semigroup is $(\mathrm{e}^{-tA})_{t>0}$. Assume that $(\mathrm{e}^{-tA})_{t>0}$ admits a negative measurable transition function $p_t(x,y)$ (the heat kernel) such that

$$\mathrm{e}^{-tA} f(x) = \int_M p_t(x,y) f(y) \, \mathrm{d}\mu(y).$$

For more details, we refer to [18, 19, 27–29].

2. Heat kernel estimates from above and below

We consider the following heat kernel estimates: the on-diagonal upper estimate

$$\text{(DUE)} \qquad p_t(x,x) \le \frac{C}{V(x,\sqrt{t})}, \qquad \forall x \in M, t > 0.$$

In fact, a Cauchy–Schwarz argument shows that (DUE) is equivalent to

$$p_t(x,y) \le \frac{C}{\sqrt{V(x,\sqrt{t})V(y,\sqrt{t})}}, \qquad \forall x, y \in M, t > 0.$$

The latter form is more adapted to situations where p_t is only assumed to be measurable, such as the metric measure space setting presented in Section 1.2.

The full Gaussian upper estimate

$$\text{(UE)} \qquad p_t(x,y) \le \frac{C}{V(x,\sqrt{t})} \exp\Big(-\frac{d^2(x,y)}{Ct}\Big), \qquad \forall x, y \in M, t > 0.$$

The on-diagonal lower estimate

$$\text{(DLE)} \qquad p_t(x,x) \ge \frac{c}{V(x,\sqrt{t})}, \qquad \forall x \in M, t > 0.$$

The full Gaussian lower estimate

$$\text{(LE)} \qquad p_t(x,y) \ge \frac{c}{V(x,\sqrt{t})} \exp\Big(-\frac{d^2(x,y)}{ct}\Big), \qquad \forall x, y \in M, t > 0.$$

(UE) plus (LE) is the Li – Yau-type estimate, i.e., $\forall x, y \in M, t > 0$,

$$\text{(LY)} \qquad \frac{c}{V(x,\sqrt{t})}\exp\Big(-\frac{d^2(x,y)}{ct}\Big) \le p_t(x,y) \le \frac{C}{V(x,\sqrt{t})}\exp\Big(-\frac{d^2(x,y)}{Ct}\Big).$$

The gradient upper estimate

$$\text{(G)} \qquad |\nabla p_t(x,y)| \le \frac{C}{\sqrt{t}V(y,\sqrt{t})}, \qquad \forall x, y \in M, t > 0.$$

Riemannian manifolds with nonnegative Ricci curvature are typical examples satisfying (LY) and (G) (see [21]).

Other forms of heat kernel estimates are interesting; for instance the fractal-like manifolds satisfy the following sub-Gaussian estimates when t is large, and also larger than $d(x, y)$:

$$p_t(x,y) \simeq \frac{1}{V(x,t^{1/m})}\exp\Bigg(-\Big(\frac{d^m(x,y)}{t}\Big)^{1/(m-1)}\Bigg),$$

for some $m > 2$ (see [3, 19]). We will not pursue this matter here.

Definition 2.1. We say that a weighted manifold (M, g, μ) satisfies the relative Faber – Krahn inequality if

$$\text{(FK)} \qquad \lambda_1(\Omega) \ge \frac{c}{r^2}\Big(\frac{V(x,r)}{\mu(\Omega)}\Big)^{2/\nu},$$

for some $c, \nu > 0$, all $x \in M, r > 0$ and $\Omega \subset B(x, r)$. Here $\lambda_1(\Omega)$ is the first eigenvalue of the Laplace operator on Ω with Dirichlet boundary conditions:

$$\lambda_1(\Omega) = \inf\Big\{\frac{\int_\Omega |\nabla u|^2\, \mathrm{d}\mu}{\int_\Omega u^2\, \mathrm{d}\mu}, u \in \mathcal{C}_0^\infty(\Omega)\Big\}.$$

Definition 2.2. We say that (M, g, μ) satisfies the Poincaré inequality if there exists $C > 0$ such that for any ball $B(x, r) \subset M$ and every f with $f, \nabla f$ locally integrable,

$$\text{(P)} \qquad \int_B |f - f_B|^2\, \mathrm{d}\mu \le Cr^2 \int_B |\nabla f|^2\, \mathrm{d}\mu,$$

where $f_A = \big(1/\mu(A)\big) \int_A f(x) d\mu(x)$.

Definition 2.3. We say that the Davies – Gaffney estimate holds if for any function f_1 supported in E and f_2 supported in F, one has

$$\langle \mathrm{e}^{-t\Delta} f_1, f_2\rangle \le \mathrm{e}^{-d^2(E,F)/(4t)} \|f_1\|_2 \|f_2\|_2.$$

Remark 2.4. One can prove that

- The Davies – Gaffney estimate is valid for Laplace – Beltrami operators on general complete Riemannian manifolds, and many more second-order self-adjoint operators.
- The Davies – Gaffney estimate is equivalent to the finite speed propagation of the associated wave equation.

For this, see [10, 25].

Theorem 2.5. *Let (M, g, μ) be a connected complete noncompact manifold, then*

$$\text{(FK)} \iff \text{(D)} + \text{(UE)} \iff \text{(D)} + \text{(DUE)}.$$

Remark 2.6. About the above theorem and the relationship between upper and lower heat kernel estimates:

(1) The characterization of (UE) in terms of (FK) is due to Grigory′an, see for instance [16].

(2) Several methods are known to obtain the full Gaussian bound (UE) from the on-diagonal bound (DUE) under the doubling condition: Davies's perturbation method (see [12]), the integrated maximum principle (see [16]) and the finite propagation speed for the wave equation (see [25]). More recently, Coulhon and Sikora ([10]) obtained the same result by using a simple complex analysis lemma.

- The integrated maximum principle ([16]) says that, for any $x \in M$, the weighted integral of the heat kernel:
$$I(x,t) = \int_M p_t^2(x,y)\exp\left(\frac{d^2(x,y)}{2t}\right) \mathrm{d}\mu(y)$$
is nonincreasing in t.
- The proof of Coulhon and Sikora: consider the polynomial case (for the doubling case, see [10, Section 4.2]). In fact, (DUE) is equivalent to
$$\|\mathrm{e}^{-t\Delta}\|_{1\to\infty} \leq Ct^{-D/2}, \quad \forall t > 0.$$
It follows that
$$\|\mathrm{e}^{-z\Delta}\|_{1\to\infty} \leq C(\Re z)^{-D/2}, \forall z \in \mathbb{C}.$$
In order to obtain (UE), the idea is to interpolate this estimate with the Davies – Gaffney estimate, and to use complex analysis via a lemma of Phragmén – Lindelöf type.

(3) We can also derive the on-diagonal lower estimate (DLE) from the upper estimates (UE), using the stochastic completeness of the manifold.

(4) One can not deduce (LE) from (DLE). A counterexample is the above-mentioned connected sum of two copies of $\mathbb{R}^D$, see [4].

(5) The lower estimate implies the upper one as well as doubling:
$$\text{(LE)} \implies \text{(D)} + \text{(UE)}.$$
The fact that
$$\text{(D)} + \text{(LE)} \implies \text{(D)} + \text{(UE)}$$
is due to Grigory′an, Hu, and Lau ([17]) in the polynomial growth case, and to my student Salahaddine Boutayeb ([5]) in the general doubling case. For the fact that
$$\text{(LE)} \implies \text{(D)},$$
see [17].

Theorem 2.7 ([23]). *Let (M, g, μ) be a connected complete noncompact manifold. Then*
$$\text{(LY)} \iff \text{(D)} + \text{(P)}.$$

Theorem 2.8 ([11]). *Assume* (D), *then* (G) *implies* (DUE) *and thus* (LY).

Finally, among the manifolds with (D) one has three distinguished subclasses, in strictly decreasing order: the ones with (UE) (or (DUE), or (FK)), and therefore (DLE), the ones with (LE) (or (P), or (LY)), the ones with (G) (and (LY)). It would be interesting to have a somewhat concrete characterization of (G). We treat this topic in the polynomial growth case in Theorem 4.3 below.

3. Riesz transform and heat kernel estimates

In this section, we mainly focus on noncompact Riemannian manifolds. Let (M, g, μ) be a weighted Riemannian manifold as before. For all $f \in C_0^\infty(M)$, one has by definition of the Laplace – Beltrami operator

$$(\Delta f, f) = \||\nabla f|\|_2^2.$$

Now, by spectral theory,

$$(\Delta f, f) = \|\Delta^{1/2} f\|_2^2.$$

Consequently,

$$\||\nabla f|\|_2 = \|\Delta^{1/2} f\|_2.$$

We can define the Riesz transform operator $\nabla \Delta^{-1/2}$. The L^p boundedness of the Riesz transform is equivalent to the existence of $C > 0$ such that

$$(\mathrm{R}_p) \qquad \||\nabla f|\|_p \leq C \|\Delta^{1/2} f\|_p.$$

It is also interesting to consider the reverse inequality:

$$(\mathrm{RR}_p) \qquad \||\nabla f|\|_p \geq c \|\Delta^{1/2} f\|_p.$$

If both (R_p) and (RR_p) hold, one has the equivalence:

$$(\mathrm{E}_p) \qquad \||\nabla f|\|_p \simeq \|\Delta^{1/2} f\|_p.$$

In this case, one can identify the first order homogeneous Sobolev spaces endowed with seminorms $\||\nabla \cdot|\|$ and $\|\Delta^{-1/2} \cdot\|$ respectively. The question of the validity of (E_p) on noncompact Riemannian manifolds has been asked by Strichartz in 1983 ([26]).

Remark 3.1. Relationship between (R_p) and (RR_p):

(1) (R_p) implies $(\mathrm{RR}_{p'})$ where p' is the conjugate exponent to p; see, for example, [9] for a proof.

(2) The converse is false, but the conjunction of $(\mathrm{RR}_{p'})$ and (Π_p) implies (R_p) (see [1]).

Here (Π_p) says that the Hodge projector of 1-forms onto exact forms is L^p bounded, i.e., for all $\omega \in C_0^\infty(M)$,

$$(\Pi_p) \qquad \||d\Delta^{-1}\delta f|\|_p \leq C_p \|\omega\|_p.$$

Theorem 3.2 ([8]). *Let (M, g, μ) be a complete noncompact Riemannian manifold. Assume that* (D) *and* (DUE) *hold. Then the Riesz transform $T = \nabla \Delta^{-1/2}$ is weak $(1,1)$ and bounded on L^p for $1 < p \leq 2$.*

Proof. The L^p boundedness for $1 < p < 2$ follows by interpolation from the weak $(1,1)$ and L^2 boundedness.

Write

$$\nabla \Delta^{-1/2} = \int_0^\infty \nabla \mathrm{e}^{-t\Delta} \frac{\mathrm{d}t}{\sqrt{t}}.$$

Denote the kernel of the Riesz transform by $K(x, y)$; then

$$K(x, y) = \int_0^\infty \nabla p_t(x, y) \frac{\mathrm{d}t}{\sqrt{t}}.$$

An integral estimate (as opposed to the much stronger pointwise estimate (G)) of the derivative of heat kernel follows from a weighted integral estimate of the gradient of the heat kernel due to to Grigor′yan (see for instance [16]). Then use the singular integral methods developed by Duong – MacIntosh [14]. □

Remark 3.3. Under the same assumptions, one can not get the L^p boundedness of the Riesz transform for $p > 2$. The counterexample given by Coulhon – Duong [8] is again the connected sum M_D of two copies of $\mathbb{R}^D$. Indeed, such a manifold satisfies

$$V(x,r) \simeq r^D,$$
$$p_t(x,y) \simeq t^{-D/2},$$

as well as, for $p > D$,

$$|f(x) - f(y)| \leq Cd(x,y)^{1-D/p}\||\nabla f|\|_p, \quad \forall f \in C_0^\infty(M_D), x, y \in M.$$

If (R_p), $p > D$, holds, we can deduce (LE), which, as we already said, is known to be wrong on M_D (see [4]). One can take $D = 2$.

The case $p > 2$ can be treated in the following way.

Theorem 3.4 ([2])**.** *Let (M, g, μ) be a complete noncompact Riemannian manifold satisfying* (D) *and* (DUE)*. If* (G) *holds, then the Riesz transform is bounded on L^p and* (E_p) *holds for $1 < p < \infty$.*

Theorem 3.5 ([2])**.** *Let (M, g, μ) be a complete noncompact Riemannian manifold satisfying* (LY)*. Let $p_0 > 2$. The following assertions are equivalent:*

(1) (R_p) *holds for $2 < p < p_0$.*
(2) *For all $2 < p < p_0$, there exists C_p such that for all $t > 0$,*

$$(\mathrm{G}_p) \qquad \||\nabla e^{-t\Delta}|\|_{p\to p} \leq \frac{C_p}{\sqrt{t}}.$$

(3) *For all $2 < p < p_0$, for all $y \in M$ and $t > 0$,*

$$(3.1) \qquad \||\nabla_x p_t(.,y)|\|_p \leq \frac{C_p}{\sqrt{t}[V(y,\sqrt{t})]^{1-1/p}}.$$

Note that (RR_p) for $1 < p < 2$ implies a tight relationship between the isoperimetric dimension and the heat kernel dimension (for a definition, see [7]). For example, the isoperimetric dimension of the Vicsek manifold equals 1, while its heat kernel dimension is $2D/(D+1) > 1$ where D is the growth exponent of the volume of large balls. As a consequence, one can show that (RR_p) does not hold on Vicsek manifolds.

Proposition 3.6 ([9])**.** *For every $p_0 \in (1,2)$, there exists a complete noncompact Riemannian manifold with bounded geometry and polynomial volume growth such that* (RR_p) *is false for every $p \in (1, p_0)$.*

Proof. Let M be a Vicsek manifold, which is such that for some $D > 1$ (arbitrarily large),

$$V(x,r) \simeq r^D, \qquad r \geq 1,$$
$$\sup_{x \in M} p_t(x,y) \simeq t^{-D/(D+1)}, \qquad t \geq 1$$

(see [3]). Let $D' = 2D/(D+1)$. Then the Nash inequality (4.3) below holds for D' instead of D, and also for $p \in (1, +\infty)$

$$\|f\|_p^{1+p/((p-1)D')} \leq C\|f\|_1^{p/((p-1)D')}\|\Delta^{1/2} f\|_p$$

(in fact, one should localize this inequality properly at infinity, unless $\dim M \le D$; for details, see [9]). Assume (RR_p), then

$$\|f\|_p^{1+p/((p-1)D')} \le C\|f\|_1^{p/((p-1)D')}\||\nabla f|\|_p. \tag{3.2}$$

We see that (3.2) also holds on the Vicsek graph Γ, the discretization of M. Taking $f = \chi_\Omega$ for $\Omega \subset \Gamma$, we finally obtain

$$|\Omega|^{1-p/D'} \le C|\partial\Omega|. \tag{3.3}$$

For $p < D'$, this contradicts the fact that the isoperimetric dimension of Γ is not better than 1. □

Theorem 3.7 ([1]). *Let M be a complete noncompact Riemannian manifold satisfying* (D) *and* (P). *Then there exists $\varepsilon > 0$ such that* (R_p) *holds for $2 < p < 2 + \varepsilon$.*

Remark 3.8. (D) and (P) self-improve to

$$|p_t(x,y) - p_t(x,z)| \le \frac{C}{V(x,\sqrt{t})}\left(\frac{d(y,z)}{\sqrt{t}}\right)^\alpha \exp\left(-\frac{d^2(x,y)}{Ct}\right),$$

for any $x \in M, t > 0$ such that $d(y,z) \le \sqrt{t}$ and some $\alpha \in (0,1)$. The limit case of this, which is in general not satisfied under (D) and (P), is the gradient estimate (G). It would be very interesting to make a quantitative connection between ε and α.

4. Heat kernel estimates and Sobolev-type inequalities

4.1. Polynomial growth case. Now we come back to manifolds with polynomial growth. Consider a metric measure space (M, d, μ) and assume, for some $D > 0$,

$$V(x,r) \simeq r^D, \quad \forall x \in M, r > 0. \tag{4.1}$$

The on-diagonal upper estimate of the heat kernel (DUE) reads

$$p_t(x,x) \le Ct^{-D/2}, \quad \forall x \in M, t > 0, \tag{$*$}$$

which is equivalent to many Sobolev-type inequalities, including the Gagliardo–Nirenberg inequality, i.e., for any $q \ge 1$, $\alpha > D/p$, there exists C such that

$$\|f\|_\infty \le C\|f\|_q^{1-D/(\alpha q)}\|\Delta^{\alpha/2}\|_q^{D/(\alpha q)}.$$

Theorem 4.1 (see for instance [7]**).** *The following conditions are equivalent:*

(1) *The upper bound* $(*)$ *holds.*
(2) *The Sobolev inequality:*

$$\|f\|_{pD/(D-\alpha p)} \le C\|\Delta^{\alpha/2} f\|_p, \quad \forall f \in C_0^\infty(M), \tag{4.2}$$

for $p > 1$ and $0 < \alpha p < D$.
(3) *The Nash inequality:*

$$\|f\|_2^{2+(4/D)} \le C\|f\|_1^{4/D}\mathcal{E}(f,f), \quad \forall f \in C_0^\infty(M) \tag{4.3}$$

where $\mathcal{E}$ is the Dirichlet form associated with the Laplace–Beltrami operator.

(4) *The Gagliardo – Nirenberg-type inequalities*:

$$\|f\|_q \le C\|f\|_2^{1-(q-2)/(2q)D}\mathcal{E}(f,f)^{(q-2)/(2q)D}, \quad \forall f \in \mathcal{C}_0^\infty(M), \tag{4.4}$$

for $q > 2$ *such that* $(q-2)/(2q)D < 1$.

Instrumental in the above equivalences is the following extrapolation lemma.

Lemma 4.2. *Let* $(T_t)_{t>0}$ *be a semigroup which is equicontinuous on* $L^1(M,\mu)$ *and* $L^\infty(M,\mu)$. *If there exist* $\alpha > 0$ *and* $1 \le p < q \le +\infty$ *such that*

$$\|T_t\|_{p\to q} \le Ct^{-\alpha}, \quad \forall t > 0,$$

then

$$\|T_t\|_{1\to\infty} \le Ct^{-\beta}, \quad \forall t > 0,$$

with $\beta = \alpha/(1/p - 1/q)$.

What is more surprising than Theorem 4.1 is that the boundedness of the Riesz transform can be expressed in the form of a Gagliardo – Nirenberg-type inequality, with the gradient in the left-hand side.

Theorem 4.3 ([11]). *Assume that* M *has Ricci curvature bounded from below and satisfies* (4.1). *Then*

$$|\nabla_x p_t(x,y)| \le Ct^{-(D+1)/2}\exp\Big(-\frac{d^2(x,y)}{Ct}\Big), \forall x, y \in M, t > 0, \tag{4.5}$$

holds for some $C > 0$ *if and only if*

$$\operatorname{Lip}(f) := \||\nabla f|\|_\infty \le C\|f\|_q^{1-(D+q)/(\alpha q)}\|\Delta^{\alpha/2}\|_q^{(D+q)/(\alpha q)}, \quad \forall f \in \mathcal{C}_0^\infty(M), \tag{4.6}$$

for some $q \in (1,\infty)$ *and some* $\alpha > D/q + 1$.

It follows from Theorems 4.3 and 3.4 that, under the assumptions of Theorem 4.3, if (4.5) or (4.6) is satisfied, then (E_p) holds on M for all $1 < p < +\infty$.

4.2. Generalization in the doubling case. Assume (D) and denote $V(x,r)$ by $V_r(x)$. We consider the following inequalities:

$$\text{(N)} \qquad \|f\|_2^2 \le C\big(\|fV_r^{-1/2}\|_1^2 + r^2\mathcal{E}(f)\big), \qquad \forall r > 0, f \in \mathcal{C}_0^\infty(M).$$

$$(\mathrm{GN}_q) \qquad \|fV_r^{1/2-1/p}\|_q^2 \le C\big(\|f\|_2^2 + r^2\mathcal{E}(f)\big), \qquad \forall f \in \mathcal{C}_0^\infty(M).$$

$$\text{(KN)} \qquad \|f\|_2^2 \le C\Big(\frac{\|f\|_1^2}{\inf_{x\in\operatorname{supp}(f)} V_r(x)} + r^2\mathcal{E}(f)\Big), \quad \forall r > 0, f \in \mathcal{C}_0^\infty(M).$$

One sees easily that, in the polynomial volume growth case, (N) and (KN) reduce to the Nash inequality (4.3) and (GN_q) reduces to the Gagliardo – Nirenberg inequality (4.4). Note that (KN) was introduced by Kigami [20].

Theorem 4.4 ([6]). *Assume that* M *satisfies* (D). *Then* (DUE) *is equivalent to* (N).

Theorem 4.5 ([6]). *Assume that* M *satisfies* (D). *Take* $q > 2$ *such that* $(q-2)/(2q)\nu < 1$, *where* ν *is as in* (D_ν). *Then* (DUE) *is equivalent to* (GN_q).

Other relations:

- (GN_q) implies (N). This is a consequence of Hölder's inequality.
- The scale-invariant Poincaré inequalities (P) imply (N).

The fact that (DUE) implies (N) and (GN_q) is easy. The converse requires the use of the following weighted $L^p - L^q$ estimates as well as a suitable generalization of Lemma 4.2.

Definition 4.6. For a function $W\colon M \to \mathbb{R}$ (or $W\colon M \times \mathbb{R}_+ \to \mathbb{R}$), denote by M_W (or $M_W(\cdot, t)$) the operator of multiplication by W.

Let $1 \leq p \leq q \leq \infty$ and γ, δ be real numbers such that $\gamma + \delta = 1/p - 1/q$. We denote

$$(\mathrm{VEV}_{p,q,r}) \qquad \|M_{V^{\gamma}_{\sqrt{t}}} e^{-tA} M_{V^{\delta}_{\sqrt{t}}}\|_p^q < \infty.$$

We abbreviate ($\mathrm{VEV}_{p,q,r}$) by ($\mathrm{VE}_{p,q}$) if $\delta = 0$ and ($\mathrm{EV}_{p,q}$) if $\gamma = 0$. The starting point of the proof of Theorems 4.4 and 4.5 is the following.

Proposition 4.7 ([6]).

$$(\mathrm{DUE}) \iff (\mathrm{EV}_{1,2}) \iff (\mathrm{VE}_{2,\infty}) \iff (\mathrm{VEV}_{1,\infty,1/2}).$$

Acknowledgement. We thank the referee for a very careful reading of the manuscript.

References

1. P. Auscher and T. Coulhon, *Riesz transform on manifolds and Poincaré inequalities*, Ann. Sc. Norm. Super. Pisa Cl. Sci. (5) **4** (2005), no. 3, 531–555.
2. P. Auscher, T. Coulhon, X. T. Duong, and S. Hofmann, *Riesz transform on manifolds and heat kernel regularity*, Ann. Sci. École Norm. Sup. (4) **37** (2004), no. 6, 911–957.
3. M. Barlow, T. Coulhon, and A. Grigor′yan, *Manifolds and graphs with slow heat kernel decay*, Invent. Math. **144** (2001), no. 3, 609–649.
4. I. Benjamini, I. Chavel, and E. A. Feldman, *Heat kernel lower bounds on Riemannian manifolds using the old ideas of Nash*, Proc. London Math. Soc. (3) **72** (1996), no. 1, 215–240.
5. S. Boutayeb, *From lower to upper estimates of heat kernels in doubling spaces*, Tbil. Math. J. **2** (2009), 61–76.
6. S. Boutayeb, T. Coulhon, and A. Sikora, *Heat kernel upper bounds and one-parameter weighted inequalities in doubling spaces*, in preparation.
7. T. Coulhon, *Heat kernel and isoperimetry on non-compact Riemannian manifolds*, Heat Kernels and Analysis on Manifolds, Graphs, and Metric Spaces (Paris, 2002) (P. Auscher, T. Coulhon, and A. Grigor′yan, eds.), Contemp. Math., vol. 338, Amer. Math. Soc., Providence, RI, 2003, pp. 65–99.
8. T. Coulhon and X. T. Duong, *Riesz transforms for* $1 \leq p \leq 2$, Trans. Amer. Math. Soc. **351** (1999), no. 3, 1151–1169.
9. ———, *Riesz transform and related inequalities on noncompact Riemannian manifolds*, Comm. Pure Appl. Math. **56** (2003), no. 12, 1728–1751.
10. T. Coulhon and A. Sikora, *Gaussian heat kernel upper bounds via the Phragmén-Lindelöf theorem*, Proc. Lond. Math. Soc. (3) **96** (2008), no. 2, 507–544.
11. ———, *Riesz meets Sobolev*, Colloq. Math. **118** (2010), no. 2, 685–704.
12. E. B. Davies, *Heat kernels and spectral theory*, Cambridge Tracts in Math., vol. 92, Cambridge Univ. Press, Cambridge, 1989.
13. N. Dungey, A. F. M. ter Elst, and D. W. Robinson, *Analysis on Lie groups with polynomial growth*, Progr. Math., vol. 214, Birkhäuser, Boston, MA, 2003.
14. X. T. Duong and A. MacIntosh, *Singular integral operators with non-smooth kernels on irregular domains*, Rev. Mat. Iberoamericana **15** (1999), no. 2, 233–265.
15. A. Grigor′yan, *Estimates of heat kernels on Riemannian manifolds*, Spectral Theory and Geometry (Edinburgh, 1998) (E. B. Davies and Yu. Safarov, eds.), London Math. Soc. Lecture Note Ser., vol. 273, Cambridge Univ. Press, Cambridge, 1999, pp. 140–225.
16. ———, *Heat kernel and analysis on manifolds*, AMS/IP Stud. Adv. Math., vol. 47, Amer. Math. Soc., Providence, RI, 2009.

17. A. Grigor′yan, J. Hu, and K.-S. Lau, *Heat kernels on metric spaces with doubling measure*, Fractal Geometry and Stochastics IV (Greifswald, 2008) (C. Bandt, P. Mörters, and M. Zähle, eds.), Progr. Probab., vol. 61, Birkhäuser, Basel, 2009, pp. 3–44.
18. P. Gyrya and L. Saloff-Coste, *Neumann and Dirichlet heat kernels in inner uniform domains*, Astérisque **336** (2011).
19. W. Hebisch and L. Saloff-Coste, *On the relation between elliptic and parabolic Harnack inequalities*, Ann. Inst. Fourier (Grenoble) **51** (2001), no. 5, 1437–1481.
20. J. Kigami, *Local Nash inequality and inhomogeneity of heat kernels*, Proc. London Math. Soc. (3) **89** (2004), no. 2, 525–544.
21. P. Li and S.-T. Yau, *On the parabolic kernel of the Schrödinger operator*, Acta Math. **156** (1986), no. 3-4, 153–201.
22. E. M. Ouhabaz, *Analysis of heat equations on domains*, London Math. Soc. Monogr. Ser., vol. 31, Princeton Univ. Press, Princeton, NJ, 2005.
23. L. Saloff-Coste, *Aspects of Sobolev-type inequalities*, London Math. Soc. Lecture Note Ser., vol. 289, Cambridge Univ. Press, Cambridge, 2002.
24. ______, *Analysis on Riemannian co-compact covers*, Surveys in Differential Geometry. Vol. IX (A. Grigor′yan and S.-T. Yau, eds.), Surv. Differ. Geom., vol. 9, Int. Press, Somerville, MA, 2004, pp. 351–384.
25. A. Sikora, *Sharp pointwise estimates on heat kernels*, Quart. J. Math. Oxford Ser. (2) **47** (1996), no. 187, 371–382.
26. R. S. Strichartz, *Analysis of the Laplacian on the complete Riemannian manifold*, J. Funct. Anal. **52** (1983), no. 1, 48–79.
27. K.-T. Sturm, *Analysis on local Dirichlet spaces.* I: *Recurrence, conservativeness and L^p-Liouville properties*, J. Reine Angew. Math. **456** (1994), 173–196.
28. ______, *Analysis on local Dirichlet spaces.* II: *Upper Gaussian estimates for the fundamental solutions of parabolic equations*, Osaka J. Math. **32** (1995), no. 2, 275–312.
29. ______, *Analysis on local Dirichlet spaces.* III: *The parabolic Harnack inequality*, J. Math. Pures Appl. (9) **75** (1996), no. 3, 273–297.

Mathematical Sciences Institute, The Australian National University, Canberra ACT 0200, Australia

E-mail address: thierry.coulhon@anu.edu.au

Centre de Recherches Mathématiques
CRM Proceedings and Lecture Notes
Volume **56**, 2013

Regularity of Minimal and Almost Minimal Sets and Cones: J. Taylor's Theorem for Beginners

Guy David

ABSTRACT. We discuss various settings for the Plateau problem, a proof of J. Taylor's regularity theorem for 2-dimensional almost minimal sets, some applications, and potential extensions of regularity results to the boundary.

1. Introduction

The main purpose of this text is to present some of the techniques used to prove the local regularity of minimal and almost minimal sets in a domain of $\mathbb{R}^n$. The notion of minimality that we shall use is a minor modification of Almgren's notion of "restricted sets" (see [3]), which seems to be very good to describe soap films or bubbles away from the boundary sets where the films are attached.

We shall first present a natural context where Almgren minimal sets arise (typically, Plateau problems for soap films). This will provide a motivation for the rest of the paper, but we shall refer to an oncoming paper in a conference proceedings [17] for more detail about the different settings for Plateau problems.

In the present text, we shall insist more on regularity properties of minimizers away from the boundary, and in particular we shall explain a proof of Jean Taylor's theorem from [59], which says that if E is a two-dimensional soap film or bubble in $\mathbb{R}^3$ (we shall say, an almost minimal set), then each interior point of E has a neighborhood where E is C^1-equivalent to a minimal cone. In this case, the list of minimal cones is known; there are only three simple types, which can be easily seen in real soap films and bubbles.

Incidentally, we give a slightly different meaning to films and bubbles here. For soap films, we think about sets that locally minimize the area, which in the case of smooth surfaces would imply that the mean curvature vanishes. For soap bubbles, there may be air trapped inside, which leads to minimizing a slightly more complicated functional and would in the smooth case yield surfaces with constant mean curvature (given by the difference between two pressures). In the case of bubbles, and also if we add smaller forces, our sets will only be almost minimal (but most of the known theorems will still be true with similar proofs).

After the initial presentation of Plateau problems (and some easier variants) in Section 2, we shall define minimal and almost minimal sets, and list the known

2010 *Mathematics Subject Classification*. 49Q20.
This is the final version of the paper.

general interior regularity results, which happen to be valid in larger classes of quasi-minimal sets: these sets (of all dimensions d) are locally Ahlfors-regular and uniformly rectifiable, the Hausdorff measure $\mathcal{H}^d$ is lowersemicontinuous along sequences of reduced (uniformly) quasi-minimal sets, and finally the various classes are stable under limits (Section 3).

Then we shall state J. Taylor's result and a partial generalization to two-dimensional almost minimal sets in higher ambient dimensions, and describe the main ingredients of a proof (a long Section 4).

In Section 5 we shall address a Bernstein problem (if E is a reduced minimal set of dimension 2 in the whole $\mathbb{R}^3$, is it a cone?), and its variant in the slightly different Mumford – Shah setting.

In Section 6 we shall mention two cases of simpler variants of a Plateau problem (but without boundary conditions) for which the regularity results presented above lead to existence theorems.

We shall also rapidly present a program of extension of some regularity results, all the way up to the boundary, for potential solutions of some type of Plateau problem, with sliding boundary conditions. See Section 7.

This text partially relies on transparencies that the author used for series of lectures in Evian (2008) and Grenoble (2010), and its content is somewhat different from the lectures in Montréal. The author wishes to thank the organizers of the seminar for their kind welcome and perfect organization, Ken Brakke for the authorization to use pictures from his site, T. De Pauw, F. Morgan, and V. Feuvrier for useful discussions and help with the references, and the referee for a careful and indulgent reading.

2. Plateau problems

Joseph Plateau (1801 – 1883) was really interested in soap films, and in particular described the typical singularities that will be mentioned below. See the book [53] from 1873, which is also very interesting for the description of lots of other funny experiments in physics. Some people also mention Lagrange (near 1760) in connection with the Plateau problem; he was clearly interested in variational problems, but I am not sure about soap films in particular.

Plateau's problem usually refers to the question of existence of sets of minimal area that are bounded by a given set (typically, a curve in $\mathbb{R}^3$). Everyone agrees on this, but the precise meaning of the words "area" and "bounded by" may differ considerably. This is why I decided to talk about Plateau problems in the plural.

In this section we shall mention a few ways to state a Plateau problem; this will provide a good motivation for the rest of the paper, but we shall not try to be exhaustive or announce a new solution to any Plateau problem. Please forgive the very schematic descriptions below; the author will try to be more precise and give more simple examples in [17].

2.1. Douglas and the parameterized surfaces. Let $\Gamma \subset \mathbb{R}^n$ be a reasonably smooth simple loop. Denote by D the unit disk in $\mathbb{R}^2$, and parameterize Γ by $g\colon \partial D \to \mathbb{R}^n$. The simplest way to state a Plateau problem is to look for Lipschitz mappings $f\colon D \to \mathbb{R}^n$, such that $f_{|\partial D} = g$, and which minimize the area

$$A(f) = \int_D J_f(x)\, dx, \tag{2.1}$$

where $J_f(x)$ denotes the positive Jacobian of f at x.

This looks nice, but there are two problems with this approach. The first one is that many functions g and even more functions f may describe the same objects Γ and $f(D)$, and it is not always clear which ones to choose. Also, and as a consequence of this too, we get the following obvious complication when we try to prove an existence result. Take a minimizing sequence $\{f_k\}$; this means that

$$\lim_{k\to+\infty} A(f_k) = \inf\{A(f) : f\colon D \to \mathbb{R}^n \text{ is Lipschitz and } f_{|\partial D} = g\}. \tag{2.2}$$

We cannot assume that the Lipschitz constant for f_k stays bounded (we would be solving a quite different problem). So, even if (usually after extracting a subsequence) the sets $f_k(D)$ converge nicely to a set E, it is unlikely a priori that the f_k will converge to an acceptable limit f, instead of having badly degenerating Lipschitz constants, or even that E will remember something useful from (2.2).

In the present case where Γ is a curve and we look for a 2-dimensional surface, we can decide to use conformal parameterizations, which have the good properties to exist in dimension 2, and to have good compactness properties. This approach was taken (among many others) by R. Garnier [32], and Tibor Radó [54] (1930). In 1931, J. Douglas [25] obtained an optimal existence theorem, using the following simple but bright idea. We decide that f will be the harmonic extension of g in D (not unrealistic given what conformal parameterizations can be), translate everything in terms of g, and get that we just need to minimize the functional

$$B(g) = \int_0^{2\pi}\int_0^{2\pi}\sum_{j=1}^{n}\frac{|g_j(\theta)-g_j(\varphi)|^2}{\sin^2((\theta-\varphi)/2)}\,d\theta\,d\varphi, \tag{2.3}$$

where the g_j are just the coordinates of g. It turns out that this is easy to do.

This is beautiful, and the paper [25] is very pleasant to go through, but the problem that Douglas solves is not exactly what Plateau intended; the surface $g(D)$ that he gets may cross itself, and near a crossing point everything happens as if $f(D)$ was composed of completely independent pieces that happen to pass through the same point. Soap films do not do that: the different pieces would interact and give different singularities. Typically, they would do one of two things. Either the two pieces would merge along some set (think about pinching two pieces of surfaces together), taking advantage that this makes the surface measure of the set $f(D)$ (which is what the soap film really minimizes) strictly smaller than $A(f)$. Or some hole would open, giving a minimal set that has a different topology and is no longer injectively parameterized by a disk.

2.2. Reifenberg and the homology groups. Even if our boundary data Γ is a nice curve, we do not necessarily want to assume that our solutions are smooth surfaces, so we shall use Hausdorff measures to define their area. Let us recall that for $d \geq 0$ and $E \subset \mathbb{R}^n$, the d-dimensional (exterior) Hausdorff measure of E is

$$\mathcal{H}^d(E) = \lim_{\delta\to 0}\mathcal{H}^d_\delta(E), \tag{2.4}$$

where

$$\mathcal{H}^d_\delta(E) = c_d \inf\Big\{\sum_{j\in\mathbb{N}} \operatorname{diam}(D_j)^d\Big\}, \tag{2.5}$$

c_d is a normalizing constant, and the infimum is taken over all coverings of E by a countable collection $\{D_j\}$ of sets, with $\mathrm{diam}(D_j) \leq \delta$ for all j.

Let us choose the normalizing constant c_d so that $\mathcal{H}^d$ coincides with the Lebesgue measure on subsets of $\mathbb{R}^d$; then $\mathcal{H}^d(E)$ is the total surface of E when E is a smooth d-dimensional surface. The advantage is that $\mathcal{H}^d(E)$ is defined for all $E \subset \mathbb{R}^n$, and its restriction to Borel sets is a Borel measure (not σ-finite, but this is all right).

Return to the Plateau problem. Let $1 \leq d < n$ and a boundary set Γ be given (for instance, $d = 2$ and Γ is a nice closed curve). We want to find a "surface" E "spanned" by Γ, and for which $\mathcal{H}^2(E)$ is minimal.

For Reifenberg [55] (1960), this last condition means that E is a compact set that contains Γ, and the boundary condition is stated in terms of Čech homology on some commutative group G. We require the inclusion $i\colon \Gamma \to E$ to induce a trivial homomorphism from $\check{H}_{d-1}(\Gamma; G)$ to $\check{H}_{d-1}(E; G)$. Or we could also take a subgroup of $\check{H}_{d-1}(\Gamma; G)$, and require that its image vanishes in $\check{H}_{d-1}(E; G)$. In other words, we select a certain number of elements in $\check{H}_{d-1}(\Gamma; G)$, and we want these elements to be boundaries inside E. In the simple case when $d = 1$ and Γ is a curve, we can take the obvious generator γ of $\check{H}_1(\Gamma; G)$, and require that E contains (the support of) a chain that fills γ.

Then we minimize the area $\mathcal{H}^d(E)$ under these constraints.

Reifenberg proves the existence of minimizers in all dimensions, but when $G = \mathbb{Z}_2$ or $G = \mathbb{R}/\mathbb{Z}$. This is a beautiful (although technical) proof by hands, with minimizing sequences and initial haircuts to make the sets look nicer.

Recently, De Pauw obtained the existence in the 2-dimensional case when the boundary is a finite union of curves and $G = \mathbb{Z}$ (using currents); he also proved that in that case the infimum for this problem is the same as for the size-minimizing currents of the next subsection. But even then it is not known whether we can define size-minimizing currents supported on the sets that he gets. See [22].

Reifenberg's solutions are nice and seem to give a good description of many soap films. Using finite groups G like $\mathbb{Z}_2$, one can even get non-orientable sets E like Möbius strips. But there are some "real-life" soap films spanned by a curve that cannot be obtained as Reifenberg solutions. The two films of Figure 1 are like this (there happens to be a problem with the orientation); see K. Brakke's home page for more examples, and maybe [17] for a short discussion.

Anyway, a more general existence result (as in [22] but in the general case) would be very welcome.

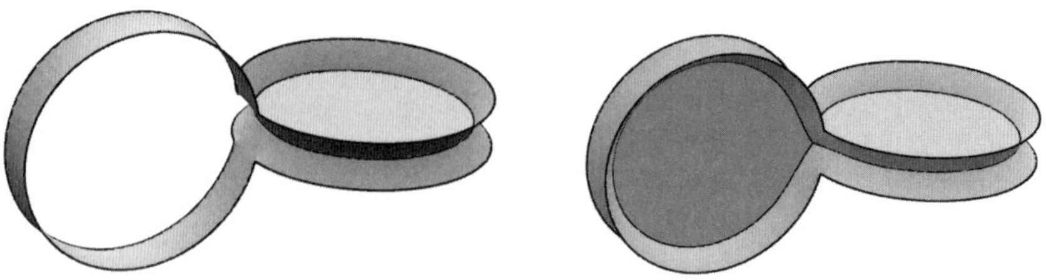

FIGURE 1. Two films that don't fit Reifenberg's setting (K. Brakke).

2.3. Plateau problems for currents. The most celebrated and successful model is probably the description of films in terms of currents, initiated by Federer, Fleming, De Giorgi, and others. See for instance [26, 28].

The most logical way to try to solve a Plateau problem would be to take a minimizing sequence of (smooth) surfaces, take a subsequence that converges in some reasonable topology, and show that the limit is a solution to our problem. The obvious difficulty is that when we choose a topology that is so rough that subsequences converge, the limit is very unlikely to be smooth.

The approach with currents is in the same spirit as for weak solutions for PDE's: first set the initial problem on a much larger class where pleasant compactness theorems exist, prove existence theorems in this class, and then show that the weak solutions that we just found are in fact much more regular than expected, and are acceptable solutions for our initial problem. We need a few definitions.

A d-dimensional *current* is a continuous linear form on the space of smooth d-forms. This is thus the same as a d-vector valued distribution. In fact, most of the distributions that will be used here are (d-vector valued) measures.

The most basic example is the current S' of integration on any smooth, oriented surface S of dimension d, which is simply defined by $\langle S', \omega\rangle = \int_S \omega$ for every d-form ω. But the point of the approach is that currents provide a much larger class of objects, with good compactness properties.

Another standard example that is relevant here is *the rectifiable current* T defined on a d-dimensional rectifiable set E such that $\mathcal{H}^d(E) < +\infty$, on which we choose a measurable orientation τ and an integer-valued multiplicity m.

Recall that a rectifiable set of dimension d is a set E such that $E \subset N \cup \bigcup_{j\in\mathbb{N}} G_j$, where $\mathcal{H}^d(N) = 0$ and each G_j is a C^1 embedded submanifold of dimension d. But we could have said that G_j is the Lipschitz image of a subset of $\mathbb{R}^d$, and obtained an equivalent definition. We shall only consider sets E such that $\mathcal{H}^d(E) < +\infty$ here. For such a set E and $\mathcal{H}^d$-almost every $x \in E$, E has what is called an approximate tangent d-plane $x + V(x)$ at x, which of course coincides with the usual tangent plane in the smooth case. A measurable orientation can be defined as the choice of a d-vector $\tau(x)$ that spans $V(x)$, which is defined $\mathcal{H}^d$-almost everywhere on E and measurable. We set

$$\langle T, \omega\rangle = \int_E m(x)\omega(x)\cdot\tau(x)\, \mathrm{d}\mathcal{H}^d(x), \tag{2.6}$$

where we let the reader guess the precise definition of the number $\omega(x)\cdot\tau(x)$ when ω is a (smooth) d-form.

The *boundary* of a d-dimensional current T is defined by duality with the exterior derivative d on forms, by

$$\langle \partial T, \omega\rangle = \langle T, d\omega\rangle \quad \text{for every } (d-1)\text{-form } \omega. \tag{2.7}$$

When S is a smooth oriented surface with boundary Γ, Green's theorem says that $\partial S' = \Gamma'$. Notice that $\partial\partial = 0$, just because $dd = 0$.

A *normal current* is a rectifiable current T such that ∂T is rectifiable too. We will like to work with normal currents here; the additional constraint on ∂T will not disturb us, because we are interested in solutions of the equation $\partial T = S$, where S defines the boundary constraint and is assumed to be nice.

There are two interesting quantities associated to a current. The first one is the *mass* of T, which is just the operator norm of T, where we put a L^∞-norm on the vector space of d-forms. That is, we see T as a measure, and compute its total mass. When T is the rectifiable current given by (2.6),

$$\text{Mass}(T) = \int_E |m(x)| \, d\mathcal{H}^d(x). \tag{2.8}$$

But we shall also consider the *size* of T, which when T is given by (2.6) is equal to

$$\text{Size}(T) = \mathcal{H}^d(\{x \in E : m(x) \neq 0\}). \tag{2.9}$$

Return to Plateau's problem. The classical way to state it in the setting of currents is to take a $(d-1)$-dimensional integral current S, with $\partial S = 0$, and minimize Mass(T) among all the d-dimensional currents T such that $\partial T = S$. Of course the condition $\partial S = 0$ is needed if we want to solve $\partial T = S$, since $\partial\partial = 0$.

With this setting, the use of currents is a great success. There is a compactness theorem that says the following. Let $\{T_k\}$ be a sequence of normal currents of dimension d, with supports in a fixed compact set in $\mathbb{R}^n$, and assume that $\text{Mass}(T_k) + \text{Mass}(\partial T_k) \leq M$ for some fixed $M < +\infty$. Then there is a subsequence that converges (in some weak norm) to a normal current T. Moreover,

$$\text{Mass}(T) \leq \liminf_{k\to+\infty} \text{Mass}(T_k) \quad \text{and} \quad \text{Mass}(\partial T) \leq \liminf_{k\to+\infty} \text{Mass}(\partial T_k). \tag{2.10}$$

We apply this to a minimizing sequence for the problem above (assuming that S is an integral current that lives on a compact set), and get a current T that solves $\partial T = S$ and minimizes Mass(T). See [26, 28].

Moreover, there are good regularity theorems for mass-minimizing currents, and the solution T that we find is quite nice. For instance, if $d = n - 1 \leq 7$, the closed support of T is a smooth manifold away from the support of S [27].

Unfortunately, mass-minimizing is not a great model for soap films. We can guess that, because we know from experience that unlike mass minimizers, 2-dimensional soap films can have singularities. The problem is the same as for the solutions of Douglas: rather than minimizing the mass of T, we should probably minimize its size Size(T). The difference is illustrated in Figure 2, where the current suggested on the right has multiplicity 2 on the central disk, and a smaller size than the sum of the two disks on the left.

But the situation for size minimizers is far from clear. Even when $d = 2$ and S is the current of integration on a smooth curve Γ, there is no general existence result for an integral current T such that $\partial T = S$ and Size(T) is minimal. The main difficulty is that if we want to apply the compactness theorem above to a minimizing sequence $\{T_k\}$, it is not clear that the masses Mass(T_k) will stay quietly bounded, because we only control the sizes.

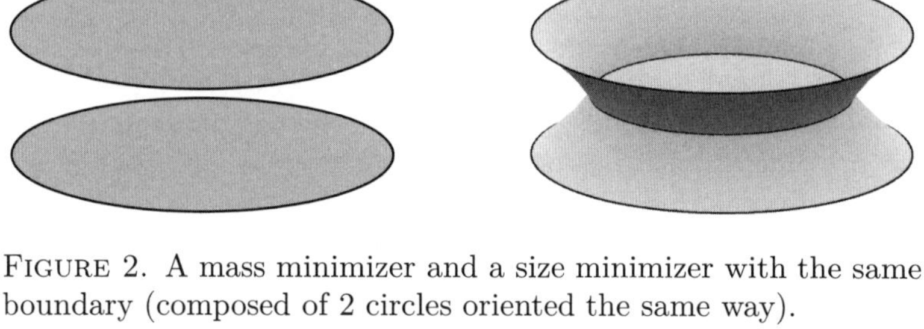

FIGURE 2. A mass minimizer and a size minimizer with the same boundary (composed of 2 circles oriented the same way).

There are some encouraging recent partial results by T. De Pauw and R. Hardt [23] where the scheme would be to minimize $\mathrm{Size}(T)+\varepsilon\,\mathrm{Mass}(T)$, get some uniform control on the minimizers, and let ε tend to 0. Also, when $d=2$ and S comes from a finite collection of smooth closed curves Γ, De Pauw showed that the infimum for the problem is the same as for Reifenberg's problem above, but for the moment failed to show that the solutions for Reifenberg's problem, which exist, give size-minimizing currents.

We should also mention that some real soap films are not orientable (they may be Möbius bands, for instance), and then the problem $\partial T = \Gamma$ does not fit well. Some ad hoc solutions to this exist (see for instance the clever constructions in [8], with covering spaces), but apparently no general scheme is available.

One can go further in the direction of weak solutions, where the idea is that maybe currents are not general enough to describe all the soap films. Almgren [2] used varifolds to give another proof of existence for the Reifenberg solutions. J. Harrison and H. Pugh state the problem and get existence results in terms of even more general objects called differential k-chains (see [33, 34]); the next question in such a context is probably a description of what the obtained minimizers look like, and hence whether they also minimize in a smaller category.

2.4. Sliding Almgren minimizers. Let us propose yet another Plateau problem, where we return to sets and a possibly more natural existence problem, but also where we have no good existence result so far.

We give ourselves a finite collection of simple compact boundary sets $\Gamma_j \subset \mathbb{R}^n$, $0 \leq j \leq j_{\max}$, and an initial candidate E_0 (a compact set such that $\mathcal{H}^d(E_0) < +\infty$).

We want to minimize $\mathcal{H}^d$ in a class $\mathcal{F}(E_0)$ of deformations of E_0 that we define now. We shall say that E is a sliding deformation of E_0, relative to the boundary pieces L_j, when $F = \varphi_1(E_0)$ for a one-parameter family $\{\varphi_t\}$, $0 \leq t \leq 1$, of mappings such that

$$(t,x) \to \varphi_t(x)\colon [0,1]\times E_0 \to \mathbb{R}^n \text{ is continuous,} \tag{2.11}$$

$$\varphi_0(x) = x \quad \text{for } x \in E_0, \tag{2.12}$$

$$\varphi_t(x) \in \Gamma_j \quad \text{when } 0 \leq j \leq j_{\max} \text{ and } x \in \Gamma_j, \tag{2.13}$$

and

$$\varphi_1 \text{ is Lipschitz.} \tag{2.14}$$

We decided to require (2.14) mostly by tradition, but this is negotiable and anyway we do not require any Lipschitz bound for φ_1. Let us denote by $\mathcal{F}(E_0)$ the class of sliding deformations of E_0 (relative to the boundary pieces L_j), and set

$$m = \inf\{\mathcal{H}^d(E) : E \in \mathcal{F}(E_0)\}. \tag{2.15}$$

Obviously $m \leq \mathcal{H}^d(E_0) < +\infty$, but we have to choose E_0 and the L_j so that $m > 0$, because otherwise the problem below is not too interesting. Then we ask for the existence of $E \in \mathcal{F}(E_0)$ such that $\mathcal{H}^d(E) = m$.

A few comments are in order. First, the use of deformations as above is not really new: the definition by Almgren of restricted sets in [3] uses competitors for E_0 that are Lipschitz images $\varphi(E_0)$, even though there is no boundary piece L_j that one wants to preserve there. A similar problem is implicitly mentioned in [59], in the proof of minimality for the cone $\mathbb{T}$ over the edges of a regular tetrahedron. The Surface Evolver software [6] also allows this as an option.

We think of our initial set E_0 as an elastic shower curtain, which is attached to the boundary pieces so that it can slide along them, but not be detached.

The most obvious case is when there is a single Γ_j, which is a nice curve, but we can also take 2-dimensional boundaries (think about a soap film leaning on a tube), or mixed problems. Incidentally, 2-dimensional boundaries in $\mathbb{R}^3$ seem interesting too, and could be easier to treat than curves.

The author likes this problem because it has a nice physical flavor, and also because it is simple and seems to give a lot of flexibility. We do not need to think too much about orienting sets, choosing a group, or defining acceptable multiplicities on them, the initial set E_0 will decide about many things for us, and various choices of E_0 will probably lead to most realistic examples.

We could also ask for local minimizers, where we would only allow as competitors the sets $E \in \mathcal{F}(E_0)$ for which $\mathcal{H}^d\big(\varphi_t(E_0)\big)$ stays below a certain threshold for all t. Of course it is fairly easy to conceive situations where a real soap film E exists and is stable, even though there is a long homotopy that deforms E into a point while preserving the boundary pieces.

But the main problem with this definition of the Plateau problem is that so far there is no good existence result. Again we are messing around with parameterizations (why should there be a nice parameterization φ_1 for the limit of a minimizing sequence?), but we can always hope that we will be able to find minimizing sequences with a good control on parameterizations. See Section 6 for a simpler example where something like this can be done.

In spite of the potential difficulties with this problem, we would like to convince the reader that it is probably interesting to study the regularity of its solutions. Firstly, it is probably a good idea to study the boundary regularity of solutions of Plateau problems in general; to the author's knowledge, very little has been done in this direction, and the sliding setting that we just described seems to be fairly appropriate, both because we can keep some control on the minimizers (because we authorize some sliding of the solutions at the boundary), and because solutions to other Plateau problems (Reifenberg solutions, or the closed supports of size-minimizing currents) may be sliding minimizers as well (when we choose them as the initial E_0). [See [17] for a rapid verification of this fact.] Also, understanding the boundary behavior is possibly an important step if we want to produce solutions, both because we will know better what to expect and because the proofs of regularity for minimizers often come with interesting ways to produce better competitors (and hence better minimizing sequences). We shall give an example of this, in a much simpler context where there is no boundary condition, in Section 6.

Let us say a little more on this topic. One nice way to try and produce existence results is to use weak solutions and an appropriate compactness result. But then it is important to see how regular these weak solutions are, and it would be nice to do this even near the boundary. It is also possible that the compactness result, or the verification of boundary conditions for the weak limit of a minimizing sequence, will be easier if we know that we can restrict to some subclass where we already have some control near the boundary. At the opposite end, it is possible (and in principle much easier) to produce what the author would like to call amnesic solutions to the Plateau problem. In the context of sets (as in this section), a amnesic solution would be a set E such that $\mathcal{H}^d(E) \leq \mathcal{H}^d(F)$ for every $F = \varphi_1(E)$, but only for all the deformations $\{\varphi_t\}$ as above that fix every point of the given boundary set

Γ, or even possibly of a neighborhood of Γ. It is easier to produce such solutions, because you just need to keep a control of a locally minimizing sequence away from Γ, and you don't even need to look precisely at what happens along Γ. But then we cannot say exactly which Plateau problem is solved by our set E, and at the same time its boundary behavior will probably be much harder to predict. In spite of this, the author does not know of any example of a amnesic minimizer that would look suspiciously irregular near the boundary, and after all the minimality away from Γ could imply some regularity at the boundary too.

See Section 7 for some first regularity results for sliding minimizers.

2.5. Other minimization problems. The Plateau problem has the advantage of being simple and being celebrated, but there are other problems with the same flavor, and which are possibly easier to solve. Here we partially repeat suggestions made in [13]. We shall return to this in Section 6, with just a little more detail.

For instance, fix a simple domain $\Omega \subset \mathbb{R}^n$, preferably closed (so that our sets E can lay along a boundary) and with some holes (so that the problem below is not trivial). Also let $M > 0$ and a continuous bounded function $g\colon \Omega \to [1, M]$ be given, and set

$$J_g(E) = \int_E g(x)\,\mathrm{d}\mathcal{H}^d(x) \quad \text{for closed sets } E \subset \Omega. \tag{2.16}$$

Then fix a class $\mathcal{F}$ of closed subsets, and look for $E \in \mathcal{F}$ that minimize $J_g(E)$.

It is reasonable to restrict to classes that are stable under small deformations in Ω. That is, if $E \subset \mathcal{F}$ and if $\{\phi_t\}$, $t \in [0,1]$, is a continuous one-parameter family of mappings $\varphi_t : E \to \Omega$ such that (2.11)–(2.14) hold with the single $\Gamma_j = \Omega$, then $\varphi_1(E) \in \mathcal{F}$.

Maybe also, if we want to restrict to localized deformations, let $r_0 > 0$ be given, and only require that $\varphi_1(E) \in \mathcal{F}$ for families $\{\phi_t\}$ such that in addition there are balls B_t of radius r_0 such that for $0 \leq t \leq 1$,

$$\varphi_t(x) = x \quad \text{for } x \in E \setminus B_t \qquad \text{and} \qquad \varphi_t(E \cap B_t) \subset B_t. \tag{2.17}$$

The additional flexibility provided by the weight g is not so much the issue here; we added it to make the existence and computation of minimal sets less trivial in the following examples. Also, we put this example in a different subsection because the only left boundary piece Ω plays a much smaller role than above, and we expect to get existence results more easily than with the standard Plateau problems.

Now there are different types of classes $\mathcal{F}$ that one can use, with possibly different answer. We expect this to depend mostly on how easy it will be to prove that the limit of a minimizing sequence in $\mathcal{F}$ will lie in $\mathcal{F}$ as well.

Example 2.1 (separation conditions). When $d = n - 1$, we can give ourselves two or more sets A_j in $\mathbb{R}^n \setminus \Omega$, and take for $\mathcal{F}$ the class of sets that separate them. A simple case is when Ω is an annulus, or the difference between a big ball and a smaller solid torus, and the A_j are the components of the complement. Here and below, we need to select Ω and the A_j so that $\inf_{E \in \mathcal{F}} J_g(E) > 0$, which is why we prefer when Ω is not trivial topologically. For such separation problems, we can state the problem in terms of the separated components, use the compactness properties of BV to find sets of finite perimeter that minimize, and obtain minimizers for our initial problem. We do this in [20], for instance, but this is very classical.

Example 2.2 (homology conditions). We try to do the same thing as before, but with possibly higher codimensions. We select smooth surfaces ω_j in $\mathbb{R}^n \setminus \Omega$, that represent non trivial elements in the $(n-d-1)$-dimensional homology of $\mathbb{R}^n \setminus \Omega$, and we let $\mathcal{F}$ be the collection of closed sets $E \subset \Omega$ for which ω_j still represents a nonzero element in the homology of $\mathbb{R}^n \setminus E$. Existence results exist, following X. Liang's thesis [40]. See Section 6 and [42].

Example 2.3 (deformations of a given set). We just proceed as in the last section: we give ourselves an initial closed set $E_0 \subset \Omega$, and call $\mathcal{F}(E_0)$ the class of continuous deformations $E = \varphi_1(E_0)$, where the φ_t are as in (2.11)–(2.13) with the single $\Gamma_j = \Omega$. That is, we are allowed to deform E_0, but points are not allowed to leave Ω. We may (or not) require that φ_1 be Lipschitz, as in (2.14), and the author does not know wether this changes the problem.

For instance, we can take $n = 2$, $d = 1$, $\Omega = \mathbb{R}^2 \setminus [B_1 \cup B_2]$ (two disjoint open balls), $g = 1$, and $E_0 = \partial B(0, R)$ (with R large). Here the minimizers are easy to guess; two cases can occur, depending on whether the balls are far from each other; see the left part of Figure 3.

Or we can take $n = 3$, $d = 2$, $\Omega = \mathbb{R}^3 \setminus A$ for some open solid torus A, $g = 1$, and $E_0 = \partial A$. Again two cases occur; when A is quite thin compared to its diameter, the minimizer should be ∂A itself; when A is quite thick, the minimizer should be the boundary of the convex hull of A, plus an additional wire, which is part of the topological problem but plays no role in the minimization. See the right part of Figure 3.

Example 2.3 may look like Examples 2.1 and 2.2, but here we do not say what topological properties of E_0 prevent it from disappearing into a lower-dimensional subset; thus the minimization problem of Example 2.3 may be more precise than in the previous example. Also, in more complicated situations, it may ve very hard for a nonspecialist to decide whether different topological conditions will lead to really different classes $\mathcal{F}$ or minimizers.

We shall return to this in Section 6 and say why in some cases, existence results can be derived from regularity results for minimizers.

3. Almost minimal sets; general regularity results and limits

We shall now focus on the interior regularity of potential solutions of Plateau problems for sets. We expect the boundary regularity to be somewhat more complicated. We shall only say a few fords about it in Section 7, and in the meantime, we stay inside.

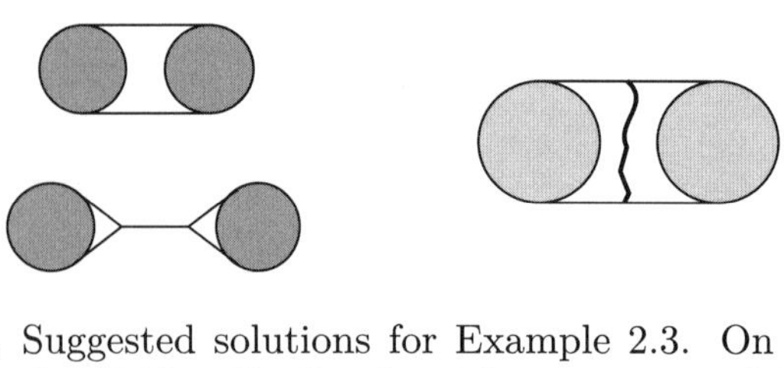

FIGURE 3. Suggested solutions for Example 2.3. On the left: 2 possibilities for E ($d = 2$). On the right: a section of the solution with a wire ($d = 3$).

3.1. Almost minimal sets, reduced sets. The best descriptions of soap films or bubbles away from the boundary seem to given by Almgren's notion of restricted sets, or variants. The definition of almost minimal sets that we give now is a minor modification of Almgren's definition, which we just simplify a little.

We work in an open set $U \subset \mathbb{R}^n$, to make the definition local and avoid boundary problems. We also use a small gauge function $h\colon \mathbb{R}_+ \to \mathbb{R}_+$ to account for perturbations (small additional forces, gently inhomogeneous space, etc.), but please feel free to take $h = 0$ and concentrate on minimal sets. For the moment we just assume that h is continuous, nondecreasing, and such that

$$\lim_{r \to 0} h(r) = 0, \tag{3.1}$$

but a standard choice would be $h(r) = Cr^\alpha$, with $\alpha > 0$ and $C \geq 0$.

We consider closed sets $E \subset U$ with locally finite $\mathcal{H}^d$-measure and (to make things simpler) only define competitors F for E in compact balls $B \subset U$.

Definition 3.1. Let $B = \overline{B}(x,r) \subset U$ be a compact ball in U. A *competitor* for E in B is a set $F = \varphi(E)$, where $\varphi\colon U \to U$ is Lipschitz (but no bounds are required), with

$$\varphi(y) = y \quad \text{for } y \in U \setminus B, \qquad \text{and} \qquad \varphi(B) \subset B. \tag{3.2}$$

And we say that the closed set $E \subset U$ is an *almost minimal set* in U, with gauge function h, when for every compact ball $B = \overline{B}(x,r) \subset U$, $\mathcal{H}^d(E \cap B) < +\infty$ and

$$\mathcal{H}^d(E \cap B) \leq \mathcal{H}^d(F \cap B) + r^d h(r) \tag{3.3}$$

for every competitor F for E in B.

Remarks. Again this is a minor variation of Almgren's definition, which also allowed competitors in other compact subsets of U, and used an accounting that was slightly different from (3.3). A few other variants exist, but they would not be significantly different for what we want to say here.

It is an important feature of the definition that φ is *not* required to be injective. That is, we are allowed to compare E with other sets obtained from E by pinching and merging pieces. Generally speaking, in these problems the definition of the competitors is as important as the accounting.

Note that since B is convex, $\varphi = \varphi_1$ for a one-parameter family of continuous mappings φ_t that satisfy (3.2) (take $\varphi_t(x) = t\varphi(x) + (1-t)x$).

It is then clear that solutions of the sliding Plateau problem mentioned in Section 2.4 are minimal sets in the complement of the Γ_j, or rather of the $\partial\Gamma_j$ (when some of the Γ_j are n-dimensional), but this stays true for other Plateau problems. For instance, the closed support of a size minimizing current T is minimal in the complement of the (closed) support of the current S that defines the boundary condition $\partial T = S$. See for instance [17].

The fact that we require φ to be Lipschitz is useful to prove this last fact, and does not make the proofs of regularity of E any harder; otherwise the Lipschitzness of φ is not so important. In the opposite direction, J. Harrison observed that it may be useful to show that we get the same class of almost minimizers if we require the functions φ in Definition 3.1 to be smooth. This seems true, but we did not write the proof yet.

A certain number of (unfortunately not so precise) regularity results hold in all dimensions and codimensions, which we shall state soon.

But let us first observe that even minimal sets of dimension 1 may have singularities. Indeed, let $Y \subset \mathbb{R}^2$ be the union of three half lines emanating from the origin, and making 120° angles. It is fairly easy to see that Y is a minimal set in $\mathbb{R}^2$, and hence also $Y \cap U$ is minimal in any open set U.

Smooth minimal surfaces are minimal in small domains, but not necessarily in big ones. For instance, let $E \subset \mathbb{R}^3$ be a catenoid. Then $E \cap B$ is minimal in B when B is a small enough ball, but this is not true when B is a very large ball with the same center as E: in very large balls $B(0,R)$, E looks a lot like the union of two parallel very large disks of radius R, that lie at distance $\log(R)$ from each other, and pinching these disks in the middle gives a significantly better competitor.

There is a minor detail that we need to address. If we add to the almost minimal set E any (closed) set of vanishing $\mathcal{H}^d$-measure, we get another almost minimal set E' because (3.3) does not change. This could make our set E look unnecessarily ugly, so we shall focus our attention to the so-called reduced almost minimal sets.

Definition 3.2. We say that the closed set $E \subset U$ is *reduced*, or *coral*, when $E = E^*$, where

$$E^* = \big\{x \in U : \mathcal{H}^d\big(E \cap B(x,r)\big) > 0 \text{ for all } r > 0\big\} \tag{3.4}$$

is the closed support of the restriction $\mathcal{H}^d_{|E}$ of $\mathcal{H}^d$ to E.

It is fairly easy to see that $\mathcal{H}^d(E \setminus E^*) = 0$, and that if E is almost minimal in U, then E^* is a reduced almost minimal set in U, with the same gauge function as E. Because of this, we shall be able to restrict to reduced sets without loss of generality. See for instance [14, Remark 2.14].

In other sources such as [11, 14, 15], we may say coral rather than reduced, because there was an earlier notion of reduction with a different definition. The name "core" for E^* was introduced in [11], and is not really widespread. But we shall mostly say reduced because there is no danger of confusion here.

Notice also that when E minimizes $\mathcal{H}^d$ or some functional like J_g above in a class of $\mathcal{F}$ of competitors, we do not say that E^* lies in the same class, but only that it is an almost minimal set too. We shall not get any information on $E \setminus E^*$, but such information would be virtually impossible to get anyway.

3.2. Ahlfors regularity and uniform rectifiability. The first important property of our almost minimal sets is their local Ahlfors regularity.

Theorem 3.3 ([3, 21]). *There exist constants $\varepsilon > 0$ and $C \geq 1$, that depend only on n and d, such that if E is a reduced almost minimal set of dimension d in $U \subset \mathbb{R}^n$, with the gauge function h, and if $x \in E$ and $r > 0$ are such that*

$$B(x, 2r) \subset U \quad \text{and} \quad h(r) \leq \varepsilon, \tag{3.5}$$

then

$$C^{-1} r^d \leq \mathcal{H}^d\big(E \cap B(x,r)\big) \leq C r^d. \tag{3.6}$$

We just gave two references here because Almgren did not explicitly state (3.6), but his statement was very close and his proof essentially works.

Note that local Ahlfors regularity is not really a regularity property, but just a size condition that says that E is neither too large, nor too small locally. But it turns out to be quite useful. The upper bound in (3.6) is not surprising (otherwise, one imagines that we could deform a part of E into a somewhat smaller d-dimensional skeleton); for the lower bound, one proves that if E gets too thin near a point, we can even deform it locally into a $(d-1)$-dimensional skeleton and make $\mathcal{H}^d(E)$ smaller. For both estimates, deformations known as Federer – Fleming projections are used.

To the author's knowledge, the strongest regularity property that holds for general almost minimal sets of any dimension is the following uniform rectifiability result.

Theorem 3.4 ([21]). *There exist constants $\varepsilon > 0$, $\theta > 0$ and $N \geq 0$, that depend only on n and d, such that if E is a reduced almost minimal set of dimension d in $U \subset \mathbb{R}^n$, with the gauge function h, and if $x \in E$ and $r > 0$ are such that* (3.5) *holds, then we can find an N-Lipschitz graph G of dimension d such that*

$$\mathcal{H}^d\big(E \cap G \cap B(x,r)\big) \geq \theta r^d. \tag{3.7}$$

Here "N-Lipschitz graph" means that Γ is the image under an isometry of $\mathbb{R}^n$ of the set $G_A = \big\{\big(x, A(x)\big) : x \in \mathbb{R}^d\big\}$, where $A\colon \mathbb{R}^d \to \mathbb{R}^{n-d}$ is such that $|A(x) - A(y)| \leq N|x-y|$ for $x, y \in \mathbb{R}^d$.

In the appropriate technical language, E is locally uniformly rectifiable, with big pieces of Lipschitz graphs (in short, BPLG). We shall not try to discuss here what it means to be uniformly rectifiable; let us just say that Lipschitz graphs and images of $\mathbb{R}^d$ by bi-Lipschitz maps are uniformly rectifiable, and that the general uniformly rectifiable set is not much less regular than that. See [19] or [16] for lots of information on uniform rectifiability. But logically we could expect more regularity from almost minimal sets than BPLG, even if we can't prove it.

Almgren [3] already knew the rectifiability, and at the time uniform rectifiability did not exist. But Theorem 3.4 is significantly more complicated to prove than the rectifiability of E.

3.3. Limits of almost minimal sets. The next result helps proving the stability almost minimality under limits. It is proved in [10], as a consequence of Theorem 3.4. But in fact, to the author's recent surprise, we could also deduce it from the rectifiability of almost minimal sets and a compactness argument.

Theorem 3.5 ([10]). *For each $\delta > 0$, there exist constants $\varepsilon > 0$, and $c > 0$, that depend only on n, d, and δ, such that if E is a reduced almost minimal set of dimension d in $U \subset \mathbb{R}^n$, with gauge function h, and if $x \in E$ and $r > 0$ are such that* (3.5) *holds, then we can find $y \in E$ and $t > 0$ such that*

$$t \geq cr, \quad B(y,t) \subset B(x,r), \tag{3.8}$$

and

$$\mathcal{H}^d\big(E \cap B(y,t)\big) \geq (1-\delta)\mathcal{H}^d\big(P \cap B(y,t)\big) \tag{3.9}$$

for any d-plane P through y.

Of course $\mathcal{H}^d\big(P \cap B(y,t)\big)$ does not depend on P. In the language of Dal Maso, Morel and Solimini, E satisfies a uniform concentration property ($\mathcal{H}^d$ is almost as concentrated on $E \cap B(y,t)$ as on $P \cap B(y,t)$). This property was introduced in [9], to prove the lowersemicontinuity of Hausdorff measure along some minimizing sequences, and then get an existence result for minimizers of the Mumford – Shah functional for image segmentation. And it is tempting to use it here for the same sort of purposes. We first need to define local Hausdorff distances and limits.

When E, F are closed sets in U and for $\overline{B}(x,r) \subset U$, we set

(3.10) $$d_{x,r}(E,F) = r^{-1} \sup\{\mathrm{dist}(y,F) : y \in E \cap B(x,r)\} + r^{-1} \sup\{\mathrm{dist}(y,E) : y \in F \cap B(x,r)\},$$

with the convention that $\sup\{\mathrm{dist}(y,F) : y \in E \cap B(x,r)\} = 0$ when $E \cap B(x,r)$ is empty, and similarly for $F \cap B(x,r)$.

Next let $\{E_k\}_{k\geq 0}$ be a sequence of closed sets in U, and E be a closed set in U. We say that $\{E_k\}$ converges to E in U when

(3.11) $$\lim_{k\to+\infty} d_{x,r}(E_k, E) = 0 \quad \text{for every choice of } x, r \text{ such that } \overline{B}(x,r) \subset U.$$

The reader may easily check that this is equivalent to various other natural definition of local convergence in E for the Hausdorff distance. For instance, we could have used an exhaustion of U by compact subsets instead of our balls $\overline{B}(x,r)$. The advantage of this notion of convergence is that for each sequence $\{E_k\}_{k\geq 0}$, we can extract a subsequence that converges to a limit E. Maybe see [11, Section 34] for additional (easy) detail. But of course we expect more difficulties when we try to prove that E inherits good properties of the E_k. Unless we start with a sequence $\{E_k\}$ with some uniform good properties.

Theorem 3.6 ([10]). *Let $\{E_k\}_{k\geq 0}$ be a sequence of d-dimensional reduced almost minimal sets in U, with the same gauge function h, and suppose that $\{E_k\}$ converges to E in U. Then*

(3.12) $$\mathcal{H}^d(E \cap V) \leq \liminf_{k\to+\infty} \mathcal{H}^d(E_k \cap V) \quad \textit{for every open set } V \subset U$$

and E is a reduced almost minimal set in U, with the same gauge function h as the E_k.

The main ingredient in the proof is the lowersemicontinuity estimate (3.12), which is a consequence of Theorem 3.12 and [9] (but see [46] for a more general statement); but some amount of local cutting and pasting is needed too. Observe that if we did not assume something on the sets E_k, (3.12) would fail miserably. For instance, E_k could be a dotted line segment of total length $\frac{1}{2}$, which converges to $E = [0,1]$; the uniform concentration property above precisely prevents this behavior.

In the world of mass minimizing currents, the analogue of Theorem 3.6 is a consequence of lowersemicontinuity of the mass, and is easy to apply because of the compactness theorem for integral currents. Almgren [2] proved a similar theorem on limits of varifolds, which may be the reason why he apparently did not try (or care) to prove Theorem 3.6.

Theorem 3.6 will be used a lot in the next section, because it allows easy compactness arguments. It also looks like a good tool to prove existence results, and this was the initial motivation in [10]. We shall see in Section 6 two examples where it can be used this way, but often the difficulty will be that E does not necessarily lie in the same class of competitors as the E_k.

In the context of existence theorems, it may be hard to find an appropriate sequence of almost minimal sets, but fortunately Theorem 3.6 also works with the following less restrictive notion of quasi-minimal sets.

3.4. Quasi-minimal sets. In some contexts it is useful to consider the larger class of quasi-minimal sets, which was introduced by Almgren [3], who called them "restricted sets."

This time let us give a definition which is essentially Almgren's. As before, E is a closed set in U, with $\mathcal{H}^d(E \cap B) < +\infty$ for every compact ball $B \subset U$. We still compare E with sets $F = \varphi(E)$, where φ is Lipschitz, but now we only require that if

$$W_\varphi = \{x \in \mathbb{R}^n : \varphi(x) \neq x\}, \tag{3.13}$$

then $W_\varphi \cup \varphi(W_\varphi) \subset\subset U$. Also, the accounting will be different.

Definition 3.7. We say that E is a quasi-minimal set with constants M and scale δ_0 if

$$\mathcal{H}^d(E \cap W_\varphi) \leq M\mathcal{H}^d\big(\varphi(E \cap W_\varphi)\big) \tag{3.14}$$

whenever $\varphi\colon U \to U$ is as above and $\operatorname{diam}\big(W_\varphi \cup \varphi(W_\varphi)\big) \leq \delta_0$.

Before, we only allowed mappings φ such that $W_\varphi \cup \varphi(W_\varphi)$ is contained in a compact ball $\overline{B} \subset U$. We could also do this here, this would give an apparently weaker notion of quasi-minimal set, but all the known regularity results would remain true with the same proofs: we only use mappings φ such that $W_\varphi \cup \varphi(W_\varphi) \subset \overline{B} \subset U$ in the proofs.

In addition, we could also add an error term and replace (3.14) with

$$\mathcal{H}^d(E \cap W_\varphi) \leq M\mathcal{H}^d\big(\varphi(E \cap W_\varphi)\big) + h(\delta)\delta^d, \tag{3.15}$$

where h is a gauge function as near (3.1), and where $\delta = \operatorname{diam}(W_\varphi \cup \varphi(W_\varphi)$. The corresponding class of generalized quasi-minimal sets is introduced in [14], and the main point is that the usual proofs go through in this case, with only minor modifications.

The accounting in (3.14) may seem a little strange, but at least it fits well with functionals like J_g in (2.16). If E minimizes such a functional in a class of sets that is stable under mappings φ as above, then E is a quasi-minimal set with constant M as soon as $1 \leq g(x) \leq M$ everywhere. This is easy to see: if $F = \varphi(E)$ is an acceptable competitor,

$$\begin{aligned}
\mathcal{H}^d(E \cap W_\varphi) &\leq \int_{E\cap W_\varphi} g(x)\, \mathrm{d}\mathcal{H}^d(x) = J_g(E) - \int_{E\setminus W_\varphi} g(x)\, \mathrm{d}\mathcal{H}^d(x) \\
&\leq J_g\big(\varphi(E)\big) - \int_{E\setminus W_\varphi} g(x)\, \mathrm{d}\mathcal{H}^d(x) \\
&= \int_{\varphi(E)} g(x)\, \mathrm{d}\mathcal{H}^d(x) - \int_{E\setminus W_\varphi} g(x)\, \mathrm{d}\mathcal{H}^d(x) \\
&\leq \int_{\varphi(E\cap W_\varphi)} g(x) d\mathcal{H}^d(x) \leq M\mathcal{H}^d\big(\varphi(E \cap W_\varphi)\big)
\end{aligned} \tag{3.16}$$

by (2.16), because E is minimal, and because

$$\varphi(E) = \varphi(E \cap W_\varphi) \cup \varphi(E \setminus W_\varphi) = \varphi(E \cap W_\varphi) \cup (E \setminus W_\varphi) \tag{3.17}$$

(recall that $\varphi(x) = x$ on $E \setminus W_\varphi$). The same argument shows that E is almost minimal when g is continuous, with a relation between the gauge function h and the modulus of continuity of g.

Theorems 3.3, 3.4, 3.5, and 3.6 are still true for generalized quasi-minimal sets. We just need to let $\varepsilon, C, \theta, N$, and c depend on M as well, and in Theorem 3.6 the conclusion is that E is quasi-minimal with the same M and h as the elements E_k of the sequence. See [10, 14, 21].

The advantage of quasi minimality is a greater flexibility, which will be used in Section 6. Also see [20] for a use of this flexibility in codimension 1 and in a slightly different context. Finally observe that Theorem 3.4 (the uniform rectifiability of E) is a little less far from optimality in the context of quasi-minimal sets, because Lipschitz graphs and bi-Lipschitz images of d-planes are easily shown to be quasi-minimal (the class of quasi-minimal sets is closed under bi-Lipschitz mappings).

4. Jean Taylor's regularity theorem

Recall that one of the main goals of this paper is to present a proof of Jean Taylor's celebrated theorem on the local regularity of 2-dimensional almost minimal sets.

4.1. Minimal cones. A minimal cone is just a cone which is also a minimal set. As before, we restrict to reduced sets to avoid ugly additional sets of vanishing measure. But we shall also accept the cones that are not necessarily centered at the origin. As we shall see soon, a good knowledge of the minimal cones is important if we want to understand the local behavior of the minimal sets of the same dimension.

It is easy to check that the (nonempty) minimal cones of dimension 1 in $\mathbb{R}^n$ are the lines, and the sets Y composed of three half lines with the same origin and that make 120° angles at that point. Thus Y is contained in a plane. Also, the union of two perpendicular lines in $\mathbb{R}^2$ is not minimal (pinch near the center to make a better competitor). Similarly, the union in $\mathbb{R}^3$ of four half lines with the same origin is not minimal either (pinch gain). See for instance [50] or [14, Section 10] for the fairly easy description above.

The list of 2-dimensional minimal cones in $\mathbb{R}^3$ is also known, and there are exactly three types of (nonempty) minimal cones. These are the planes (which we shall also call cones of type $\mathbb{P}$), the cones of type $\mathbb{Y}$ obtained as unions of three half planes with a common boundary line L and that make 120° angles along L, and the cones of type $\mathbb{T}$ obtained as the (positive) cone over the union of the edges of a regular tetrahedron, centered at the center of the tetrahedron (see the left of Figure 4). Thus a cone of type $\mathbb{T}$ has six faces, that meet by sets of three and with 120° angles along four edges (half lines emanating from the center).

We can observe these minimal cones as tangent objects in soap films and bubbles; this is not a surprise because we shall see soon that every blow-up limit of an almost minimal set at one of its points is a minimal cone (see the right part of Figure 4).

The fact that cones of types $\mathbb{P}$ and $\mathbb{Y}$ are minimal is easy; for cones of type $\mathbb{T}$, the best is to use a calibration argument [48] (see also [38]) (integrate against a carefully chosen vector field and use Green's theorem to prove that any competitor (that separates) has a large measure).

The fact that no other cone in $\mathbb{R}^3$ is minimal is more complicated. It is not too hard to see that the intersection of such a cone with the unit sphere would be composed of arcs of great circles that make 120° angles, but there are quite a few possible configurations, and one needs to describe them and remove the ones for which the corresponding cone is not minimal. For instance, the cone over the union

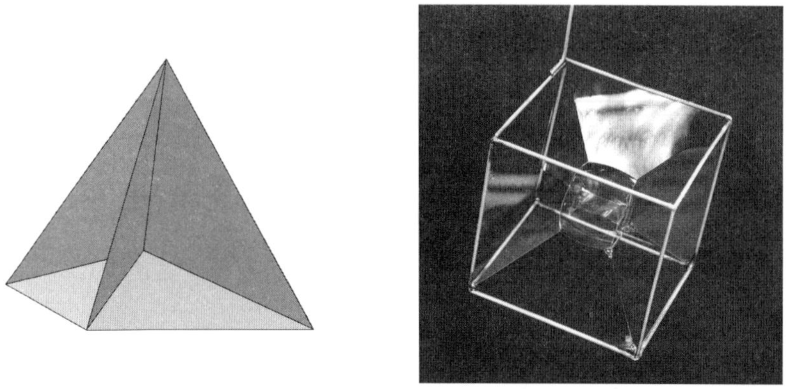

FIGURE 4. Left: A piece of cone of type $\mathbb{T}$. Right: A soap bubble where the two types of singularities are visible.

of the edges of a cube is not minimal, but again there are other examples. This discussion was done by Plateau, Lamarle [36], Heppes [35], and Taylor [59]. Also see Ken Brakke's home page (`http://www.susqu.edu/brakke/`) for pictures of the nonminimal cones and better competitors.

The list of 2-dimensional minimal cones in $\mathbb{R}^n$ is not known yet. It is fairly easy to see that the cones of type $\mathbb{P}, \mathbb{Y}$, and $\mathbb{T}$ are still minimal in $\mathbb{R}^n$, but many more may exist.

The first new candidates are unions $P_1 \cup P_2$ of two planes of dimension 2. If P_1 is orthogonal to P_2, a small projection argument shows that $P_1 \cup P_2$ is minimal. If P_1 and P_2 make a small angle (and in particular if they are not transverse), we can pinch $P_1 \cup P_2$ and get a better competitor, so it is not minimal. A recent theorem of X. Liang [40,41] says that if P_1 and P_2 are almost orthogonal (i.e., assuming that $0 \in P_1 \cup P_2$, if for some small constant $\varepsilon > 0$, $|u \cdot v| \leq \varepsilon |u||v|$ for $u \in P_1$ and $v \in P_2$), then $P_1 \cup P_2$ is minimal. There is a conjecture of Morgan [49] on the precise values of the angles that d-planes P_1 and P_2 need to make when $P_1 \cup P_2$ is minimal, but (when $d = 2$) only the necessary condition was proved [37].

The last known new candidate is the product $Y \times Y$ of two one-dimensional sets Y contained in orthogonal 2-planes. This cone is composed of 9 faces (that correspond to 9 arcs of circles in $\partial B(0,1)$), which meet along 6 half lines (that correspond to the vertices where the arcs meet). At the time of the lectures it was not known whether it is minimal, but this was recently proved by Liang [43].

Other than that, we only have a general knowledge on what the minimal cones of dimension 2 in $\mathbb{R}^n$, $n \geq 4$, look like. Let E be such a cone (centered at 0), and set $K = E \cap \partial B(0,1)$. Then K is a finite union of great circles (circles centered at 0), and of arcs of great circles. The great circles do not meet each other or the rest of K, and the arcs of great circles meet at their endpoints, by sets of three and with 120° angles. And in addition we have a lower bound on the length of each arc of circle. See [14].

For instance, $\mathbb{Y}$ corresponds to 3 half circles, $\mathbb{T}$ to 6 shorter arcs, and $P_1 \cup P_2$ to 2 disjoint great circles. But even in $\mathbb{R}^4$ we do not have a full list of candidates for K and E.

Even less is known for higher-dimensional minimal cones. Almgren [1] proved that if K is a smooth 2-dimensional surface in the 3-sphere, then the cone over K is not minimal, except if it is a hyperplane. Also, the cone over the union of the $(n-2)$-dimensional faces of the hypercube in $\mathbb{R}^n$ is minimal when $n \geq 4$ [5].

4.2. A statement of Jean Taylor's theorem (1976), and an extension.

Theorem 4.1 ([59]). *Let E be a reduced local almost minimal set of dimension 2 in some open set $U \subset \mathbb{R}^3$, with gauge function $h(r) \leq Cr^\alpha$ (for some choice of $\alpha > 0$ and $C \geq 0$). Then for each $x \in E$, there is a ball $B(x,r)$ where E is the image of a minimal cone centered at x by a C^1-diffeomorphism of $\mathbb{R}^3$ that fixes x.*

See Definitions 3.1 and 3.2 for reduced almost minimal sets. Also recall that there are only three types of minimal cones: the planes and the cones of type $\mathbb{Y}$ or $\mathbb{T}$.

We can make the conclusion a little more precise. There is a minimal cone Z centered at x and a $C^{1+\beta}$-diffeomorphism $\phi\colon \mathbb{R}^3 \to \mathbb{R}^3$, such that $\phi(x) = x$, $D\phi(x)$ is the identity mapping, and $\phi(Z) \cap B(x,r) = E \cap B(x,r)$. Here $\beta > 0$ is a small constant, that can be computed in terms of α, but, as far as the author knows, the optimal β is not known. We have uniform bounds on $r\phi(r^{-1}x)$, but an unfortunate point is that we do not have a good lower bound for r; in particular, we cannot prove yet that if E is very close to a minimal cone of type $\mathbb{T}$ in $B(x,r) \subset U$, and if $h(r) = 0$, then E is C^1-equivalent to a $\mathbb{T}$ in $B(x, r/100)$.

The reader should probably not pay too much attention to the C^1 diffeomorphism ϕ, the main point of the conclusion is that near x, E is composed of C^1 faces, that meet along C^1 curves with 120° angles and with the same combinatorics as for a minimal cone; the rest would follow from the implicit function theorem anyway.

The two types of singularities ($\mathbb{Y}$ and $\mathbb{T}$) really occur in soap films and bubbles, where they are very easy to produce. See the right side of Figure 4 again.

Our assumption on h is not optimal; if

$$\int_0^1 h(r)\frac{dr}{r} < +\infty \tag{4.1}$$

we get a bi-Hölder equivalence, and if we ask a little more (for instance, that $h(r) \leq C[\log(A/r)]^{-b}$ for some $b > 30$) we obtain the C^1 equivalence, but the conditions that we get are probably not optimal. See [15, Section 13].

The author does not know what is the optimal regularity near the singularities of E, but away from them (at points where E is locally equivalent to a plane), E is a standard minimal surface, and hence real-analytic!

Let us mention a slightly more recent generalization of Theorem 4.1 to higher ambient dimensions.

Theorem 4.2 ([14]). *] Let E be a reduced local almost minimal set of dimension 2 in some open set $U \subset \mathbb{R}^n$, with a gauge function h such that (4.1) holds. Then for each $x \in E$, there is a ball $B(x,r)$ in which E coincides with the image, under a bi-Hölder diffeomorphism of $\mathbb{R}^n$, of a minimal cone.*

This looks like Theorem 4.1, and certainly many of the estimates are similar, but the author suspects that there are also differences in the proof, and does not really know wether J. Taylor's proof can be made to work in $\mathbb{R}^n$.

The statements too are different, and Theorem 4.2 is far from being as perfect as Theorem 4.1. Its main defect is that when $n \geq 4$, we do not know the list of minimal cones. Our statement is less precise because of this: we get that E is close to a minimal cone, and we can use the general description of minimal cones given in the last subsection to say that E decomposes into faces, with some combinatorial and angle constraints, but this is not such a precise information after all.

The bi-Hölder exponent can be taken to be as close to 1 as we want, but even if $h(r) \leq Cr^{\alpha}$ as above we do not get a C^1 equivalence in general. We only prove the C^1 equivalence when some blow-up limit of E at x is a minimal cone with the so-called "full length" property (see [15]). This property may fail for some minimal cones of dimension 2; the first candidate for failure is the product $Y \times Y \subset \mathbb{R}^4$.

Let us give a definition of the full length property, just to give an idea. Write $K = X \cap \partial B(0,1)$ as a union of great circles or arcs of great circles. Cut them in 2 or 3 pieces when their length is larger than $9\pi/10$, so as to get a collection of arcs γ of lengths less than $9\pi/10$; thus $K = \bigcup_\gamma \gamma$. Call V the set of vertices where the different arcs meet.

Consider any mapping $\varphi\colon V \to \partial B(0,1)$, with $\sup_{x\in V} |\varphi(x) - x|$ small. When γ is an arc of K with endpoints x and $y \in V$, call $\varphi_*(\gamma)$ the geodesic from $\varphi(x)$ to $\varphi(y)$; it is unique because $\mathcal{H}^1(\gamma) \leq 9\pi/10$. Set $\varphi_*(K) = \bigcup_\gamma \varphi_*(\gamma)$ and call $\varphi_*(X)$ the cone over $\varphi_*(K)$.

We say that the cone E satisfies the full length property when there exists $c > 0$ such that, whenever φ above is such that $\mathcal{H}^1(\varphi_*(K)) > \mathcal{H}^1(K)$, there is a competitor $\widetilde{X}$ for $\varphi_*(X)$ in $B(0,1)$ (as in Definition 3.1) such that

$$\mathcal{H}^2(\widetilde{X} \cap B(0,1)) \leq \mathcal{H}^2(\varphi_*(X) \cap B(0,1)) - c\big[\mathcal{H}^1(\varphi_*(K)) - \mathcal{H}^1(K)\big]. \tag{4.2}$$

There are slightly simpler sufficient conditions, but anyway it seems complicated to check (4.2) without knowing more precisely what K is. The minimal cones in $\mathbb{R}^3$ and the known minimal cones in $\mathbb{R}^n$ have the full length property (and the verification is not hard), except for $Y \times Y$ for which it may well fail.

4.3. The density is (almost) monotone. From now on we describe the main points of the proof of Theorem 4.1 that was given in [14], which again is probably not so different from the original proof. As we shall see, the details are sometime a little long, but the scheme is fairly easy to understand.

The first main ingredient is a standard of minimal surfaces: the monotonicity of density.

Proposition 4.3. *Let E be a reduced minimal set of dimension d in $U \subset \mathbb{R}^n$, and set*

$$\theta(x,r) = r^{-d}\mathcal{H}^d(E \cap B(x,r)) \tag{4.3}$$

for $x \in E$ and $0 < r \leq \operatorname{dist}(x, \mathbb{R}^n \setminus U)$. Then

$$\theta(x,\cdot) \text{ is nondecreasing.} \tag{4.4}$$

The idea of the proof is simple. Observe that $r \to \mathcal{H}^d(E \cap B(x,r))$ is nondecreasing, so it is the integral of its derivative (seen as a Stiljes measure), which is at least as large as its almost-everywhere derivative. Because of this (and a small computation), it is enough to check that for a.e. $r \leq \operatorname{dist}(x, \mathbb{R}^n \setminus U)$,

$$r^{-d}\frac{\partial}{\partial r}\big(\mathcal{H}^d(E \cap B(0,r)\big) \geq d\,r^{-d-1}\mathcal{H}^d(E \cap B(0,r)). \tag{4.5}$$

But for almost every r,

$$\frac{\partial}{\partial r}\big(\mathcal{H}^d(E\cap B(0,r)\big) \geq \mathcal{H}^{d-1}\big(E\cap \partial B(0,r)\big) \tag{4.6}$$

[we know that E is rectifiable, so we can reduce to C^1 surfaces, where eventually we just need to compute the contribution of a small element of surface to both sides of (4.6)], so it is enough to show that

$$\mathcal{H}^d\big(E\cap B(0,r)\big) \leq \frac{r}{d}\mathcal{H}^{d-1}\big(E\cap \partial B(0,r)\big). \tag{4.7}$$

Then we observe that

$$\frac{r}{d}\mathcal{H}^{d-1}\big(E\cap \partial B(0,r)\big) = \mathcal{H}^d\big(\Gamma\cap B(0,r)\big), \tag{4.8}$$

where Γ denotes the cone over $E\cap \partial B(0,r)$.

Now $[\Gamma\cap B(0,r)]\cup E\setminus B(0,r)]$ is not directly a competitor for E, but we can approximate it by Lipschitz deformations of E in $B(0,r)$. [Expand a lot a small annulus near $\partial B(0,r)$ and contract most of $B(0,r)$ to the origin, as suggested by Figure 5.] The comparison and a limiting argument yield (4.7) and the monotonicity of θ. □

Proposition 4.3 has two useful extensions. If E is merely almost minimal in E, but (4.1) holds, a similar argument yields that $\theta(x,\cdot)$ is almost nondecreasing, in the sense that

$$\theta(x,r)\exp\Big\{C\int_0^r h(2t)\frac{dt}{t}\Big\} \text{ is a nondecreasing function of } r\leq \mathrm{dist}(x,\mathbb{R}^n\setminus U). \tag{4.9}$$

Notice that by (4.1) the exponential tends to 1 when r tends to 0, so (4.9) really means that $\theta(x,\cdot)$ is almost nondecreasing. Also, the nondecreasing function in (4.9) has a limit when r tends to 0, and so

$$\theta(x) = \lim_{r\to 0}\theta(x,r) \text{ exists for every } x\in E \tag{4.10}$$

when (4.1) holds.

Besides, the case when $\theta(x,\cdot)$ is locally constant in Proposition 4.3 is under control:

(4.11) if $x\in E$ and E is a (reduced) minimal set in $B(x,r_0)$ and $\theta(x,\cdot)$ is constant on $]0,r_0[$, then E coincides in $B(x,r_0)$ with a (reduced) minimal cone centered at x.

Surprisingly (and maybe out of clumsiness) this requires some additional work in [14]; it is not hard to see that equality almost-everywhere in (4.6) implies that

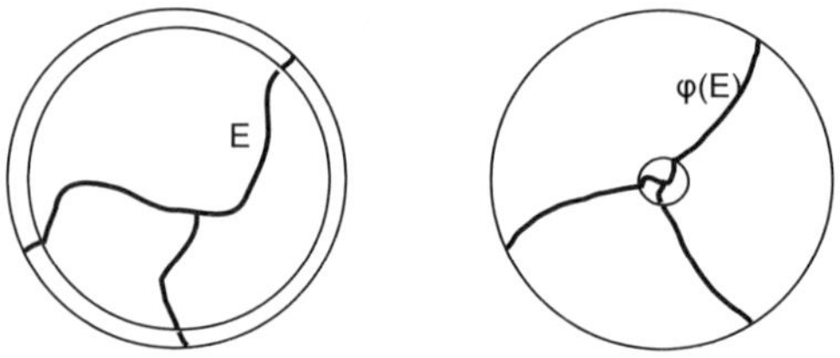

FIGURE 5. How to find a deformation of E (on the left) that looks like the cone over $E\cap\partial B(0,r)$. The thin annulus is mapped to the thick one.

for almost every $y \in E \cap B(x, r_0)$, the tangent plane to E at y goes through x, but then we still need to show that E contains the line segment $[x, y]$, and this takes some time and the complicated construction of a competitor.

We shall often use the following consequence of (4.11).

Lemma 4.4. *For each small $\delta > 0$, we can find $\varepsilon > 0$ (that depends on δ and the dimensions n and d) such that the following holds. Still assume that (4.1) holds, and that E is a reduced almost minimal set in U, with gauge function h. Let $x \in E$ and $B(x, 2r) \subset U$ be such that $h(2r) \leq \varepsilon$ and*

$$\theta(x) - \varepsilon \leq \theta(x, \rho) \leq \theta(x) + \varepsilon \quad \text{for } 0 < \rho \leq 2r, \tag{4.12}$$

where $\theta(x) = \lim_{t \to 0} \theta(x, t)$ as above. Then there is a minimal cone Z centered at x such that

$$d_{x,r}(E, Z) \leq \delta \tag{4.13}$$

and, for $y \in \mathbb{R}^n$ and $t > 0$ such that $B(y, t) \subset B(x, r)$,

$$\left|\mathcal{H}^d(E \cap B(y, t)) - \mathcal{H}^d(Z \cap B(y, t))\right| \leq \delta r^d. \tag{4.14}$$

See (4.3), (4.10), and (3.10) for the definitions. When E is minimal, instead of (4.12) we just need to say that $\theta(x, 2r) \leq \theta(x) + \varepsilon$, because of (4.4), but in the almost minimal case, it is simpler to require (4.12). Also, do not be shocked by (4.14); when t/r is very small, $\mathcal{H}^d(E \cap B(y, t))$ is much smaller than r^d anyway, and (4.14) does not say much.

We shall sometimes refer to Lemma 4.4 as the almost-constant density property. Lemma 4.4 is deduced from (4.11) and Theorem 3.6 (about limits) by a standard compactness argument; let us just sketch the proof, and refer to Proposition 7.24 (which follows from Proposition 7.1) in [14] for details.

We proceed by contradiction and suppose that for some small $\delta > 0$, Lemma 4.15 does not hold with $\varepsilon_k = 2^{-k}$. Pick an almost minimal set E_k in a domain $U_k \subset \mathbb{R}^n$, and a ball $B(x_k, 2r_k) \subset U_k$ which satisfy the hypotheses of the lemma, but not the conclusion. By dilation invariance, we can assume that $x_k = 0$, $r_k = 1$, and $U_k \supset B(0, 2)$. Then replace $\{E_k\}$ with a subsequence that converges to some limit E in $B(0, 2)$. By Theorem 3.6, E is minimal in $B(0, 2)$. More precisely, for each k_0 we can apply Theorem 3.6 with the gauge function $\widetilde{h}_{k_0} = \sup_{k \geq k_0} h_k$; we use the fact that $h_k(2) \leq \varepsilon_k = 2^{-k}$ to show that $\widetilde{h}_{k_0}$ is a gauge function, and apply Theorem 3.6 to show that E is almost minimal with the gauge $\widetilde{h}_{k_0}$. Then we observe that $\widetilde{h}_{k_0}(2) \leq 2^{-k_0}$ and get that in fact E is minimal in $B(0, 2)$.

By (3.12),

$$\mathcal{H}^d(E \cap B(y, t)) \leq \liminf_{k \to +\infty} \mathcal{H}^d(E_k \cap B(y, t)) \quad \text{when } B(y, t) \subset B(0, 2). \tag{4.15}$$

But the proof of Theorem 3.6 also gives the uppersemicontinuity estimate

$$\mathcal{H}^d(E \cap \overline{B}(y, t)) \geq \limsup_{k \to +\infty} \mathcal{H}^d(E_k \cap \overline{B}(y, t)) \quad \text{when } \overline{B}(y, t) \subset B(0, 2); \tag{4.16}$$

the idea of the proof is that if (4.16) did not hold, some Hausdorff measure would drop when we pass to the limit, and we could use this to construct a strictly better competitor for E_k. But the proof relies on the construction of Theorem 3.6 and the fact that the E_k are asymptotically minimal.

Let θ denote the density function associated to E (as in (4.3)) and θ_k its analogue for E_k. Notice that for almost every $t \in (0, 2)$, $\mathcal{H}^d(E \cap \partial B(0, t)) = 0$

(the set of t where this fails is at most countable, because the sets $E \cap \partial B(0,t)$ are disjoint and $\mathcal{H}^d(E \cap B(0,r)) < +\infty$ for $r < 2$). For such t,

$$\begin{aligned}(4.17)\qquad \limsup_{k\to+\infty} \theta_k(0,t) &= t^{-d} \limsup_{k\to+\infty} \mathcal{H}^d(E_k \cap B(0,t)) \leq t^{-d}\mathcal{H}^d(E \cap \overline{B}(0,t)) \\ &= t^{-d}\mathcal{H}^d(E \cap B(0,t)) = \theta(0,t) \leq \liminf_{k\to+\infty} \theta_k(0,t) \\ &\leq \liminf_{k\to+\infty} \theta_k(0)\end{aligned}$$

by (4.16), (4.15), and (4.12). Set $L = \liminf_{k\to+\infty} \theta_k(0)$. By (4.16) again, $L \leq \limsup_{k\to+\infty} \theta_k(0,t)$; then (4.17) says that $\limsup_{k\to+\infty} \theta_k(0,t) = L$, and (looking at the intermediate inequalities)

$$(4.18)\qquad \lim_{k\to+\infty} \theta_k(0,t) = L \quad \text{and} \quad \theta(0,t) = L.$$

This holds for almost every $t \in (0,2)$, hence $\theta_k(0,t) = L$ for all $t \in (0,2)$, by (4.4) and because E is minimal. Then (4.11) says that E coincides, in $B(0,2)$, with a minimal cone Z centered at 0.

Since (our subsequence of) $\{E_k\}$ converges to E, we get that (4.13) holds for k large. We can also use (4.15) and (4.16) to get that (4.14) holds for some large k; the verification takes a little time, but is not too complicated (see the proof of Proposition 7.1 in [14] for details); the desired contradiction and Lemma 4.4 then follow. □

4.4. Blow-up limits. Let E be a reduced almost minimal set in U, and pick $x \in E$. A blow-up limit of E at x is any set

$$(4.19)\qquad E_\infty = \lim_{k\to+\infty} \frac{1}{r_k}[E - x],$$

where $\{r_k\}$ is a sequence in $(0,+\infty)$ that tends to 0. Here the definition of convergence is slightly different, because the domains vary. That is, $U_k = (1/r_k)[U - x]$ tends to $\mathbb{R}^n$, in the sense that $\mathrm{dist}(0, \mathbb{R}^n \setminus U_k) = (1/r_k)\,\mathrm{dist}(x, \mathbb{R}^n \setminus U)$ tends to $+\infty$. Then (4.19) is defined by requiring that

$$(4.20)\qquad \lim_{k\to+\infty} d_{0,R}\left(E_\infty, \frac{1}{r_k}[E-x]\right) = 0 \quad \text{for every } R > 0.$$

By (the proof of) the standard fact that the set of compact subsets of any $\overline{B}(0,R)$, with the Hausdorff distance, is itself compact, it is easy to see that from any sequence $\{r_k\}$ that tends to 0, we can extract a subsequence such that the $(1/r_k)[E-x]$ converge. So some blow-up limits of E at a given $x \in E$ exist. But, without additional information, we have to expect that E has more than one blow-up limit at x. We claim that if E is a reduced almost minimal set and E_∞ is a blow-up limit of E at some $x \in E$,

(4.21) E_∞ is a reduced minimal cone in $\mathbb{R}^n$, centered at 0, and $\mathcal{H}^d(E_\infty \cap B(0,1)) = \theta(x)$.

Indeed, let $\{r_k\}$ be such that $E_k = (1/r_k)[E-x]$ converges to E_∞. Notice that E_k is almost minimal in the domain U_k above, and with the gauge function $h_k(r) = h(rr_k)$ (just apply Definition 3.1 and conjugate with a dilation). Then apply Theorem 3.3 in any fixed ball $B(0,R)$, and observe that the h_k tend to 0 uniformly in $[0,2R]$. We get that E_∞ is minimal in $\mathbb{R}^n$ (because, as above (4.15), it is almost minimal in any $B(0,R)$ and with arbitrarily small gauge functions). By (3.12) and as for (4.15), we get that for $t > 0$,

$$\begin{aligned} \mathcal{H}^d\big(E_\infty \cap B(0,t)\big) &\leq \liminf_{k\to+\infty} \mathcal{H}^d\big(E_k \cap B(0,t)\big) \\ &= \liminf_{k\to+\infty} r_k^{-d}\mathcal{H}^d\big(E \cap B(x,tr_k)\big) \\ &= t^d \lim_{k\to+\infty} \theta(x,tr_k) = t^d\theta(x). \end{aligned} \tag{4.22}$$

Similarly, the same uppersemicontinuity estimate that is hidden in the proof of Theorem 3.3 and that we used for (4.16) shows that

$$\begin{aligned} \mathcal{H}^d\big(E_\infty \cap \overline{B}(0,t)\big) &\geq \limsup_{k\to+\infty} \mathcal{H}^d\big(E_k \cap \overline{B}(y,t)\big) \\ &= \limsup_{k\to+\infty} r_k^{-d}\mathcal{H}^d\big(E \cap \overline{B}(x,tr_k)\big) \\ &\geq t^d \limsup_{k\to+\infty} \theta(x,tr_k) = t^d\theta(x). \end{aligned} \tag{4.23}$$

So $\mathcal{H}^d\big(E_\infty \cap B(0,t)\big) = t^d\theta(x)$ for a.e. $t > 0$, hence for all $t > 0$. That is, the density associated to E_∞ at the origin is constant on $(0,+\infty)$, and (4.11) says that E_∞ is a reduced minimal cone. We already checked that $\mathcal{H}^d\big(E_\infty \cap B(0,t)\big) = t^d\theta(x)$ for all t, so (4.21) holds.

Of course we expect the minimal cones to be significantly easier to study than the minimal sets of the same dimension. Thus the almost monotonicity of density implies a better control on blow-up limits of minimal sets, that in turn we can try to use to get a good local control of the minimal sets themselves. One could even dream of using this control to get regularity results for minimal cones of one more dimension, and prove regularity results by induction. To some extent, this is what we will do for the 2-dimensional minimal sets, but the program seems hard to continue, because when we try to go from 3-dimensional minimal cones to 3-dimensional minimal sets, we apparently need topological information that we can't get easily. See Remark 4.7.

Let us say two words about blow-in limits. Suppose that E is a reduced minimal set in the whole $\mathbb{R}^n$. A blow-in limit of E is any limit of a sequence $\{(1/r_k)E\}$, where this time $\lim_{k\to+\infty} r_k = +\infty$. We could have used $(1/r_k)[E-x]$ for some x, but it is easy to see that we would have obtained the same limit. The same proof as for (4.21) yields that if E_∞ is a blow-in limit of E,

(4.24) $\quad E_\infty$ is a minimal cone in $\mathbb{R}^n$ centered at 0, and $\mathcal{H}^d\big(E_\infty \cap B(0,1)\big) = \lim_{r\to+\infty} \theta(0,r)$.

4.5. Reifenberg's topological disk theorem. The following extension of Reifenberg's topological disk theorem [55] will allow us to parameterize our set E, as in the statements of Theorems 4.1 and 4.2, once we have significant information on its geometry. It is nice to know that Reifenberg initially proved this theorem in connection to minimal sets. But, at least in the author's opinion, we should not put too much stress on the use of this result, especially for the C^1 regularity theorem; it is natural that a good control on the distance between E and minimal cones, at all scales and locations, should yield the existence of a nice parameterization. On the other hand, Reifenberg's theorem is quite nice in itself, and will be quite agreeable to use, because it will spare us the tedious construction of a parameterization. See [24, 60] for some other applications and generalizations of Reifenberg's argument.

Theorem 4.5 ([18]). *For each $\tau > 0$, we can find $\varepsilon_0 > 0$, that depends only on τ and the dimension n, such that if $E \subset \mathbb{R}^n$ is a closed set, with $0 \in E$ and such that for each $x \in E \cap B(0,2)$ and $r \in (0,2]$, we can find a minimal cone $Z = Z(x,r)$ of dimension 2 such that*

$$d_{x,r}(E,Z) \leq \varepsilon_0, \tag{4.25}$$

then there is a minimal cone Z_0 of dimension 2 and a bi-Hölder homeomorphism $f\colon \mathbb{R}^n \to \mathbb{R}^n$ such that

$$|f(x) - x| \leq \tau \qquad \text{for } x \in B(0,2), \tag{4.26}$$

$$(1-\tau)|x-y|^{1+\tau} \leq |f(x)-f(y)| \leq (1+\tau)|x-y|^{1-\tau} \qquad \text{for } x,y \in B(0,2), \tag{4.27}$$

and

$$E \cap B(0,1-\tau) \subset f\big(Z_0 \cap B(0,1)\big) \subset E \cap B(0,1+\tau). \tag{4.28}$$

In the sketch of proof below, we shall just use the special case when $n = 3$. Then the minimal cones are known to be of type $\mathbb{P}, \mathbb{Y}$, or $\mathbb{T}$. But the general case is used for Theorem 4.2.

In Reifenberg's original theorem, one only uses affine d-planes, but on the other hand all the dimensions and codimensions are allowed.

Notice that $Z(x,r)$ is not necessarily centered at x, (which is better because otherwise the assumptions would not even hold when E is a cone of type $\mathbb{Y}$ centered near the origin).

We would have preferred to have a quasiconformal mapping f, but even in the original theorem of Reifenberg this is not always possible, for instance, when E is the product of a line with a flat snowflake.

Theorem 4.5 has a C^1 variant where one assumes that $d_{x,r}(E,Z) \leq \varepsilon_0 r^\alpha$ for some $\alpha > 0$, and one gets a C^1 parameterization at the end. This is the version that we use in [15], although it is obtained in [18] as a corollary of the proof of Theorem 4.5, it could probably be obtained directly and with somewhat less work.

Reifenberg's theorem is a nice example where some small local information on E (namely (4.25)), but known at every scale and location, gives a much better global control on the set that one would expect a priori. For instance, (4.25) for the pair (x,r) seems to allow small holes in E, of size $\varepsilon_0 r$, but in fact, due to the fact that potential holes have to be small at every scale, they just don't exist. A similar theorem that the author likes a lot is the John and Nirenberg theorem on the exponential integrability of functions in BMO.

An important information that we use many times in the proof is that (4.25) also forces $Z(x,r)$ to depend fairly nicely on x and r. Because of this, we can rapidly decompose $E \cap B(0,2)$ into the sets of points of type $\mathbb{P}$, $\mathbb{Y}$, and $\mathbb{T}$, depending on the closeness of $Z(x,r)$ to a cone of type $\mathbb{P}, \mathbb{Y}$, and $\mathbb{T}$ at small scales r, and one can show, for instance, that $B(0,1)$ contains at most one point of type $\mathbb{T}$.

The proof in [18] uses the same ideas as in Reifenberg's original theorem. It is a typical "top-down" algorithm. We start from an initial cone Z_0 (typically, $Z_0 = Z(0,2)$), and construct the final mapping $f\colon Z_0 \to \mathbb{R}^n$ as an infinite composition of mappings g_k that we construct at geometrically decaying scales r_k. At the scale r_k, E is locally approached by minimal cones, and we construct g_k so that it pushes the points of the image of Z_0 in the direction of E (and of these minimal cones). Technical details (with partitions of unity and distortion estimates) are needed because we need to make sure that our mappings are injective, and make the proof a little long, but the idea is fairly simple.

4.6. J. Taylor's theorem near a point of type $\mathbb{P}$. Let us now try to describe the proof of the bi-Hölder part of J. Taylor's theorem that was given in [14].

Of course many technical details will be ignored, and we should also say that getting the local C^1 equivalence is much more difficult than the mere bi-Hölder equivalence, but at least the scheme of the proof, and some of the challenges for future generalizations, are reasonably clear in this context. Also, one can argue that the bi-Hölder equivalence should be enough for many purposes.

So let E be a reduced almost minimal set in U, and let $x_0 \in E$ be given. We want to find a small neighborhood of x_0 where Theorem 4.5 can be used, and for this we need to show that E is well approximated by minimal cones near x_0. Let us state this more precisely. For $x \in E$ and $r > 0$, set

$$\beta_c(x,r) = \inf\{d_{x,r}(E,Z) : Z \text{ is a minimal cone}\}; \tag{4.29}$$

we want to find $r_0 > 0$ such that

$$\beta_c(x,r) \leq \varepsilon_0 \quad \text{for } x \in E \cap B(x_0, 2r_0) \text{ and } 0 < r \leq 2r_0, \tag{4.30}$$

with ε_0 as in Theorem 4.5. Once this is done, we just have to apply Theorem 4.5 to $(1/r_0)[E - x_0]$ and conclude.

Now something needs to be done. It is true that

$$\lim_{r \to 0} \beta_c(x,r) = 0 \quad \text{for each } x \in E. \tag{4.31}$$

Indeed, otherwise we could find $\varepsilon > 0$ and a sequence $\{r_k\}$ that tends to 0, such that $\beta_c(x, r_k) \geq \varepsilon$. Then we could extract a subsequence for which the sets $E_k = (1/r_k)[E - x]$ converge to a limit E_∞, and by (4.21) E_∞ is a minimal cone. By definition, $d_{0,1}(E_\infty, E_k)$ tends to 0 for this sequence, but since $\beta_c(x,r_k) \leq d_{x,r}(E, x + r_k E_\infty) = d_{0,1}(E_\infty, E_k)$ by definitions, we get a contradiction that proves (4.31).

But (4.31) is not enough for (4.30); we also need some uniform estimates, and this is the whole point of the arguments that we shall describe below.

To make things simpler, we shall assume that E is actually minimal near x; the difference is not enormous, but this way we won't have to deal with error terms. We shall only describe the argument in $\mathbb{R}^3$; the proof extends to $\mathbb{R}^n$, with some modifications in the topological parts, but the general idea is the same. Finally, we may assume that $x_0 = 0$.

Recall from (4.21) that all the blow-up limits of E at 0 are minimal cones, with the same density $\theta(0) = \lim_{r \to 0} r^{-2}\mathcal{H}^2(E \cap B(0,r))$. There are three types of minimal cones ($\mathbb{P}, \mathbb{Y}$, or $\mathbb{T}$), with different densities π, $3\pi/2$, and some number $d_T > 3\pi/2$, so all the blow-up limits of x_0 are of the same type, determined by the density $\theta(0)$.

In this subsection we just deal with the easiest case when 0 is of type $\mathbb{P}$, which means that $\theta(0) = \pi$ and all the blow-up limits of E at 0 are planes.

Let $\varepsilon_1 > 0$ be small, to be chosen later, and chose a small radius $r_1 > 0$ such that E is minimal in $B(0, 2r_1) \subset U$ and

$$\theta(0, 2r_1) \leq \theta(0) + \varepsilon_1 = \pi + \varepsilon_1. \tag{4.32}$$

Let ε_2 be very small, to be chosen later. If ε_1 is small enough (depending on ε_2), the almost constant density property Lemma 4.4, applied with $\delta = \varepsilon_2$, says that we can find a minimal cone Z centered at 0 such that

$$d_{0,r_1}(E, Z) \leq \varepsilon_2 \tag{4.33}$$

and

$$\left|\mathcal{H}^d(E \cap B(y,t)) - \mathcal{H}^d(Z \cap B(y,t))\right| \leq \varepsilon_2 r_1^d. \tag{4.34}$$

for $y \in \mathbb{R}^n$ and $t > 0$ such that $B(y,t) \subset B(0,r_1)$. We apply this with $B(y,t) = B(0,r_1)$ and get that

$$\begin{aligned}\mathcal{H}^d(Z \cap B(0,r_1)) &\leq \mathcal{H}^d(E \cap B(0,r_1)) + \varepsilon_2 r_1^d = [\theta(0,r_1) + \varepsilon_2]\, r_1^d \\ &\leq [\theta(0,2r_1) + \varepsilon_2] r_1^d \leq [\pi + \varepsilon_1 + \varepsilon_2] r_1^d\end{aligned} \tag{4.35}$$

because $\theta(0,\cdot)$ is nondecreasing (by (4.4)), and by (4.32). Hence Z is a plane (cones of type $\mathbb{Y}$ and $\mathbb{T}$ have a larger density).

Now let $y \in B(0, r_1/2)$ be given. Apply (4.34) with $t = r_1/2$ to get that

$$\begin{aligned}\theta(y, r_1/2) &= (r_1/2)^{-d}\mathcal{H}^d(E \cap B(y, r_1/2)) \\ &\leq (r_1/2)^{-d}\mathcal{H}^d(Z \cap B(y, r_1/2)) + 2^d\varepsilon_2 \leq \pi + 2^d\varepsilon_2\end{aligned} \tag{4.36}$$

because $\mathcal{H}^d(Z \cap B(y, r_1/2)) \leq \pi(r_1/2)^d$ (recall that Z is a plane). By monotonicity (i.e., (4.4)),

$$\theta(y,r) \leq \pi + 2^d\varepsilon_2 \quad \text{for } 0 < r \leq r_1/2. \tag{4.37}$$

Then $\theta(y) = \lim_{r\to 0}\theta(y,r) \leq \pi + 2^d\varepsilon_2$ by (4.4). If ε_2 is small enough, $\theta(y) < 3\pi/2$ (the density of cone of type $\mathbb{Y}$), y is a point of type $\mathbb{P}$, and $\theta(y) = \pi$.

Now apply Lemma 4.4 (the almost constant density property) with $\delta = \varepsilon_0$. If ε_2 is small enough, we deduce from that lemma that there exists a minimal cone $Z = Z(y, r/2)$, such that $d_{y,r/2}(E, Z) \leq \varepsilon_0$. Therefore, $\beta_c(y, r/2) \leq \varepsilon_0$ for $0 < r \leq r_1/2$. Since y was any point of $E \cap B(0, r_1/2)$, we just proved (4.30) with $r_0 = r_1/8$.

Notice that we could fairly easily show that $Z(y, r/2)$ is a plane, as we did for Z near (4.35), and so the standard Reifenberg theorem would have been enough in this case.

Anyway, this concludes our description of Theorem 4.1, but with a bi-Hölder equivalence only and when E is locally minimal and x is a point of type $\mathbb{P}$.

4.7. A topological lemma for the existence of $\mathbb{Y}$-points. In the previous argument, we were lucky with the densities, because we found approximating cones Z which had the minimal possible density, and this allowed us to apply the almost constant density property (see below (4.37)). Let us see what happens when we assume that 0 is a point of type $\mathbb{Y}$, i.e., when

$$\theta(0) = \lim_{r\to 0} r^{-d}\mathcal{H}^d(E \cap B(0,r)) = \frac{3\pi}{2}. \tag{4.38}$$

We start as above, choose $r_1 > 0$ such that E is minimal in $B(0, 4r_1) \subset U$ and $\theta(0, 4r_1) \leq \theta(0) + \varepsilon_1 \leq 3\pi/2 + \varepsilon_1$. Since $\theta(0,\cdot)$ is monotone, we also get that

$$\frac{3\pi}{2} = \theta(0) \leq \theta(0,r) \leq \theta(0,4r_1) \leq \frac{3\pi}{2} + \varepsilon_1 \quad \text{for } 0 < r \leq 4r_1. \tag{4.39}$$

Let ε_2 be small, as above, and apply the almost constant density property to the pair $(0, 2r_1)$. If ε_1 is small enough (depending on ε_2, not on r_1), we obtain a minimal cone Z centered at 0 such that

$$d_{0,2r_1}(E, Z) \leq \varepsilon_2 \tag{4.40}$$

and

$$\left|\mathcal{H}^d(E \cap B(y,t)) - \mathcal{H}^d(Z \cap B(y,t))\right| \leq \varepsilon_2 2^d r_1^d \tag{4.41}$$

for $y \in \mathbb{R}^n$ and $t > 0$ such that $B(y,t) \subset B(0, 2r_1)$. Apply (4.41) with $y = 0$ and $t = 2r_1$ to get that

$$\begin{aligned} \mathcal{H}^d(Z \cap B(0,2r_1)) &\leq \mathcal{H}^d(E \cap B(0,2r_1)) + \varepsilon_2 2^d r_1^d \\ &= \theta(0, 2r_1)\, r_1^d + \varepsilon_2 2^d r_1^d \leq \left(\frac{3\pi}{2} + \varepsilon_1 + 2^d \varepsilon_2\right) r_1^d \end{aligned} \tag{4.42}$$

by (4.39). If ε_1 and ε_2 are small enough, $3\pi/2 + \varepsilon_1 + 2^d\varepsilon_2 < d_T$ (the density of a cone of type $\mathbb{T}$), and Z, which is centered at the origin, cannot be of type $\mathbb{T}$. By the same argument,

$$\begin{aligned} \mathcal{H}^d(Z \cap B(0,2r_1)) &\geq \mathcal{H}^d(E \cap B(0,2r_1)) - \varepsilon_2 2^d r_1^d \\ &= \theta(0, 2r_1) 2^d r_1^d - \varepsilon_2 2^d r_1^d \\ &\geq \left(\frac{3\pi}{2} - 2^d \varepsilon_2\right) 2^d r_1^d > \pi 2^d r_1^d \end{aligned} \tag{4.43}$$

by (4.41) and (4.39), and if ε_2 is small enough, so Z is not a plane either. That is, Z is of type $\mathbb{Y}$. Then

$$\mathcal{H}^d(Z \cap B(y,t)) \leq \frac{3\pi}{2} t^d \tag{4.44}$$

for every ball $B(y,t)$; the easiest way to see this is to observe that since Z is a minimal set, $t^{-d}\mathcal{H}^d(Z \cap B(y,t))$ is a nondecreasing function of t (in fact, (4.4) is also true with the same proof when the center x lies out of E), whose limit at $+\infty$ is $3\pi/2$. Return to (4.41), and apply it with $y \in E \cap B(0, r_1)$ and $t = r_1$; this yields

$$\mathcal{H}^d(E \cap B(y,r_1)) \leq \mathcal{H}^d(Z \cap B(y,r_1)) + \varepsilon_2 2^d r_1^d \leq \frac{3\pi}{2} r_1^d + \varepsilon_2 2^d r_1^d, \tag{4.45}$$

by (4.44). Then, by (4.4),

$$\theta(y, \rho) \leq \theta(y, r_1) = r_1^{-d}\mathcal{H}^d(E \cap B(y,r_1)) \leq \frac{3\pi}{2} + 2^d \varepsilon_2 \tag{4.46}$$

for $y \in B(0, r_1)$ and $0 < \rho \leq r_1$. Denote by

$$E_Y = \left\{ y \in E : \theta(y) = \frac{3\pi}{2} \right\} \tag{4.47}$$

the set of points of type $\mathbb{Y}$ in E; (4.46) will be easier to use when $y \in E_Y$, because since $\theta(y) = 3\pi/2$, we can apply the almost constant density property again, to the pair $(y, \rho/2)$. We get that for

$$y \in E_Y \cap B(0, r_1) \quad \text{and} \quad 0 < \rho \leq r_1/2, \tag{4.48}$$

we can find a minimal cone $Z(y, \rho)$, centered at y, such that

$$d_{y,\rho}(E, Z(y,\rho)) \leq \varepsilon_3 \tag{4.49}$$

and, as in (4.14),

$$\big|\mathcal{H}^d(E\cap B(x,r)) - \mathcal{H}^d(Z(y,\rho)\cap B(x,r))\big| \leq \varepsilon_3\rho^d \tag{4.50}$$

for $x \in \mathbb{R}^n$ and $r > 0$ such that $B(x,r) \subset B(y,\rho)$, with an $\varepsilon_3 > 0$ which is as small as we want (provided that we take ε_2 small enough).

This is essentially as good as before, except that it only applies to $y \in E_Y$. If we want to continue the argument, we need to be able to find sufficiently many points of E_Y. The following existence lemma will help.

Lemma 4.6. *Let E be a reduced and minimal set of dimension 2 in $B(0,4)$, and assume that there is a cone Y of type $\mathbb{Y}$, centered at 0, such that $d_{0,4}(E,Y) \leq \varepsilon$. Then (if ε is small enough) $E \cap B(0,1)$ contains at least a point of type $\mathbb{Y}$.*

Let us sketch the proof. Incidentally, this is one of the places where it is clear that J. Taylor used similar arguments. We first need to check that

$$\theta(x,1) \leq \frac{3\pi}{2} + \eta \quad \textit{for } x \in B(0,2), \tag{4.51}$$

where $\eta > 0$ is as small as we want.

We prove this by contradiction and compactness. If this fails, we can find E_k and Y_k, with $d_{0,4}(E_k, Y_k) \leq \varepsilon_k = 2^{-k}$ but for which (4.51) fails. By rotation invariance, we may assume that $Y_k = Y$ for a fixed Y, and then $\{E_k\}$ converges to Y in $B(0,4)$.

Since (4.51) fails, we can find $x_k \in B(0,2)$ such that $\theta_k(x_k,1) \geq 3\pi/2 + \eta$, where θ_k is the density function for E_k. We may replace $\{E_k\}$ with a subsequence, so that $\{x_k\}$ converges to some $x \in \overline{B}(0,2)$.

Let $\rho > 1$ be very close to 1, to be chosen soon. By the analogue of (4.44) for Y and the uppersemicontinuity estimate that follows from the proof of Theorem 3.6 (and as for (4.16)),

$$\begin{aligned}\frac{3\pi\rho^d}{2} &\geq \mathcal{H}^d(Y\cap \overline{B}(x,\rho)) \geq \limsup_{k\to+\infty} \mathcal{H}^d(E_k \cap \overline{B}(x,\rho)) \\ &\geq \limsup_{k\to+\infty} \mathcal{H}^d(E_k\cap B(x_k,1)) \geq \frac{3\pi}{2} + \eta\end{aligned} \tag{4.52}$$

because $B(x_k,1) \subset B(x,\rho)$ for k large, and by definition of x_k. This is impossible if ρ is close enough to 1; (4.51) follows.

Return to the sets E and Y of the lemma. Choose η so small that $3\pi/2+\eta < d_T$, where d_T denotes the density of a cone of type $\mathbb{T}$. Then $\theta(x) \leq \theta(x,1) < d_T$ for $x \in E\cap B(0,2)$, by (4.4) and (4.51), and so $E \cap B(0,2)$ contains no point of type $\mathbb{T}$.

Let us proceed by contradiction, and assume that $E \cap B(0,1)$ contains no point of type $\mathbb{Y}$. Then every point $x \in E\cap B(0,1)$ is of type $\mathbb{P}$, and the part of Theorem 4.1 that we already proved implies that x has a neighborhood where E is bi-Hölder equivalent to a plane.

We want to find a curve in $B(0,1)$ which intersects E three times transversally; then we will deform this curve to a point inside $B(0,1)$, say that the number of intersections stays odd, and get the desired contradiction.

Let us assume for simplicity that Y is composed of three vertical half planes. Denote by P the horizontal plane through 0, set $S = P \cap \partial B(0,\frac{1}{2})$, and denote by a_1, a_2, a_3 the three points of $S \cap Y$. Since $d_{0,4}(E,Y) \leq \varepsilon$, we can find $x_i \in E$ such that $|x_i - a_i| \leq 4\varepsilon$.

Let P_i denote the (vertical) plane parallel to the face of Y that contains a_i, and which passes through x_i. Notice that

$$d_{x_i,10^{-1}}(E, P_i) \leq 100\varepsilon \tag{4.53}$$

by elementary geometry and because $d_{0,4}(E, Y) \leq \varepsilon_k$. Because of (4.53), and by the same proof by compactness as for (4.51), we get that

$$\theta\left(x_i, \frac{1}{20}\right) \leq \pi + \eta. \tag{4.54}$$

Then (if η is chosen small enough) our proof of Theorem 4.1 for points of type $\mathbb{P}$ shows that in $B(x_i, 10^{-4})$, E is equivalent to a plane, through a bi-Hölder mapping f_i such that $|f_i(z) - z| \leq 10^{-5}$ for $z \in \mathbb{R}^n$, as in (4.26).

We then use the three f_i to replace three short arcs of S near the a_i with three short curves, obtained as images of line segments by the f_i, and that cut E transversally (in the local coordinates given by the f_i). The rest of S does not meet E (recall that $d_{0,4}(E, Y) \leq \varepsilon$). So we get a curve γ that meets E exactly three times.

We contract γ inside $B(0, \frac{1}{2})$, to a point of $B(0, \frac{1}{2}) \setminus E$. With a little bit of topology, we show that along the contraction, the number of intersections with E stays odd, which contradicts the fact that it is null at the end. For the topological verification, which we shall not do here, we manage to reduce to the situation where we just need to compare two curves that coincide out of a small neighborhood of a point $x \in E \cap B(0, 1)$, where we have seen that E is equivalent to a plane through a bi-Hölder mapping. In addition, we choose our contraction so that in these local coordinates, both curves are piecewise linear and cross E transversally; the fact that the parity of the number of intersections is preserved is then easy.

This completes our description of the proof of Lemma 4.6. In higher codimensions, the lemma stays true, but we need to replace the final argument with curves by a little bit of degree theory. □

4.8. How to end the proof at a point of type $\mathbb{Y}$ or $\mathbb{T}$. We return to the the proof of the bi-Hölder variant of Theorem 4.1 near a point of type $\mathbb{Y}$. As we shall see, nothing very exciting happens after we use Lemma 4.6 to find points in E_Y, but we describe the argument with some detail to convince the reader of precisely that.

Recall that we chose a small radius r_1 below (4.38); we want to take $r_0 = 10^{-3} r_1$, and prove that

$$\beta_c(x, r) \leq \varepsilon_0 \quad \text{for } x \in E \cap B(x_0, 2r_0) \text{ and } 0 < r \leq 2r_0, \tag{4.55}$$

which means that we want to find a cone $Z(x, r)$ such that $d_{x,r}\big(E, Z(x, r)\big) \leq \varepsilon_0$.

For points of E_Y, we already found $Z(x, r)$, by (4.48) and (4.49) (and if we choose $\varepsilon_3 < \varepsilon_0$). So let us consider $x \in E \cap B(x_0, 2r_0) \setminus E_Y$, and set

$$\delta(x) = \mathrm{dist}(x, E_Y) \leq |x| \leq 2r_0 = 2 \cdot 10^{-3} r_1, \tag{4.56}$$

where the inequalities come from the fact that 0 is of type $\mathbb{Y}$ and $x \in B(0, 2r_0)$. Choose $y \in E_Y$ such that

$$|y - x| \leq 2\delta(x) \leq 4 \cdot 10^{-3} r_1. \tag{4.57}$$

Let $0 < r \leq 2r_0$ be as in (4.55), and set

$$\rho = 10\delta(x) + 10r < 10^{-1} r_1; \tag{4.58}$$

then (4.48) holds, and we can find $Z(y,\rho)$ such that (4.49) and (4.50) holds. In particular, $d_{y,\rho}\big(E, Z(y,\rho)\big) \le \varepsilon_3$.

If $r \ge \delta(x)/20$, we can use $Z(x,r) = Z(y,\rho)$, because then

$$\tag{4.59} \begin{aligned} d_{x,r}\big(E, Z(x,r)\big) &= d_{x,r}\big(E, Z(y,\rho)\big) \le \frac{\rho}{r} d_{y,\rho}\big(E, Z(y,\rho)\big) \\ &\le \frac{\rho\varepsilon_3}{r} = \frac{10\delta(x)+10r}{r}\varepsilon_3 \le 300\varepsilon_3 \le \varepsilon_0 \end{aligned}$$

since $B(x,r) \subset B(y,\rho)$ and if ε_3 is small enough. So we may assume that $0 < r \le \delta(x)/20$.

Denote by L the singular set of $Z(y,\rho)$. Thus L is the line at the intersection of the faces of $Z(y,\rho)$ if it is of type $\mathbb{Y}$, the union of four half lines emanating from the center of $Z(y,\rho)$ if it is of type $\mathbb{T}$, and the empty set if $Z(y,\rho)$ is a plane (which is unlikely). Let us check that

$$\tag{4.60} \mathrm{dist}(x, L) \ge \frac{\delta(x)}{4}.$$

Otherwise, we can find $\xi \in L$ such that $|\xi - x| \le \delta(x)/2$ and, if $Z(y,\rho)$ is of type $\mathbb{T}$, such that the distance from ξ to the center of $Z(y,\rho)$ is larger than $\delta(x)/5$. Because of this, $Z(y,\rho)$ coincides in $B(\xi, \delta(x)/20)$ with a minimal cone Y of type $\mathbb{Y}$, centered at ξ. Notice that

$$\tag{4.61} |y-\xi| + \frac{\delta(x)}{40} \le |y-x| + |x-\xi| + \frac{\delta(x)}{40} \le 2\delta(x) + \frac{\delta(x)}{2} + \frac{\delta(x)}{40} < \frac{\rho}{3}$$

by (4.57) and (4.58), so $B(\xi, \delta(x)/40) \subset B(y,\rho)$ and hence

$$\tag{4.62} \begin{aligned} d_{\xi,\delta(x)/40}(E, Y) &= d_{\xi,\delta(x)/40}\big(E, Z(y,\rho)\big) \le \frac{40\rho}{\delta(x)} d_{y,\rho}\big(E, Z(y,\rho)\big) \\ &\le \frac{400(\delta(x)+r)\varepsilon_3}{\delta(x)} \le 500\,\varepsilon_3 \end{aligned}$$

by definition of Y, (4.58), (4.49), and because $r \le \delta(x)/20$. If ε_3 is small enough, (4.62) will allow us to apply Lemma 4.6 to $E' = \big(160/\delta(x)\big)(E-\xi)$. Indeed E is minimal in $U \supset B(0,4r_1)$, and $B(0,4r_1) \supset B(\xi,\delta(x)/40)$ because $|\xi| + \delta(x)/40 \le |x| + |\xi - x| + \delta(x)/40 \le 2r_0 + \delta(x)/2 + \delta(x)/40 \le 2r_0 + \delta(x) \le 4\cdot 10^{-3} r_1$ by (4.56). Then E' is minimal in $B(0,4)$, and $d_{0,4}(E', Y-\xi) = d_{\xi,\delta(x)/40}(E,Y) \le 500\varepsilon_3$ (recall that Y is centered at ξ, and use (4.62)). So Lemma 4.6 gives a point of type $\mathbb{Y}$ in $E' \cap B(0,1)$, and so $E \cap B(\xi, \delta(x)/40)$ contains a point of type $\mathbb{Y}$. But then

$$\tag{4.63} \delta(x) = \mathrm{dist}(x, E_Y) \le |x-\xi| + \frac{\delta(x)}{40} \le \frac{\delta(x)}{2} + \frac{\delta(x)}{40} < \delta(x);$$

this contradiction proves (4.60). Then $Z(y,\rho)$ coincides with a plane in $B = B(x, \delta(x)/10)$. In addition, $B \subset B(y,\rho)$ by (4.57) and (4.58), so we can apply (4.50) to it. We get that

$$\tag{4.64} \begin{aligned} \mathcal{H}^d\left(E \cap B\left(x, \frac{\delta(x)}{10}\right)\right) &\le \mathcal{H}^d\left(Z(y,\rho) \cap B\left(x, \frac{\delta(x)}{10}\right)\right) + \varepsilon_3\rho^d \\ &\le \left(\frac{\delta(x)}{10}\right)^d \pi + \varepsilon_3\rho^d. \end{aligned}$$

By the monotonicity property (4.4), we obtain that

$$\theta(x,t) \leq \theta\Big(x, \frac{\delta(x)}{10}\Big) \leq \pi + \varepsilon_3 \Big(\frac{10\rho}{\delta(x)}\Big)^d \tag{4.65}$$

or $0 < t \leq \delta(x)/10$. Obviously $\theta(x) \geq \pi$ because $x \in E$, so if ε_3 is small enough, we can apply the almost constant density property (Lemma 4.4) one last time, and get that for $0 < r \leq \delta(x)/20$ we can find a minimal cone $Z(x,r)$ such that $d_{x,r}\big(E, Z(x,r)\big) \leq \varepsilon_0$. This is exactly what we needed for (4.55) in our last remaining case, and this completes our description of the proof when the origin is a point of type $\mathbb{Y}$.

Let us finally say a few words about the last case when 0 is a point of type $\mathbb{T}$. We do not need an existence lemma for points of type $\mathbb{T}$, because we already have the origin, and we expect no other point of type $\mathbb{T}$.

Let us only sketch the argument, because it is almost the same as when 0 is a point of type $\mathbb{Y}$, but longer. We start as before: we choose r_1 so small that $\theta(0,4r_1) \leq \theta(0) + \varepsilon_1$, and hence also $\theta(0,r) \leq \theta(0) + \varepsilon_1$ for $0 < r \leq 4r_1$. By the almost constant density property, we can find minimal cones $Z(0,r)$, $0 < r \leq 2r_1$, centered at 0, and such that

$$d_{0,r}\big(E, Z(0,r)\big) \leq \varepsilon_2 \tag{4.66}$$

and

$$\big|\mathcal{H}^d\big(E \cap B(y,t)\big) - \mathcal{H}^d\big(Z(0,r) \cap B(y,t)\big)\big| \leq \varepsilon_2 r^d \tag{4.67}$$

for $y \in \mathbb{R}^n$ and $t > 0$ such that $B(y,t) \subset B(0,r)$.

When we apply this with $y = 0$ and $t = r$, we see that $\mathcal{H}^d\big(Z(0,r) \cap B(y,t)\big) > (3\pi/2)t^d$ (because $\theta(0,r) \geq \theta(0) = d_T$), hence $Z(0,r)$ is of type $\mathbb{T}$.

When we apply (4.67) to pairs (y,t) such that $r/10 \leq |y| \leq r/2$ and $t = |y|/10$, we get that

$$\mathcal{H}^d\big(E \cap B(y,t)\big) \leq \mathcal{H}^d\big(Z(0,r) \cap B(y,t)\big) + \varepsilon_2 r^d \leq \frac{3\pi t^d}{2} + \varepsilon_2 r^d \leq \frac{3\pi t^d}{2} + C\varepsilon_2 t^d \tag{4.68}$$

because $Z(0,r)$ coincides with a cone of type $\mathbb{P}$ or $\mathbb{Y}$ in $B(y,t)$. Thus $\theta(y,|y|/10) \leq 3\pi/2 + C\varepsilon_2$.

If in addition $y \in E_Y$, this allows us to apply the almost constant density property and find, for $0 < \rho \leq |y|/20$, a minimal cone $Z(y,t)$ such that

$$d_{y,\rho}\big(E, Z(y,\rho)\big) \leq \varepsilon_3 \tag{4.69}$$

and

$$\big|\mathcal{H}^d\big(E \cap B(x,t)\big) - \mathcal{H}^d\big(Z(y,\rho) \cap B(x,t)\big)\big| \leq \varepsilon_3 \rho^d \tag{4.70}$$

for $x \in \mathbb{R}^n$ and $t > 0$ such that $B(x,t) \subset B(y,\rho)$.

Now we claim that we control the balls centered at the origin (by (4.66)), the balls $B(y,\rho)$ centered on E_Y and for which $\rho \leq |y|/20$ (by (4.69)), but also for which $\rho > |y|/20$, because in this last case we can use $Z(y,\rho) = Z\big(0, 2(|y|+\rho)\big)$, just because

$$d_{y,\rho}\big(E, Z(y,\rho)\big) \leq \frac{2(|y|+\rho)}{\rho} d_{0,2(|y|+\rho)}\big(E, Z(y,\rho)\big) \leq C\varepsilon_2 \tag{4.71}$$

by (4.66).

We are thus left with the balls $B(x,t)$ centered on $E \setminus [E_Y \cup \{0\}]$. For these we proceed as before. We set $\delta(x) = \min\big(|x|, \mathrm{dist}(x, E_Y)\big)$, and notice that we have a control on $B\big(x, 10\delta(x)\big)$ because of one of the balls that we already treated. So we get a nearby cone Z, and we can take $Z(x,t) = Z$ if $t \geq 10^{-5}\delta(x)$, say. For $t < 10^{-5}\delta(x)$, we first show that $B(x, \delta(x)/100)$ does not meet the singular set of Z, as we did before (4.64), use this to show that $\mathcal{H}^d\big(Z \cap B(x, \delta(x)/1000)\big) \leq \pi(\delta(x)/1000)^d + C\varepsilon_3\delta(x)^d$, deduce from this that $\theta(x, 2t) \leq \theta(x, \delta(x)/1000) \leq \pi + C\varepsilon_3$, and use the almost constant density property to find the good cone $Z(x,t)$. The details are a little long and boring, but no difficulty appears.

This completes our description of Theorem 4.1, in the simpler case of minimal sets, and with the easier bi-Hölder conclusion. □

4.9. Last comments, C^1 regularity, and epiperimetry.

Remark 4.7. At this date we do not have an existence lemma for points of type $\mathbb{T}$, that would be similar to Lemma 4.6. Such a lemma would say that if E is a reduced almost minimal set in $B(0, 4r)$, with a sufficiently small gauge function h, and if $d_{0,4}(E, T) \leq \varepsilon$ for some minimal cone T of type $\mathbb{T}$ centered at the origin, then there is a point of type $\mathbb{T}$ in $B(0,1)$.

We did not need this for J. Taylor's result, but it would be enough to prove that every reduced minimal set in the whole $\mathbb{R}^3$ is a minimal cone (see Section 5). Also, if we want to establish a J. Taylor theorem for 3-dimensional sets in $\mathbb{R}^4$, we will probably need a lemma that finds points of type $\mathbb{T} \times \mathbb{R}$ when E looks a lot like the minimal cone $\mathbb{T} \times \mathbb{R}$ in the unit ball.

The corresponding extension of Lemma 4.6 for points of type $\mathbb{Y} \times \mathbb{R}$ in $\mathbb{R}^4$ is true, and leads to a Hölder regularity result for 3-dimensional almost minimal sets near points of type $\mathbb{Y} \times \mathbb{R}$; see [44].

Remark 4.8 (on C^1 regularity). Theorem 4.1 is significantly more delicate than its bi-Hölder variant. Here we just needed to check that the numbers $\beta_c(x,r)$ that control the Hausdorff distances to minimal cones stay small. For the C^1 result, we would need to prove that they have some definite decay.

Almost equivalently, we proved that the density $\theta(x,r)$ is a nondecreasing function of r, and for the C^1 result, it seems that we need to prove that the density excess $f(r) = \theta(x,r) - \theta(x)$ decays when r tends to 0, at some definite speed.

As we have seen, the monotonicity of $\theta(x,\cdot)$ is obtained rather easily, essentially by comparing E with the cone over $E \cap \partial B(x,r)$. The necessary decay for the density excess is harder to get. In [59] it is obtained as the consequence of some anterior epiperimetry result [58], which I am not sure I understand.

In [15], the decay for f is obtained in the following way. Because of Theorem 4.2 (the bi-Hölder estimate), we can reduce to situations where E is quite close to a minimal cone X. We suppose that E does not coincide with X, measure the difference in terms of the density excess $f(r) = \theta(x,r) - \theta(x)$, use this difference to construct a competitor which is even better than the cone over $E \cap \partial B(x,r)$, make sure that the improvement over the cone is large enough compared to the density excess, and use this to get a differential inequality on f that implies the desired decay. Finally we check that f controls other geometric quantities like the $\beta_c(x,r)$.

Somehow, the fact that we can suppose that E is close to a the minimal cone X acts as a linearization of our problem: a situation where E is completely wild would be much harder to control at once.

We use X in the following way. Since X is a minimal cone, $X \cap \partial B(x,r)$ is a collection of geodesics g_i on $\partial B(x,r)$ that may meet by sets of three at vertices (see the description in Section 4.1). We use the fact that E is close to X to find a net of nearby curves γ_i, with almost the same vertices, and such that the symmetric difference between $E \cap \partial B(x,r)$ and $\bigcup_i \gamma_i$ has small $\mathcal{H}^1$ measure. We can even modify the γ_i so that they are Lipschitz graphs with small constants.

Recall that we want to construct good competitors for E, better than the cone over $E \cap \partial B(x,r)$, and after a first deformation that is not too costly, we are reduced to constructing good competitors for the cone over $\bigcup_i \gamma_i$. This is much easier now. For instance, we can proceed independently for each i, find a competitor for the cone Γ_i over γ_i with the same boundary (i.e., γ_i and the two segments from x to the extremities of γ_i), and then glue.

We then observe that for each i, Γ_i is the graph of a radial extension, and the graph of the corresponding harmonic extension does significantly better, with a good control on the difference in terms of how close γ_i is to a geodesic. This is what we mean by a linearization: we are able to reduce to a situation that is so close to a flat case that we can use asymptotic expansions.

This works well when (some of) the γ_i are not too close to geodesics, but we also need to do something when the γ_i are almost geodesics, but not with the same endpoints as that g_i. This situation should be simpler now because we can compute better with the cone over a collection of geodesics, but this is the place where additional information on the metric properties of our cone X comes into play. We want to know how the sum of the lengths of the g_i changes when we move the vertices, and the full length property discussed near (4.2) is the notion that comes out from the computations.

It is probable that the two different ways to get the desired decay estimates for f are similar; in both cases, some specific knowledge of metric properties of the minimal cone X are needed.

5. Mumford – Shah minimal sets and a Bernstein problem

We start with a variant, for minimal sets, of the classical Bernstein problem where you take a graph over the plane, assume that it is a minimal surface, and ask whether it must be a plane.

(5.1) Is every 2-dimensional reduced minimal set in $U = \mathbb{R}^3$ equal to a cone?

See Definitions 3.1 and 3.2 for reduced minimal sets. We expect this to happen because this is the only way for the set not to get worse at infinity. The question would also make sense in higher dimensions, but it should be harder and we don't even know the answer in the present case.

The author discovered this question (and Almgren minimal sets at the same time) because of the Mumford – Shah functional in image segmentation, where a positive answer helps describe the blow-up limits of minimal segmentations, and then get some regularity properties for the minimal segmentations themselves.

Fortunately the answer to this question is easier in the Mumford – Shah context. Let us define the notion of Mumford – Shah minimal set only in the domain $U =$

$\mathbb{R}^n$ (this is simpler), and refer the reader to [4, 11, 12, 46] for information on the Mumford – Shah functional and image processing.

Definition 5.1. Let E be a closed set in $\mathbb{R}^n$, and let $B \subset \mathbb{R}^n$ be a closed ball. An MS-competitor for E in B is a closed set $F \subset \mathbb{R}^n$ such that

$$F \setminus B = E \setminus B \tag{5.2}$$

and, for $x, y \in \mathbb{R}^n \setminus [B \cup E]$,

$$F \text{ separates } x \text{ from } y \text{ if } E \text{ separates } x \text{ from } y. \tag{5.3}$$

Here we say that E separates x from y when x and y lie in different connected components of $\mathbb{R}^n \setminus E$. Clearly, if F is an MS-competitor for E in B, then it is also an MS-competitor for E in any larger ball B. In $\mathbb{R}^2$, for example, $\mathbb{R} \cup \partial B(0,1)$ is a competitor for $\mathbb{R}$ in (any ball that contains) $\overline{B}(0,1)$, while $\mathbb{R} \setminus (-1,1)$ is not a competitor for $\mathbb{R}$ (in any ball).

Definition 5.2. An MS-minimal set in $\mathbb{R}^n$ is a closed set $E \subset \mathbb{R}^n$ such that $\mathcal{H}^{n-1}(E \cap B) < +\infty$ for every ball B, and

$$\mathcal{H}^{n-1}(E \cap B) \leq \mathcal{H}^{n-1}(F \cap B) \tag{5.4}$$

whenever $B \subset \mathbb{R}^n$ is a closed ball and F is an MS-competitor for E in B.

Thus $d = n - 1$ here. Sets of dimensions smaller than $n - 1$ do not separate points in $\mathbb{R}^n$, so the notion would not be relevant for $d < n - 1$. Notice also that (5.4) stays true if we replace B with a larger ball.

In terms of Alexis Bonnet's definitions relative to the Mumford – Shah functional, $E \subset \mathbb{R}^n$ is an MS-minimal set if and only if the pair (u, E), where u is constant on every component of $\mathbb{R}^n \setminus E$, is a global Mumford – Shah minimizer. Of course we shall not use this here.

It is not so hard to check that if E is a closed subset of $\mathbb{R}^n$ and if F is an Almgren competitor for E (i.e., if $F = \varphi(E)$ for some Lipschitz mapping $\varphi\colon \mathbb{R}^n \to \mathbb{R}^n$ such that $\varphi(x) = x$ for $|x|$ large, as in Definition 3.1), then F is an MS-competitor for E. That is, if x, y lie far from the origin and in different connected components of $\mathbb{R}^n \setminus E$, then $\varphi(E)$ also separates them.

Because of this, every MS-minimal set in $\mathbb{R}^n$ is a minimal set as in Definition 3.1, with $h = 0$ (we shall say Al-minimal to distinguish). And it is true that

$$\text{every 2-dimensional reduced MS-minimal set in } \mathbb{R}^3 \text{ is a cone.} \tag{5.5}$$

Here reduced is still meant as in Definition 3.2. Of course, our set is then a plane or a cone of type $\mathbb{Y}$ or $\mathbb{T}$, as in Section 4.1, because it is Al-minimal. We don't know whether something like this is true in higher dimensions.

Let us describe the proof of (5.5) given in [14] because it is amusing. Its idea (the amusing part) was known from F. Morgan; see [47], even though he was working in a different context. If we had an existence lemma for points of type $\mathbb{T}$ (like Lemma 4.6), we would easily get a positive answer to (5.1) by the same argument.

So let E be a reduced Al-minimal set of dimension 2 in $\mathbb{R}^3$ (we shall mention it explicitly when we use the MS-minimality). Pick any origin $x \in E$, and recall from (4.4) that $\theta(x,r) = r^{-2}\mathcal{H}^2\big(E \cap B(x,r)\big)$ is nondecreasing. Then it has limits $\theta(x) = \lim_{r\to 0} \theta(x,r)$ and $\theta_\infty = \lim_{r\to+\infty} \theta(x,r)$, and the last one is easily seen not to depend on x.

If for some $x \in E$, $\theta(x) = \theta_\infty$, then $\theta(x,\cdot)$ is constant and (4.11) says that E is a minimal cone. So we may assume that

$$\theta(x) < \theta_\infty \quad \text{for all } x \in E. \tag{5.6}$$

Recall that we can find blow-in limits of E (as above (4.24)), and by (4.24) they are minimal cones with density θ_∞. Let us distinguish cases, depending on the value of θ_∞.

If $\theta_\infty = \pi$, i.e., if the blow-in limits of E are planes, we pick any $x \in E$, and any blow-up limit of E at x. Such blow-up limits exist, and they are cones of type $\mathbb{P}, \mathbb{Y}$, or $\mathbb{T}$. Then $\theta(x)$, which is the density of such a cone, cannot be smaller than $\theta_\infty = \pi$. This contradicts (5.6) and we get the result.

Next suppose that $\theta_\infty = 3\pi/2$, i.e., that the blow-in limits of E are cones of type $\mathbb{Y}$. Select a sequence $\{r_k^{-1}E\}$, with $\lim_{k\to+\infty} r_k = +\infty$, that tends to a cone Y of type $\mathbb{Y}$. Recall that this implies that for $R > 0$,

$$\lim_{k\to+\infty} d_{0,R}(r_k^{-1}E, Y) = 0, \tag{5.7}$$

as in (4.20). We apply this with $R = 4$, and choose k so large that $d_{0,4}(r_k^{-1}E, Y) \leq \varepsilon$, where ε is as in Lemma 4.6. Then this lemma says that $r_k^{-1}E$ contains a point y of type $\mathbb{Y}$, and of course $w = x + r_k y$ is a point of type $\mathbb{Y}$ in E. This is impossible again, because (5.6) says that $\theta(w) < \theta_\infty = 3\pi/2$.

We are left with the case when $\theta_\infty = d_T$, where d_T is the density at the origin of a cone of type $\mathbb{T}$. If we had an existence lemma for points of type $\mathbb{T}$, we could proceed as above, find a point $w \in E$ such that $\theta(w) = d_T$, and get a contradiction with (5.6). And in fact we will essentially prove an existence lemma in the present situation with an MS-minimal set.

So let us assume that E contains no point of type $\mathbb{T}$ (otherwise, we get a contradiction with (5.6)). Let T be a blow-in limit of E; it is a minimal cone of type $\mathbb{T}$, centered at the origin. As before, we can find a radius r_k such that if we set $F = r_k^{-1}E$,

$$\delta_{0,4}(F, T) \leq \varepsilon, \tag{5.8}$$

where $\varepsilon > 0$ is as small as we want. Of course, F is also a minimal set, and it has no point of type $\mathbb{T}$. Denote by F_Y the set of points of type $\mathbb{Y}$ in F.

Denote by L the singular set of T; thus L is the union of four half lines L_i, $1 \leq i \leq 4$. Denote by a_i, $1 \leq i \leq 4$, the point of $L_i \cap \partial B(0,2)$ (you may start to look at Figure 6). If ε is small enough, we can apply Lemma 4.6 to $B(a_i, 10^{-10})$ to find a point y_i of type $\mathbb{Y}$ in F, with $|y_i - a_i| \leq 10^{-10}$.

We shall use Theorem 4.2 (the bi-Hölder equivalence) rather than the full theorem of J. Taylor (Theorem 4.1), because it is easier to prove. But the reader is free to apply Theorem 4.1 secretly and work with C^1 curves below. Near each y_i, F is bi-Hölder equivalent to a minimal cone of type Y, and F_Y is bi-Hölder equivalent to a line. In fact, the proof of Theorem 4.2 also says that this happens in the ball $B(y_i, 10^{-3})$, say, and that in this ball F_Y coincides with a simple curve that really crosses $\partial B(0,2)$, in the sense that one of its ends lies out of $B(0, 2+10^{-6})$ and the other one lies inside $B(0, 2-10^{-6})$.

We also need to know that outside of the $B(y_i, 10^{-6})$, $\partial B(0,2)$ never gets close to F_Y. This is true because near such points of $\partial B(0,2)$, T and F are very close to a plane, and Theorem 4.2 implies that F is equivalent to a plane and contains no point of type $\mathbb{Y}$.

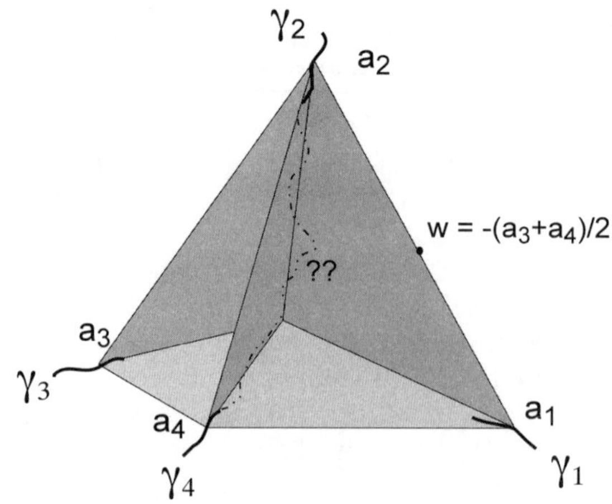

FIGURE 6. The set T and the curves γ_i (imagine that F is very close to T).

Return to F_Y. We know that near y_j, F_Y coincides with a simple (bi-Hölder) curve γ_i. We want to follow this curve as it gets inside $B(0,2)$, and follow it as long as we can. That is, we try to continue γ_i into a simple curve contained in F_Y, as long as possible and as suggested by Figure 6.

By Theorem 4.2 again, we know that every point $y \in F_Y$ has a small neighborhood where F_Y is a simple curve. We use this to continue γ_i, little by little. We never hit a point x where we cannot continue γ_i, because $x \in F$ (which is closed), x cannot be of type $\mathbb{T}$ (there is no such point), nor a point of type $\mathbb{P}$ (because such points lie at positive distance from F_Y, by Theorem 4.2), and because at a point of type $\mathbb{Y}$ we can continue the curve. Similarly, γ_i does not return to itself (the first point where this happens would not be of type $\mathbb{Y}$), nor accumulate on a complicated set (any point of this set would be of type $\mathbb{Y}$, a contradiction), so the only thing that it can do is eventually leave $B(0,2)$. In addition, this does not happen near the same a_i (as above, the curve does not return to itself). This means that γ_i goes from y_i to some other y_j.

Let y be any point of F_Y. By Theorem 4.2, there is a small neighborhood of y where $\mathbb{R}^n \setminus E$ is equivalent to the complement of a cone of type $\mathbb{Y}$. This means that there are at most three connected components of $B(0,3) \setminus F$ that touch y. Some of these three components may coincide (because we don't know what happens away from y), but this is all right. Let us denote by $H(y)$ this set of at most three components. By definitions and the description of F near y, $H(y') = H(y)$ for $y \in F_Y$ close to y. That is, H is locally constant along F_Y.

Denote by O_k the connected component of $B(0,3) \setminus F$ that contains $-a_k$. Recall that we have a bi-Hölder equivalence in the large ball $B(y_i, 10^{-3})$, and because of this it is fairly easy to check that $H(y_i)$ is composed of the three O_k, $k \neq i$. Again two of these components could be equal.

But when we go from y_i to y_j along γ_i, $H(y)$ stays constant. So $H(y_i) = H(y_j)$. But $H(y_j)$ is composed of the three components O_k, $k \neq j$. Altogether, each of the four components O_k lies in $H(y_i)$, and since $H(y_i)$ has at most three elements, we find that there exist $k \neq l$ such that $O_k = O_l$.

In the context of Al-minimal sets, this would not yet give a topological contradiction: one can produce sets that are locally C^1 equivalent to a cone of type $\mathbb{Y}$ or $\mathbb{P}$ at each point, and that coincide with a cone of type $\mathbb{T}$ on $\mathbb{R}^n \setminus B(0,1)$. We only proved that for such a set F, $\mathbb{R}^n \setminus F$ has at most 3 components (in fact, at most 2 if we use another curve γ_k in F_Y). See [14] for an example of such a set, which was probably also known before by B. Hardt and F. Morgan. See for instance [51, p. 110]. But Liang [40] showed that an Al-minimal set which is not a cone cannot be modeled on this example, because the two tubes in $\mathbb{R}^3 \setminus F$ that connect opposite components are not knotted.

In the context of MS-minimal sets, we are happy because we find out that the big wall between $-a_k$ and $-a_l$ is not needed. That is, set $w = -(a_k + a_l)/2$(see Figure 6 again). This point lies on the middle of a face of T, the face that separates $-a_k$ from $-a_l$. Theorem 4.2 (applied to a point of F near w) shows that $B(w, 10^{-3})$ does not meet any other component of $B(0,3) \setminus F$ than O_k and O_l. Then set $G = F \setminus B(w, 10^{-3})$. It is not hard to see that G is a competitor for F in $B = \overline{B}(0,4)$ (the only components that could possibly become equal when we replace F with G are O_k and O_l, which happen to be equal before we start). Also, $\mathcal{H}^2(G \cap B) < \mathcal{H}^2(F \cap B)$, so (5.4) fails, F is not minimal, and we get the desired contradiction. This completes our description of the proof of (5.5); see [14] for details. □

Since the situation is already fairly complicated in $\mathbb{R}^3$, the author does not expect an analogue of (5.5) in higher dimensions to be easy.

Remark 5.3 (about topological minimal sets in higher codimensions). The notion of MS-minimal is nice, because of its connections to the Mumford–Shah functional, and also because in this context existence results could be easier. The definition above only makes sense in codimension 1, but a generalization of the notion was recently proposed by Xiangyu Liang [40, 42]. Let us discuss this rapidly. As above, we only define her topological minimal sets in $\mathbb{R}^n$, because this is simpler.

Definition 5.4. Let E be a closed set in $\mathbb{R}^n$, and let $B \subset \mathbb{R}^n$ be a closed ball. A topological competitor of dimension d for E in B is a closed set $F \subset \mathbb{R}^n$ such that

$$F \setminus B = E \setminus B \tag{5.9}$$

and, for each Euclidean sphere S of dimension $n-d-1$ that is contained in $\mathbb{R}^n \setminus [B \cup E]$, if S represents a nonzero element in the singular homology group $H_{n-d-1}(\mathbb{R}^n \setminus E; \mathbb{Z})$, then S also represents a nonzero element in $H_{n-d-1}(\mathbb{R}^n \setminus F; \mathbb{Z})$.

When $d = n-1$, S is composed of two points, and it represents a nonzero element in $H_0(\mathbb{R}^n \setminus E; \mathbb{Z})$ if and only if these two points lie in different connected components of $\mathbb{R}^n \setminus E$, so Definition 5.4 is really a generalization of Definition 5.1. The corresponding notion of topological minimal set is as follows (we just change the class of competitors).

Definition 5.5. A topological minimal set of dimension d in $\mathbb{R}^n$ is a closed set $E \subset \mathbb{R}^n$ such that $\mathcal{H}^d(E \cap B) < +\infty$ for every ball B, and for which

$$\mathcal{H}^d(E \cap B) \le \mathcal{H}^d(F \cap B) \tag{5.10}$$

whenever $B \subset \mathbb{R}^n$ is a closed ball and F is a topological competitor of dimension d for E in B.

There is an analogue of the Bernstein problems (5.1) and (5.5) for this notion, but even the case of 2-dimensional topological minimal sets in $\mathbb{R}^4$ is unclear.

Topological minimal sets have some nice properties. First, one can show that they are Al-minimal sets. For this, using homology groups was a good idea, because it turns out that the nonvanishing of S in $H_{n-d-1}(\mathbb{R}^n \setminus E; \mathbb{Z})$ is stable under deformations of E in $\mathbb{R}^n$ (away from S); a corresponding attempt with homotopy groups would fail.

Liang shows that the almost orthogonal union of two planes of dimension d in $\mathbb{R}^n$ is topologically minimal, and that the product by $\mathbb{R}$ of a topologically minimal set is topologically minimal; note that the corresponding property for Al-minimal sets is not known. A consequence of this is that some non transversal unions of 3-planes in $\mathbb{R}^5$, for instance, are topological, and hence also Al-minimal sets. See [40, 42] for details.

Another nice feature of the notion is that, as for the MS-minimality, related existence theorem can be proved. See the next section.

6. Existence results for simpler problems

Simpler means, compared to the Plateau problems above. In this section we describe how a construction of V. Feuvrier [29, 30] can be used to prove some existence results for minimizers.

We start with notation for a general situation. Let Ω be a simple "domain," which we often want to be closed, so that limits of sets in Ω stay in Ω. Or Ω can be a manifold. Let $\mathcal{F}$ be a class of closed subsets of Ω. Also let $g\colon \Omega \to [0, +\infty)$ be a given continuous function, and assume that $1 \leq g(x) \leq N$ for some $N \geq 1$. We want to minimize the functional

$$J_g(E) = \int_E g(x)\, \mathrm{d}\mathcal{H}^d(x) \tag{6.1}$$

over the class $\mathcal{F}$. Of course we expect many things to depend on the specific choice of $\mathcal{F}$, and for our construction to have a chance to work we will need $\mathcal{F}$ to be closed under some class of deformations.

It is better to choose Ω and $\mathcal{F}$ so that

$$0 < m_g(\mathcal{F}) := \inf\{J_g(E) : E \in \mathcal{F}\} < +\infty, \tag{6.2}$$

because otherwise nothing interesting happens. This is why we like to take domains Ω with some topology, to make sure that sets $E \in \mathcal{F}$ cannot be deformed to one point, for instance.

Here is the most obvious way to try to get a minimizer. Take a minimizing sequence, i.e., a sequence $\{E_k\}$ in $\mathcal{F}$ such that

$$\lim_{k \to +\infty} J_g(E_k) = m_g(\mathcal{F}), \tag{6.3}$$

and extract a subsequence (which will also be denoted by $\{E_k\}$) which converges to a limit E_∞. If we want to be sure that this is possible, we usually have no choice on the notion of convergence: we take the convergence for the Hausdorff distance (locally in Ω if Ω is open). That is, we could also hope to work in a subclass of $\mathcal{F}$ for which we have better compactness properties, but this will probably happen because in that subclass some apparently stronger notion of convergence is implied by the Hausdorff convergence, so let us not bother yet.

There are two main problems now. The first one is that it is not always true that

$$J_g(E_\infty) \leq \lim_{k \to +\infty} J_g(E_k), \tag{6.4}$$

i.e., J_g is not lowersemicontinuous. For instance, dotted lines segments $E_k \subset [0,1]$ may converge to $E_\infty = [0,1]$, while $\mathcal{H}^1(E_k) = 1/2$ for all k. This is the main issue that the Feuvrier construction addresses. The other problem is that we cannot be sure that

$$E_\infty \in \mathcal{F}. \tag{6.5}$$

This part will depend on $\mathcal{F}$; some classes will be easier to deal with than others, and we shall just give two examples where we can get something like (6.5). But in general, we can expect serious difficulties. For instance, if $\mathcal{F}$ is the class of all the sliding deformations of an initial set E_0, as in Example 2.3 in Section 2.4, we have the usual problem relative to parameterizations: each E_k'' is a obtained through a deformation by mappings $\varphi_{t,k}$, but we have no reason to believe that the $\varphi_{t,k}$ will converge to nice mappings $\varphi_{t,\infty}$.

Example 6.1. This example is related to Example 2.3, but we will need extra assumptions. We try to follow [29, 31], but where unfortunately the details of this precise example are not carried out; the reader will thus have to trust the sketchy proof below. Here Ω will be a simple manifold without boundary obtained by pasting faces of dyadic cubes in $\mathbb{R}^n$. That is, we start from a finite collection of dyadic cubes Q_j of unit length, and glue some faces together so as to get a manifold without boundary $\mathbf{M}$; it does not matter whether the manifold is orientable or not. For instance, we could get a torus $\mathbb{R}/\mathbb{Z} \times \mathbb{R}/\mathbb{Z} \simeq \mathbb{S}^1 \times \mathbb{S}^1$ from a single square, by gluing the opposite faces correctly. The fact that we have no boundary left is convenient because they would restrict our definition of deformations. The precise form of rigid manifold is also convenient, because polyhedral grids are much easier to construct there.

We also have to assume that $d = 2$, and that the function $g\colon \mathbf{M} \to [1,N]$ is Hölder-continuous, say. This is because we intend to use Theorem 4.2.

Let us proceed as in Example 2.3. Select a closed set $E_0 \in \Omega$, and denote by $\mathcal{F}(E_0)$ the class of continuous deformations of E_0 inside Ω. That is, $F \in \mathcal{F}(E_0)$ if $F = \varphi_1(E_0)$, for some mappings $\varphi_t\colon E_0 \to \mathbf{M}$, $0 \leq t \leq 1$, such that $(x,t) \to \varphi_t(x)\colon E_0 \times [0,1] \to \mathbf{M}$ is continuous, and $\varphi_0(x) = x$ on E_0.

Claim 6.2. *Under all these assumptions, we can find $F \in \mathcal{F}(E_0)$ such that*

$$J_g(F) = \inf\{J_g(E) : E \in \mathcal{F}(E_0)\}. \tag{6.6}$$

Let us explain how we can prove this. The construction may be more general, but it is simpler under our assumptions. Its main point is that we want to have (6.4), for a minimizing sequence that we shall need to select carefully. The idea is not new, this is what Reifenberg did in [55], already in the context of the Plateau problem (with homology).

We start from any minimizing sequence $\{E_k\}$ as above, and our first idea (that will not work) is to replace each E_k with its Federer – Fleming projection on a dyadic net. Recall that in codimension 1, this would be done in one step, as follows. We cut $\mathbf{M}$ into dyadic cubes Q of sidelength 2^{-m}, for some large integer $m = m(k)$. For each such Q, we select a center $\xi_Q \in Q \setminus E_k$, and use the radial projection centered at ξ_Q to project $E \cap Q$ on ∂Q. (Notice that we do not change E on the boundaries ∂Q.) When $d < n-1$, we have to iterate this procedure, project again on faces of dimension $n-2$, and so on. After a few steps, we get a new candidate which is contained in the d-dimensional skeleton of our dyadic net. Usually, when we get to faces of dimension d, we cannot continue because our set covers the full face. But if by chance it meets a face without containing it, we can iterate once more in this face. Eventually we get a new competitor E'_k, which is a finite union of faces of dimensions at most d.

Unfortunately, $J_g(E'_k)$ may be significantly larger than $J_g(E_k)$ and $m_g(\mathcal{F}(E_0))$; even in the 2-plane, we may have projected the diagonal of a square onto a $\sqrt{2}$ times longer pair of sides. But given E_k, V. Feuvrier constructs a polyhedral net, similar to the usual dyadic net, but which near the flat rectifiable parts of E_k has faces that are almost parallel to E_k, so that after the Federer – Fleming projection on the net, $J_g(E'_k) - J_g(E_k)$ is as small as we want. This also uses the fact that g is continuous (and then we work at scales that are so small that g acts like a constant). The construction of the polyhedral nets is the main part of [29] and [30], and the main difficulty is to be able to complete a collection of non parallel dyadic nets living in disjoint regions, in such a way as to keep a lower bound on all the angles in the various faces of all the dimensions. The reason why we decided to stick to our rigid manifold $\mathbf{M}$ is because Feuvrier's construction of polyhedral nets has not yet been adapted to manifolds. [There are difficulties even with the definition of polyhedral nets, because of the non-affine changes of charts.]

The fact that this lower bound on angles is independent of E_k and the small scale 2^{-m} at which we decide to work is important for the next part of the argument. We replace E'_k (the Federer – Fleming projection of E_k on our net) with another finite union E''_k of faces of the same net, obtained as a deformation of E'_k in $\mathbf{M}$ (and hence also of E_k, since Federer – Fleming projections are deformations), and for which $J_g(E''_k)$ is minimal. The existence is easy, because there are only finitely many combinations of faces.

When we do this, we do not increase $J_g(E'_k)$, and we gain something because can use the minimality property of E''_k (among unions of faces) to prove that it is quasi-minimal (as in Section 3.4) in our manifold $\mathbf{M}$. The quasi minimality constant M that we get depends only on N and the lower bound for the angles in the grid, and not on k or $m = m(k)$. The proof of quasi minimality uses more Federer – Fleming projections, and is not really difficult (see below (6.34) for some hints); again the main difficulty was the construction of the grid.

There is another issue that needs to be addressed: E''_k is a union of d-dimensional faces, and also of lower-dimensional faces where we were able to contract more, and we can only control the d-dimensional part correctly. That is, set

$$F_k = (E''_k)^* = \{x \in E''_k : \mathcal{H}^d(E''_k \cap B(x,r)) > 0 \text{ for all } r > 0\} \tag{6.7}$$

as in (3.4). This is the d-dimensional part that we control, and for the rest we only know that

$$\mathcal{H}^{d-1}(E''_k \setminus F_k) < +\infty, \tag{6.8}$$

but with no control on precisely where $E''_k \setminus F_k$ may lie. So we only want to take a limit of the sets F_k.

Notice that F_k is a reduced quasi-minimal set (it is easy to see that E^* is quasi-minimal when E is quasi-minimal). We replace $\{F_k\}$ with a subsequence for which $\{F_k\}$ converges to a limit F, and use the fact that Theorem 3.6, and in particular the lowersemicontinuity estimate (3.12), stays true for sequences of reduced quasi-minimal sets (with uniform estimates, as here). We get that

$$J_g(F) \leq \liminf_{k\to+\infty} J_g(F_k) = \liminf_{k\to+\infty} J_g(E_k'') = \inf\{J_g(E) : E \in \mathcal{F}(E_0)\}. \tag{6.9}$$

[It is not hard to see that the continuous function g in (6.1) does not disturb the lowersemicontinuity estimate.]

So this takes care of (6.4) and (6.6), but we still need to show that some minor modification of F lies $\mathcal{F}(E_0)$, and this is where we shall use Theorem 4.2 and need more assumptions.

By the proof of Theorem 3.6, and if g was identically equal to 1, we would get that F is locally minimal in $\mathbf{M}$. Here g may vary, and we only get, by an argument similar to what we did near (3.16), that F is almost minimal, with a gauge function h that depends on the modulus of continuity of g. If g is Hölder-continuous, $h(r) \leq Cr^\alpha$ for some $\alpha > 0$, which is more than enough to apply Theorem 4.2.

We get that near each point $x \in F$, F is bi-Hölder equivalent to a minimal cone. Then, near x, there is a Hölder retraction onto F, defined on a small neighborhood of F. We claim that we can make this global, i.e., that there is an ε-neighborhood V_ε of F, and a Hölder retraction $R\colon V_\varepsilon \to F$, so that $R(x) = x$ on F.

Let us check this in a probably too brutal way. For each $x \in F$, there is a radius $r(x) > 0$ such that the conclusion of Theorem 4.2 holds in $B\big(x, 10r(x)\big)$, with uniform estimates. We use the bi-Hölder equivalence provided by Theorem 4.2 to get a Hölder retraction $R_x\colon B\big(x, 5r(x)\big) \to F$, which we obtain by conjugating a Lipschitz retraction on the minimal cone provided by Theorem 4.2 with the bi-Hölder mapping also provided by that theorem. Thus

$$R_x(y) = y \quad \text{for } y \in F \cap B\big(x, 5r(x)\big), \tag{6.10}$$

and

$$|R_x(y) - R_x(z)| \leq C|y-z|^\alpha \quad \text{for } y, z \in B\big(x, 5r(x)\big), \tag{6.11}$$

where the exponent $\alpha < 1$ can be chosen as close to 1 as we want, because we can take the exponent in Theorem 4.2 arbitrarily close to 1.

We don't like the fact that the radius $r(x)$ depends on x, so we cover the compact set F by a finite number of balls $B\big(x_i, r(x_i)\big)$, and then set $r = \inf_i r(x_i)$. For each $x \in F$, we now pick i such that $x \in B\big(x_i, r(x_i)\big)$, and denote by S_x the restriction of R_{x_i} to $B(x, 4r) \subset B\big(x_i, 5r(x_i)\big)$. Obviously

$$S_x(y) \in F \qquad \text{for } y \in B(x, 4r), \tag{6.12}$$

$$S_x(y) = y \qquad \text{for } y \in F \cap B(x, 4r), \tag{6.13}$$

and

$$|S_x(y) - S_x(z)| \leq C|y-z|^\alpha \qquad \text{for } y, z \in B(x, 4r). \tag{6.14}$$

We need mappings defined on the whole $\mathbf{M}$, so we define $T_x\colon \mathbf{M} \to \mathbf{M}$ by

$$T_x(y) = y \qquad \text{for } y \in \mathbf{M} \setminus B(x, 3r), \tag{6.15}$$

$$T_x(y) = S_x(y) \quad \text{for } y \in B(x, 2r), \tag{6.16}$$

and, for $y \in B(x, 3r) \setminus B(x, 2r)$,

$$T_x(y) = \frac{3r - |y-x|}{r} S_x(y) + \frac{|y-x| - 2r}{r} y. \tag{6.17}$$

There is a small abuse of notation here, because in principle $\mathbf{M}$ is a manifold, but we can choose r so small that the convex combination in (6.17) makes sense.

Pick a (new) maximal collection of points $x_i \in F$, $i \in I$, so that $|x_i - x_j| \geq r/2$ when $i \neq j$. Then the $B(x_i, r)$ cover F and the $B(x_i, 5r)$ have bounded overlap. Set $T_i = T_{x_i}$ for $i \in I$. We take for R the composition of all the T_i (taken in any order that we like), restricted to $V_\varepsilon = \{y \in \mathbf{M} : \mathrm{dist}(y, F) \leq \varepsilon\}$, and where ε is very small, to be chosen soon.

Let us verify that R is the desired retraction. Let $y \in V_\varepsilon$ be given, and let us see what happens when we apply the successive T_i to it. To simplify the notation, assume that $I = \{1, \dots, m\}$, and that we apply the T_i in the natural order. Set $y_0 = y$ and, by induction, $y_i = T_i(y_{i-1})$ for $i \geq 1$. We want to check that

$$\mathrm{dist}(y_i, F) \leq \varepsilon_i, \tag{6.18}$$

for some small numbers ε_i that we shall compute along the way. This is the case when $i = 0$, with $\varepsilon_0 = \varepsilon$. Suppose $i \geq 1$ and this is true for $i - 1$. If $y_{i-1} \in \mathbf{M} \setminus B(x_i, 3r)$, then $y_i = T_i(y_{i-1}) = y_{i-1}$ by (6.15), and (6.18) holds (if $\varepsilon_i \geq \varepsilon_{i-1}$). So we may assume that $y_{i-1} \in B(x_i, 3r)$. Pick $z_{i-1} \in F$ such that $|z_{i-1} - y_{i-1}| \leq \varepsilon_{i-1}$. Then $z_{i-1} \in B(x_i, 4r)$ (if $\varepsilon_{i-1} \leq r$; we shall take care of this later). Notice that $T_i(y_{i-1}) \in [S_{x_i}(y_{i-1}), T_i(y_{i-1})]$ by (6.15) – (6.17); then

$$\begin{aligned} |y_i - y_{i-1}| &= |T_i(y_{i-1}) - y_{i-1}| \leq |S_{x_i}(y_{i-1}) - y_{i-1}| \\ &\leq |S_{x_i}(y_{i-1}) - S_{x_i}(z_{i-1})| + |z_{i-1} - y_{i-1}| \\ &\leq C|z_{i-1} - y_{i-1}|^\alpha + |z_{i-1} - y_{i-1}| \leq C\varepsilon_{i-1}^\alpha + \varepsilon_{i-1} \end{aligned} \tag{6.19}$$

by (6.18), because $S_{x_i}(z_{i-1}) = z_{i-1}$ by (6.13), and by (6.14). Then

$$\mathrm{dist}(y_i, F) \leq \mathrm{dist}(y_{i-1}, F) + C\varepsilon_{i-1}^\alpha + \varepsilon_{i-1} \leq C\varepsilon_{i-1}^\alpha + 2\varepsilon_{i-1} \tag{6.20}$$

by (6.18) for $i - 1$. We choose

$$\varepsilon_i = C\varepsilon_{i-1}^\alpha + 2\varepsilon_{i-1} \tag{6.21}$$

and get (6.18). This gives a definition of all ε_i in terms of $\varepsilon_0 = \varepsilon$, and now we choose ε so small that $\sum_{i=0}^m \varepsilon_i \leq r/2$, say. This way we can be sure that

$$|y_i - y| \leq \sum_{j \leq i} |y_j - y_{j-1}| \leq r/2 \quad \text{for } 1 \leq i \leq m \tag{6.22}$$

and (as was used before) $\varepsilon_i < r$ for $0 \leq i \leq m$.

A consequence of (6.22) is that $y_i = y_{i-1}$ most of the time. Even more, let $w \in V_\varepsilon \cap B(y, r/2)$ and define $w_i = T_i \circ \cdots \circ T_1(w)$ as we did for y_i. By (6.15), $w_i = w_{i-1}$ as soon as w_{i-1} lies out of $B(x_i, 3r)$; by (6.22) (applied to w) this happens as soon as $|w - x_i| > 7r/2$, hence also as soon as $|y - x_i| > 4r$. This means that in $V_\varepsilon \cap B(y, r/2)$, R is also the composition of the mappings T_i, $i \in I(y)$, where $I(y) = \{i : |x_i - y| \leq 4r\}$. Because of this (and the fact that the balls $B(x_i, 5r)$ have bounded overlap), R is still Hölder-continuous, with a new exponent that is as close to 1 as we want. We don't care about the Hölder constant for that exponent, just the exponent. Also, it appears that we have been too conservative in our choice of ε (because we could have counted how many times y_i changes), but this is all right.

The fact that $R(y) \in F$ for $y \in V_\varepsilon$ is easy. Since the $B(x_i, r)$ cover F, we can find $i \geq 1$ such that $y \in B(x_i, r + \varepsilon)$, and then $y_{i-1} \in B(x_i, 2r)$ by (6.22). Then $y_i = T_i(y_{i-1}) = S_{x_i}(y_{i-1}) \in F$ by (6.16) and (6.12). The ulterior mappings T_j leave y_j alone, by (6.13) and (6.15) – (6.17), so $R(y) = y_m \in F$. Finally, $R(y) = y$

for $y \in F$, for the same reason using (6.13). This completes our definition of the Hölder retraction R.

We want to use R to define a one parameter family of deformations g_t. Set $d_F(x) = \mathrm{dist}(x, F)$, and let $\tau > 0$ be much smaller than ε. We want to keep

$$g_t(x) = x \quad \text{when } d(x) \geq 2\tau, \tag{6.23}$$

and otherwise we set

$$g_t(x) = \psi_t(x)R(x) + \big(1 - \psi_t(x)\big)x \quad \text{when } d(x) \leq 2\tau, \tag{6.24}$$

where

$$\psi_t(x) = t\,\mathrm{Min}\Big(1, \frac{2\tau - d(x)}{\tau}\Big) \quad \text{when } d_F(x) \leq 2\tau. \tag{6.25}$$

We abuse notation slightly again, because $\mathbf{M}$ is a manifold. But it is locally the same as $\mathbb{R}^n$, which is enough to define the intermediate points $\psi_t(x)R(x) + \big(1 - \psi_t(x)\big)x$ and compute with them locally. Notice that $g_0(x) = x$ for $x \in \Omega_\varepsilon$, and $g_t(x)$ is a continuous function of x and t.

Return to the sets E_k''. By construction, E_k, E_k', and finally E_k'' are deformations of E_0, which means that $E_k'' = \varphi_1(E_0)$, for some acceptable family of continuous mappings φ_t, $0 \leq t \leq 1$, as above. Now we define a new family $\{f_t\}$, $0 \leq t \leq 1$, by

$$f_t(x) = \varphi_{2t}(x) \quad \text{for } x \in E_0 \text{ and } 0 \leq t \leq \tfrac{1}{2} \tag{6.26}$$

and

$$f_t(x) = g_{2t-1}\big(\varphi_1(x)\big) \quad \text{for } x \in E_0 \text{ and } 0 \leq t \leq \tfrac{1}{2}. \tag{6.27}$$

Again f_t is a continuous function of $x \in E_0$ and $t \in [0, 1]$, and $f_0(x) = x$ for $x \in E_0$. Set

$$G_k = f_1(E_0) = g_1 \circ \varphi_1(E_0) = g_1(E_k'') \tag{6.28}$$

by definitions; then $G_k \in \mathcal{F}(E_0)$. We want to check that for k large,

$$J_g(G_k) \leq \inf\{J_g(E) : E \in \mathcal{F}(E_0)\}, \tag{6.29}$$

because as soon as we do this, we will know that G_k is a minimizer, and Claim 6.2 will follow.

The main piece of E_k'' is $F_k = (E_k'')^*$; since F is the limit of the F_k, we get that for k large, $d_F(y) \leq \tau$ for $y \in F_k$. Then (6.25) and (6.24) yield $\psi_1(y) = 1$ and $g_1(y) = R(x) \in F$. Thus

$$g_1(F_k) \subset F. \tag{6.30}$$

The remaining part is $Z_k = E_k'' \setminus F_k$. By (6.8), $\mathcal{H}^{d-1}(Z_k) < +\infty$. But g_1 is Hölder-continuous with the same exponent as R, which can be taken as close to 1 as we want. Then $\mathcal{H}^d\big(g_1(Z_k)\big) = 0$. Altogether,

$$\begin{aligned} J_g(G_k) &= \int_{G_k} g(x)\,\mathrm{d}\mathcal{H}^d(x) = \int_{g_1(F_k \cup Z_k)} g(x)\,\mathrm{d}\mathcal{H}^d(x) \\ &= \int_{g_1(F_k)} g(x)\,\mathrm{d}\mathcal{H}^d(x) \leq \int_F g(x)\,\mathrm{d}\mathcal{H}^d(x) \\ &= J_g(F) \leq \inf\{J_g(E) : E \in \mathcal{F}(E_0)\} \end{aligned} \tag{6.31}$$

because $G_k = g_1(E_k'') = g_1(F_k \cup Z_k)$ by (6.28), and by (6.30) and (6.9). Thus (6.29) holds, G_k is a minimizer, and our proof of existence is complete. □

Remarks 6.3. The reader noticed that our assumptions were probably not optimal. Perhaps we can allow simple domains Ω (rather than our flat manifolds without boundary), but at least the argument used for the next example would have to be modified.

When $n > 2$, we could be in trouble if our definition of deformations demanded that φ_1 be Lipschitz, because we only know that R is Hölder. When $n = 2$, we can apply Theorem 4.1 to get a Lipschitz retraction, and conclude as above. Otherwise we would need to prove a better regularity or a better retraction theorem.

When $d > 2$, we do not know the existence of neighborhood retractions near almost minimal sets, even though this would look like a weak regularity property to prove, so we don't have an existence theorem.

Example 6.4. We return to Example 2.2 and its homology conditions (which also includes simpler separation conditions, as in Example 2.1). Let Ω be our closed domain. We shall assume that for some $\varepsilon > 0$, there is a Lipschitz retraction $R\colon \Omega_\varepsilon \to \Omega$. Here $\Omega_\varepsilon = \{x \in \mathbb{R}^n : \operatorname{dist}(x, \Omega) \leq \varepsilon\}$, and retraction means that $R(x) = x$ for $x \in \Omega$ and $R(x) \in \Omega$ for $x \in \Omega_\varepsilon \setminus \Omega$.

Theorem 6.5 ([42]). *Let $\Omega \subset \mathbb{R}^n$ be compact, and suppose that for some $\varepsilon > 0$ there is a Lipschitz retraction $R\colon \Omega_\varepsilon \to \Omega$. Let $g\colon \Omega \to [1, N]$ be continuous, and let ω_j, $j \in J$, be a collection of smooth $(n-d-1)$-dimensional surfaces in $\mathbb{R}^n \setminus \Omega_\varepsilon$, such that ω_j represents a nonzero element in the singular homology group $H_{n-d-1}(\mathbb{R}^n \setminus \Omega; \mathbb{Z})$. Set*

$$\mathcal{F} = \{E \subset \Omega : E \text{ is closed and each } \omega_j \text{ represents a nonzero element in } H_{n-d-1}(\mathbb{R}^n \setminus E; \mathbb{Z})\}. \tag{6.32}$$

Then there is a set $E \in \mathcal{F}$ such that $J_g(E) = \inf\{J_g(F) : F \in \mathcal{F}\}$.

The proof is based on the same program as for Example 6.1, but there are minor complications because Ω has a boundary. We explain one way to fix this; it could be that the organization in [42] is a little more efficient, but probably not too much.

First extend g to Ω_ε in a continuous way, so that we can talk about $J_g(E)$ when $E \subset \Omega_\varepsilon$. The simplest is to take $g \circ R$ on Ω_ε, and observe that this does not change g on Ω.

Now let $\{E_k\}$ be a minimizing sequence as above, and for each k choose a Feuvrier polyhedral net $\mathcal{R}_k$ with two constraints. The first one is that $\operatorname{diam}(Q) \leq \alpha_k$ for every polyhedron Q of the net, where $\alpha_k << \varepsilon$ will tend to 0. The main one is that $\mathcal{R}_k$ is adapted to E_k, in the sense that if E_k' denotes the Federer – Fleming projection of E_k on this net, then $J_g(E_k') \leq J_g(E_k) + 2^{-k}$. The fact that g is continuous is used here, because somewhere in the argument we compare $\mathcal{H}^d(E_k')$ and $J_g(E_k')$ in small balls. The details are as in Example 6.1.

Notice that $E_k' \subset \Omega_{n\alpha_k} \subset \Omega_\varepsilon$ by construction (we never move a point by more than α_k when we do a step of the Federer – Fleming projection). Thus we shall not change $J_g(E_k')$ when we replace g with the larger g_k defined by

$$g_k(x) = g(x) \quad \text{for } x \in \Omega_{n\alpha_k} \qquad \text{and} \qquad g_k(x) = N_1 \quad \text{for } x \in \Omega_\varepsilon \setminus \Omega_{n\alpha_k}, \tag{6.33}$$

where N_1 is a very large constant, that depends on N and the Lipschitz constant for the retraction R.

Choose a (fixed) union Ω_1 of (standard) dyadic cubes, that contains a neighborhood of Ω and is contained in Ω_ε. It will be more convenient to work in Ω_1. For k large, $\Omega_{2n\alpha_k} \subset \Omega_1$; we use this and Feuvrier's result to modify our network $\mathcal{R}_k$ so that it stays the same in $\Omega_{n\alpha_k}$, and coincides with a dyadic refinement of the usual dyadic net in $\Omega_1 \setminus \Omega_{2n\alpha_k}$.

Now replace E'_k with E''_k, which is a finite union of faces of our new network $\mathcal{R}_k$ and a deformation of E'_k in Ω_1, and minimizes $J_k(E''_k) = \int_{E''_k} g_k(x) d\mathcal{H}^d(x)$ under these constraints. The existence is easy because the total number of competitors is finite. Then

$$J_k(E''_k) \leq J_k(E'_k) = J_g(E'_k) \leq J_g(E_k) + 2^{-k} \tag{6.34}$$

by definitions.

Now E''_k is a quasi-minimal set in Ω_1, with (bad) constants that depend on N_1 and the angles of our networks, but do not depend on k. The verification requires some computations, but the idea is not too complicated. If $F = \varphi_1(E''_k)$ is a deformation of E''_k in Ω_1, we cannot compare F directly to E''_k, because it is not a union of faces in $\mathcal{R}_k$, but we can replace F with its Federer – Fleming projection G on $\mathcal{R}_k$. Then $J_k(E''_k) \leq J_k(G)$ by definition of E''_k. When we replace F with G, we only move the points of $Z = F \setminus \mathcal{S}_d$, where $\mathcal{S}_d$ denotes the d-dimensional skeleton of $\mathcal{R}_k$. Because of the angle conditions on the faces, when we apply the projection we can manage to multiply the contribution of Z to $J_k(F)$ by no more than CN_1. At the same time $Z \subset \varphi_1(W)$, where $W = \{x \in E''_k : \varphi_1(x) \neq x\}$, because $E''_k \subset \mathcal{S}_d$ by definition. Now if (3.14) fails, i.e., if $\mathcal{H}^d(\varphi_1(W))$ is much smaller than $\mathcal{H}^d(W)$, then $\mathcal{H}^d(\varphi_1(Z)) \leq \mathcal{H}^d(\varphi_1(W)) \ll \mathcal{H}^d(W)$, and the contribution of its Federer – Fleming projection to $J_g(G)$ is still small compared to $\mathcal{H}^d(W)$. The rest of $\varphi_1(W)$ stays the same, so its contribution stays very small too. When we sum up, we find that G is strictly better than E''_k, a contradiction.

So the E''_k are quasi-minimal with bad, but uniform constants. Set $F_k = (E''_k)^*$; the F_k are also quasi-minimal with uniform constants. Replace $\{F_k\}$ with a subsequence for which F_k tends to a limit F. We can apply (3.12), which yields that

$$\begin{aligned} J_g(F) &\leq \liminf_{k\to+\infty} J_g(F_k) \leq \liminf_{k\to+\infty} J_g(E''_k) \\ &\leq \liminf_{k\to+\infty} [J_g(E_k) + 2^{-k}] = \inf\{J_g(F) : F \in \mathcal{F}\} \end{aligned} \tag{6.35}$$

because the fact that we multiplied $\mathcal{H}^d$ with the continuous function g does not matter for the first inequality, then because $F_k \subset E''_k$, by (6.34), and because $\{E_k\}$ is a minimizing sequence.

We now want to check that $F \in \mathcal{F}$. We first need to see that F_k stays close to Ω. We claim that

$$F_k \subset \Omega_{\beta_k}, \tag{6.36}$$

where β_k can be computed from α_k, and tends to 0 when k tends to $+\infty$. Again the computation is a little long, but the idea is simple: since g_k is so much larger on $\Omega \setminus \Omega_{n\alpha_k}$ (see (6.33)), it is not interesting for F_k to get far from Ω. The existence of the Lipschitz retraction R will be used here, to provide competitors that have a substantial piece in Ω where g_k is so much smaller. But we'll have to try different competitors before we find one for which the thin annulus where we distort things without winning is small.

Set $a_l = 2^l n\alpha_k$ for $l \geq 0$, and

$$(6.37)\qquad \begin{aligned} V_l^- &= \{x \in \Omega_1 : \operatorname{dist}(x,\Omega) \leq a_l\},\\ V_l &= \{x \in \Omega_1 : a_l < \operatorname{dist}(x,\Omega) \leq 2a_l\},\\ V_l^+ &= \{x \in \Omega_1 : \operatorname{dist}(x,\Omega) > 2a_l\}.\end{aligned}$$

We define a competitor $\varphi(E_k'')$ for E_k'' by

$$(6.38)\qquad \begin{aligned} &\varphi(x) = x && \text{for } x \in V_l^-,\\ &\varphi(x) = \frac{2a_l - \operatorname{dist}(x,\Omega)}{a_l}\, x + \frac{\operatorname{dist}(x,\Omega) - a_l}{a_l}\, R(x) && \text{for } x \in V_l,\\ &\varphi(x) = R(x) && \text{for } x \in V_l^+.\end{aligned}$$

It would be easy to write down a one-parameter family $\{\varphi_t\}$ of mappings that ends with φ, but let us not bother the reader with this. Notice that we can expect to win a lot when we apply φ in the last region, because $g_k(x) = N_1$ in the last two regions (by (6.33)), while $g_k\big(R(x)\big) = g\big(R(x)\big) \leq N$ (which is much smaller) because $R(x) \in \Omega$. That is,

$$(6.39)\qquad \begin{aligned} J_k\big(\varphi(E_k'')\big) &\leq \int_{\varphi(E_k''\cap V_l^-)} g_k + \int_{\varphi(E_k''\cap V_l)} g_k + \int_{\varphi(E_k''\cap V_l^+)} g_k\\ &\leq \int_{E_k''\cap V_l^-} g_k + N_1\mathcal{H}^d\big(\varphi(E_k''\cap V_l)\big) + N\mathcal{H}^d\big(\varphi(E_k''\cap V_l^+)\big)\\ &\leq \int_{E_k''\cap V_l^-} g_k + CN_1\mathcal{H}^d(E_k''\cap V_l) + CN\mathcal{H}^d(E_k''\cap V_l^+)\end{aligned}$$

by (6.38) and because φ is C-Lipschitz (including on V_l where we can use the fact that $|R(x) - x| \leq C\operatorname{dist}(x,\Omega) \leq Ca_l$). At the same time,

$$(6.40)\qquad \begin{aligned} J_k(E_k'') &= \int_{E_k''\cap V_l^-} g_k + \int_{E_k''\cap V_l} g_k + \int_{E_k''\cap V_l^+} g_k\\ &= \int_{E_k''\cap V_l^-} g_k + N_1\mathcal{H}^d(E_k''\cap V_l) + N_1\mathcal{H}^d(E_k''\cap V_l^+)\end{aligned}$$

by (6.33), so

$$(6.41)\qquad J_k\big(\varphi(E_k'')\big) - J_k(E_k'') \leq CN_1\mathcal{H}^d(E_k''\cap V_l) - \frac{N_1}{2}\mathcal{H}^d(E_k''\cap V_l^+)$$

if N_1/N is large enough. If we were allowed to compare $\varphi(E_k'')$ directly to E_k'', we would deduce from (6.41) that

$$(6.42)\qquad \mathcal{H}^d(E_k''\cap V_l^+) \leq 2C\mathcal{H}^d(E_k''\cap V_l).$$

Now we cannot do that, but we can compose φ with a Federer – Fleming projection that sends $\varphi(E_k'')$ back to our net, and by the same computations we eventually obtain (6.42) anyway (but with a larger constant).

Observe that (6.42) shows that $\mathcal{H}^d(E_k''\cap V_l^+)$ decays exponentially fast. Indeed, V_l^+ is the disjoint union of V_{l+1} and V_{l+1}^+ by (6.42) and because $a_{k+1} = 2a_k$, so

$$(6.43)\quad \mathcal{H}^d(E_k''\cap V_l^+) = \mathcal{H}^d(E_k''\cap V_{l+1}) + \mathcal{H}^d(E_k''\cap V_{l+1}^+) \geq \Big(1 + \frac{1}{2C}\Big)\mathcal{H}^d(E_k''\cap V_{l+1}^+)$$

by (6.42) and, after some iterations,

$$\mathcal{H}^d(E_k'' \cap V_l^+) \le \left(\frac{2C}{2C+1}\right)^k \mathcal{H}^d(E_k''). \tag{6.44}$$

Now we could modify the definition of the a_l to make the sequence $\{a_l/\alpha_k\}$ bounded (but with very large values), and still get, with the same computations as above, that $\lim_{l\to+\infty} \mathcal{H}^d(E_k'' \cap V_l^+) = 0$. Let us cheat slightly instead, and use the local Ahlfors-regularity of $F_k = (E_k'')^*$ to get a less precise estimate which is enough. If F_k meets V_l at some point x, then

$$\mathcal{H}^d\big(F_k \cap (V_{l-1} \cup V_l \cup V_{l+1})\big) \ge \mathcal{H}^d\big(F_k \cap B(x, a_l/2)\big) \ge C^{-1} a_l^d. \tag{6.45}$$

We are now ready to prove (6.36). Let $\beta > 0$ be small, and denote by l_k the largest integer such that $a_l = 2^l n\alpha_k < \beta$. Obviously l_k is well defined for k large, and tends to $+\infty$ when k tends to $+\infty$. Then (6.44) says that $\mathcal{H}^d\big(F_k \cap (V_{l_k-1} \cup V_{l_k} \cup V_{l_k+1})\big)$ tends to 0 (recall that $\mathcal{H}^d(E_k'')$ stays bounded). But the right-hand side of (6.45) is comparable to β^d, so it does not tend to 0. This contradiction proves that for k large, $F_k \cap V_{l_k} = \varnothing$. Then $\mathcal{H}^d(E_k'' \cap V_{l_k}) = 0$ by definition of $F_k = (E_k'')^*$, (6.42) implies that $\mathcal{H}^d(E_k'' \cap V_{l_k}^+) = 0$, and (by local Ahlfors-regularity of F_k again) F_k does not meet $V_{l_k}^+$ either. In other words, $F_k \subset V_{l_k}^- \subset \Omega_\beta$ for k large; (6.36) follows.

Recall that we want to show that $F \in \mathcal{F}$. By (6.36), we know that $F \subset \Omega$; we still need to check that for $j \in J$, ω_j represents a nonzero element in $H_{n-d-1}(\mathbb{R}^n \setminus F;\mathbb{Z})$. We know this with E_k (instead of F), just because $E_k \in \mathcal{F}$. This is then true for E_k' and E_k'', because they are deformation of E_k in Ω_1, and it is known that such deformations away from ω_j preserve the fact that $\omega_j \ne 0$ in homology. This is also true for F_k, because removing pieces of $(d-1)$-dimensional faces from E_k'' does not change the fact that ω_j represents a nonzero element in $H_{n-d-1}(\mathbb{R}^n \setminus E_k'';\mathbb{Z})$, this time by a general position argument (we could move the support of a $n-d$ chain that closes ω_j so that it does not meet the faces). Finally, ω_j represents a nonzero element in $H_{n-d-1}(\mathbb{R}^n \setminus F;\mathbb{Z})$, because if the support of a chain that closes ω_j does not meet F, it does not meet F_k for k large. This completes the verification of the fact that $F \in \mathcal{F}$, and the description of the proof of Theorem 6.5. We refer to [42] for details. □

7. Boundary regularity

The author believes that the logical way to try to prove existence results under sliding boundary conditions (as in Section 2.4) is to first prove regularity properties for potential solutions. Apparently there has not been so much work on the subject of boundary regularity, but one can hope that the sliding context of Section 2.4 and Definition 7.1 below will provide sufficient flexibility to allow some positive results.

Figure 7 (probably the best time of the author's Montréal lectures), which comes from the link on soap films on the Borromean rings in K. Brakke's homepage, is supposed to illustrate the fact that it may be hard to study precisely the boundary behavior of soap films. See also [7; 39; 51, Figure 13.9.3 on p. 137] for information and conjectures about boundary regularity (maybe of slightly different objects).

Here we just give definitions, announce some partial results (that still need to be written down or proofread) and mention plans for the future.

We give ourselves a finite collection of simple compact boundary sets $\Gamma_j \subset \mathbb{R}^n$, $0 \le j \le j_{\max}$; for instance, each Γ_j can be a finite union of (closed) faces (of various dimensions) of dyadic cubes. Or we can choose the Γ_j like this, and consider their images under a (same) bi-Lipschitz mapping. Other options are possible, but the author decided to restrict to these for the moment.

We shall also proceed locally, in an open ball B. Let $E \subset B$ be closed in B, and assume that $\mathcal{H}^d(E \cap K) < +\infty$ for every compact subset K of B. A sliding competitor for E in B is a set $F = \varphi_1(E)$, where the one-parameter family $\{\varphi_t\}$, $0 \leq t \leq 1$, has the following properties (similar to what we asked in Section 2.4):

$$(t,x) \to \varphi_t(x) \colon [0,1] \times E \to \mathbb{R}^n \text{ is continuous,} \tag{7.1}$$

$$\varphi_0(x) = x \quad \text{for } x \in E_0, \tag{7.2}$$

$$\varphi_t(x) \in \Gamma_j \quad \text{when } 0 \leq j \leq j_{\max} \text{ and } x \in \Gamma_j, \tag{7.3}$$

$$\varphi_1 \text{ is Lipschitz} \tag{7.4}$$

(which we again keep out of respect for traditions); for the locality, we set

$$W_t = \{x \in E : \varphi_t(x) \neq x\}, \quad W = \bigcup_{t \in [0,1]} (W_t \cup \varphi_t(W_t)), \tag{7.5}$$

and demand that W be contained in a compact subset of B (in short, $W \subset\subset B$).

Notice that if the first set Γ_0 contains E, it may play the role of a closed domain where everything happens (the set Ω that we used in examples).

Definition 7.1. We say that E is a *sliding almost minimal* set with gauge function h (and relative to the choice of boundaries Γ_j) when

$$\mathcal{H}^d(E \cap W) \leq \mathcal{H}^d(\varphi(E) \cap W) + h(\mathrm{diam}(W))\,\mathrm{diam}(W)^d \tag{7.6}$$

when the φ_t and W are as above.

As usual, we would like to write $\mathcal{H}^d(E) \leq \mathcal{H}^d(\varphi_1(E)) + h(\mathrm{diam}(W))\,\mathrm{diam}(W)^d$, but $\mathcal{H}^d(E)$ may be infinite, and anyway E and $\varphi_1(E)$ coincide out of W. There is also a notion of sliding quasi-minimal set, as in Definition 3.7 but with the sliding competitors, but let us not write the definition here.

Of course we set things so that the minimizers of the sliding Plateau problem of Section 2.4, if they exist, are sliding almost minimizers with this definitions.

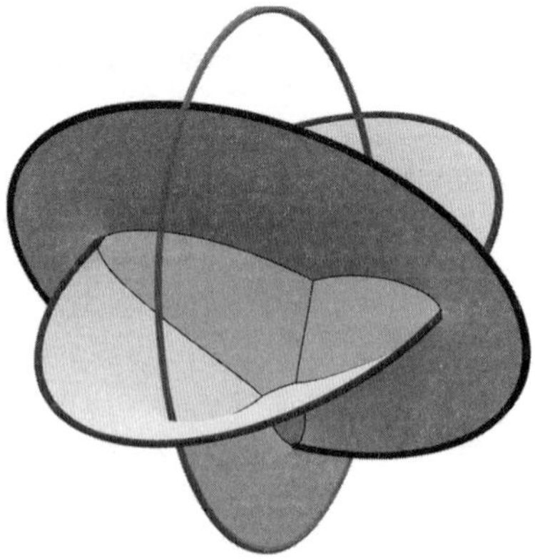

FIGURE 7. A soap film which leaves one of the boundary curves (K. Brakke). Guess how it looks like at small scales; is its behavior stable when the width of the wires tend to 0?

This may also be the case for the solutions of some other problems (for instance, minimize the size of a current under a boundary constraint), but the author did not take the time to check yet.

So we want to see to which extent the interior regularity theory described above extends to the sliding context. Here are a few things that seem to work.

The local Ahlfors-regularity of E for almost minimal and quasi-minimal sets (as in Theorem 3.3) seems to extend without real trouble.

The situation for the uniform rectifiability of E (Theorem 3.4) is a little more complicated; the main stopping time argument in the proof does not go through, and we only seem to be only able to prove that E is uniformly rectifiable under additional dimension conditions. The case when $d = 2$ and the boundaries Γ_j are at most 2-dimensional is still all right, but in fact the uniform rectifiability up to the boundary, when we can prove it, does not mean much more than what we knew anyway from the interior regularity.

Fortunately, sliding quasi-minimal sets are still rectifiable, and even every Hausdorff limit of reduced quasi-minimal sets (with uniform estimates) is rectifiable. This is somewhat easier, and it is still enough to prove the concentration property (Theorem 3.5), and then (with some work) an analogue of Theorem 3.6 (the lowersemicontinuity of $\mathcal{H}^d$ and the stability under limits), which therefore still hold in the sliding context.

The monotonicity of the density $r^{-d}\mathcal{H}^d\big(E \cap B(0,r)\big)$, when all the Γ_j are cones centered at the origin and E is minimal, seems to be true, and maybe also the fact that E coincides with a cone when the density is constant, yielding the fact that blow-up limits at the boundary are sliding minimal cones.

What should be done next is still a long program. First, we should try to get a list of the sliding minimal cones at a boundary point, depending on the list of boundary cones Γ_j. But also, some more precise understanding of how the set converges to a blow-up limit will be welcome, starting with the simple blow-up limits that we already know.

We should mention that we cannot hope the monotonicity of $r^{-d}\mathcal{H}^d\big(E \cap B(x,r)\big)$ at some point x near the boundaries to be true as before, so we should then estimate and quantify more precisely the lack of monotonicity (or equivalently define something like a density profile).

Figure 7 (probably the best time of the author's Montréal lectures), which comes from the link on soap films on the Borromean rings in K. Brakke's homepage, is supposed to illustrate the fact that it may be hard to study precisely the boundary behavior of soap films. See also [7; 39; 51, Figure 13.9.3 on p. 137] for information and conjectures about boundary regularity (maybe of slightly different objects).

References

1. F. J. Almgren Jr., *Some interior regularity theorems for minimal surfaces and an extension of Bernstein's theorem*, Ann. of Math. (2) **84** (1966), 277–292.
2. ______, *Existence and regularity almost everywhere of solutions to elliptic variational problems among surfaces of varying topological type and singularity structure*, Ann. of Math. (2) **87** (1968), 321–391.
3. ______, *Existence and regularity almost everywhere of solutions to elliptic variational problems with constraints*, Mem. Amer. Math. Soc. **4** (1976), no. 165.
4. L. Ambrosio, N. Fusco, and D. Pallara, *Functions of bounded variation and free discontinuity problems*, Oxford Math. Monogr., Oxford Univ. Press, New York, 2000.
5. K. A. Brakke, *Minimal cones on hypercubes*, J. Geom. Anal. **1** (1991), no. 4, 329–338.

6. ———, *The surface evolver*, Experiment. Math. **1** (1992), no. 2, 141–165.
7. ———, *Minimal surfaces, corners, and wires*, J. Geom. Anal. **2** (1992), no. 1, 11–36.
8. ———, *Soap films and covering spaces*, J. Geom. Anal. **5** (1995), no. 4, 445–514.
9. G. Dal Maso, J.-M. Morel, and S. Solimini, *A variational method in image segmentation: existence and approximation results*, Acta Math. **168** (1992), no. 1-2, 89–151.
10. G. David, *Limits of Almgren quasiminimal sets*, Harmonic analysis at Mount Holyoke (South Hadley, MA, 2001) (W. Beckner, A. Nagel, A. Seeger, and H. F. Smith, eds.), Contemp. Math., vol. 320, Amer. Math. Soc., Providence, RI, 2003, pp. 119–145.
11. ———, *Singular sets of minimizers for the Mumford – Shah functional*, Progr. Math., vol. 233, Birkhäuser, Basel, 2005.
12. ———, *Open questions on the Mumford – Shah functional*, Perspectives in Analysis (Stockholm, 2003) (M. Benedicks, P. W. Jones, and S. Smirnov, eds.), Math. Phys. Stud., vol. 27, Springer, Berlin, 2005, pp. 37–49.
13. ———, *Quasiminimal sets for Hausdorff measures*, Recent Developments in Nonlinear Partial Differential Equations (West Lafayette, IN, 2004) (D. Danielli, ed.), Contemp. Math., vol. 439, Amer. Math. Soc., Providence, RI, 2007, pp. 81–99.
14. ———, *Hölder regularity of two-dimensional almost-minimal sets in* $\mathbb{R}^n$, Ann. Fac. Sci. Toulouse Math. (6) **18** (2009), no. 1, 65–246.
15. ———, $C^{1+\alpha}$*-regularity for two-dimensional almost-minimal sets in* $\mathbb{R}^n$, J. Geom. Anal. **20** (2010), no. 4, 837–954.
16. ———, *Uniformly rectifiable sets*, available at `http://math.u-psud.fr/~gdavid/`. Lecture notes from a course in Park City, UT, 2003.
17. ———, *Should we solve Plateau's problem again?*, Analysis and Applications — A Conference in Honor of Elias M. Stein (Princeton, NJ, 2011).
18. G. David, T. De Pauw, and T. Toro, *A generalization of Reifenberg's theorem in* $\mathbb{R}^3$, Geom. Funct. Anal. **18** (2008), no. 4, 1168–1235.
19. G. David and S. Semmes, *Analysis of and on uniformly rectifiable sets*, Math. Surveys Monogr., vol. 38, Amer. Math. Soc., Providence, RI, 1993.
20. ———, *Quasiminimal surfaces of codimension* 1 *and John domains*, Pacific J. Math. **183** (1998), no. 2, 213–277.
21. ———, *Uniform rectifiability and quasiminimizing sets of arbitrary codimension*, Mem. Amer. Math. Soc. **144** (2000), no. 687.
22. T. De Pauw, *Size minimizing surfaces*, Ann. Sci. Éc. Norm. Supér. (4) **42** (2009), no. 1, 37–101.
23. T. De Pauw and R. Hardt, *Size minimization and approximating problems*, Calc. Var. Partial Differential Equations **17** (2003), no. 4, 405–442.
24. G. David and T. Toro, *Reifenberg parameterizations for sets with holes*, Mem. Amer. Math. Soc. **215** (2012), no. 1012.
25. J. Douglas, *Solution of the problem of Plateau*, Trans. Amer. Math. Soc. **33** (1931), no. 1, 263–321.
26. H. Federer, *Geometric measure theory*, Grundlehren Math. Wiss., vol. 153, Springer, New York, 1969.
27. ———, *The singular sets of area minimizing rectifiable currents with codimension one and of area minimizing flat chains modulo two with arbitrary codimension*, Bull. Amer. Math. Soc. **76** (1970), 767–771.
28. H. Federer and W. H. Fleming, *Normal and integral currents*, Ann. of Math. (2) **72** (1960), 458–520.
29. V. Feuvrier, *Un résultat d'existence pour les ensembles minimaux par optimisation sur des grilles polyédrales*, Ph.D. Thesis, Université de Paris-Sud, 2008.
30. V. Feuvrier, *Remplissage de l'espace euclidien par des complexes polyédriques d'orientation imposée et de rotondité uniforme*, available at `arXiv:0812.4709`.
31. V. Feuvrier, *Condensation of polyhedric structures onto soap films*, available at `arXiv:0906.3505`.
32. R. Garnier, *Le problème de Plateau*, Ann. Sci. École Norm. Sup. (3) **45** (1928), 53–144.
33. J. Harrison, *On Plateau's problem for soap films with a bound on energy*, J. Geom. Anal. **14** (2004), no. 2, 319–329.
34. ———, *Solution of Plateau's problem*, available at `arXiv:1106.5839`.

35. A. Heppes, *Isogonale sphärische Netze*, Ann. Univ. Sci. Budapest. Eötvös Sect. Math. **7** (1964), 41–48.
36. E. Lamarle, *Sur la stabilité des systèmes liquides en lames minces*, Acad. Roy. Belg. Cl. Sci. Mem. Coll. in-8o **35** (1865).
37. G. Lawlor, *Pairs of planes which are not size-minimizing*, Indiana Univ. Math. J. **43** (1994), no. 2, 651–661.
38. G. Lawlor and F. Morgan, *Paired calibrations applied to soap films, immiscible fluids, and surfaces or networks minimizing other norms*, Pacific J. Math. **166** (1994), no. 1, 55–83.
39. ———, *Curvy slicing proves that triple junctions locally minimize area*, J. Differential Geom. **44** (1996), no. 3, 514–528.
40. X. Liang, *Ensembles et cônes minimaux de dimension 2 dans les espaces euclidiens*, Ph.D. Thesis, Université de Paris-Sud, 2010.
41. ———, *Almgren-minimality of unions of two almost orthogonal planes in* $\mathbb{R}^4$, available at arXiv:1103.1468.
42. ———, *Topological minimal sets and their applications*, Calc. Var. Partial Differential Equations, to appear, available at arXiv:1103.3871.
43. ———, *Almgren and topological minimality for the set* $Y \times Y$, available at arXiv:1203.0564.
44. T. D. Luu, *Régularité des cônes et ensembles minimaux de dimension 3 dans* $\mathbb{R}^4$, Ph.D. Thesis, Université de Paris-Sud, 2011.
45. P. Mattila, *Geometry of sets and measures in Euclidean spaces*, Cambridge Stud. Adv. Math., vol. 44, Cambridge Univ. Press, Cambridge, 1995.
46. J.-M. Morel and S. Solimini, *Variational methods in image segmentation*, Progr. Nonlinear Differential Equations Appl., vol. 14, Birkhäuser, Boston, MA, 1995.
47. F. Morgan, *Size-minimizing rectifiable currents*, Invent. Math. **96** (1989), no. 2, 333–348.
48. ———, *Minimal surfaces, crystals, shortest networks, and undergraduate research*, Math. Intelligencer **14** (1992), no. 3, 37–44.
49. ———, *Soap films and mathematics*, Differential Geometry: Partial Differential Equations on Manifolds (Los Angeles, CA, 1990) (R. Greene and S.-T. Yau, eds.), Proc. Sympos. Pure Math., vol. 54, Amer. Math. Soc., Providence, RI, 1993, pp. 375–380.
50. ———, *$(\mathbf{M}, \epsilon, \delta)$-minimal curve regularity*, Proc. Amer. Math. Soc. **120** (1994), no. 3, 677–686.
51. ———, *Geometric measure theory: A beginner's guide*, 2nd ed., Academic Press, San Diego, CA, 1995.
52. D. Mumford and J. Shah, *Optimal approximations by piecewise smooth functions and associated variational problems*, Comm. Pure Appl. Math. **42** (1989), no. 5, 577–685.
53. J. Plateau, *Statique expérimentale et théorique des liquides soumis aux seules forces moléculaires*, Gauthier-Villars, Paris, 1873.
54. T. Radó, *The problem of the least area and the problem of Plateau*, Math. Z. **32** (1930), no. 1, 763–796.
55. E. R. Reifenberg, *Solution of the Plateau Problem for m-dimensional surfaces of varying topological type*, Acta Math. **104** (1960), 1–92.
56. ———, *An epiperimetric inequality related to the analyticity of minimal surfaces*, Ann. of Math. (2) **80** (1964), 1–14.
57. ———, *On the analyticity of minimal surfaces*, Ann. of Math. (2) **80** (1964), 15–21.
58. J. E. Taylor, *Regularity of the singular sets of two-dimensional area-minimizing flat chains modulo 3 in* R^3, Invent. Math. **22** (1973), 119–159.
59. ———, *The structure of singularities in soap-bubble-like and soap-film-like minimal surfaces*, Ann. of Math. (2) **103** (1976), no. 3, 489–539.
60. T. Toro, *Geometric conditions and existence of bi-Lipschitz parameterizations*, Duke Math. J. **77** (1995), no. 1, 193–227.

Département de Mathématiques d'Orsay, Université Paris-Sud, Bâtiment 425, 91405 Orsay Cedex, France
E-mail address: guy.david@math.u-psud.fr

Centre de Recherches Mathématiques
CRM Proceedings and Lecture Notes
Volume **56**, 2013

Lectures on Ma – Trudinger – Wang Curvature and Regularity of Optimal Transport Maps

Young-Heon Kim

ABSTRACT. Optimal transportation concerns the phenomena when the cost of matching two mass distributions is minimized. Regarding the regularity of such optimal transport maps, a new notion of curvature, called MTW curvature, was found recently by Ma, Trudinger and Wang. In these lectures, we discuss MTW curvature and regularity of optimal transport, focusing on the case when the transportation cost is given by the Riemannian distance squared.

1. Introduction

The present article is based on four lectures given in the SMS 2011 Summer School (50th edition) "Metric Measure Spaces: Geometric and Analytic Aspects" June 27 – July 8, 2011, in CRM, Montréal. In these lecture notes, the goal is to explain the new curvature notion, called Ma – Trudinger – Wang curvature (or simply MTW curvature) that was discovered by Ma, Trudinger and Wang [52], in the context of the regularity theory of optimal transportation maps. We study the geometry and analysis involving this curvature, explaining why this is relevant to the optimal transportation theory, and giving some key examples, and discussing how to prove regularity, in particular, Hölder continuity of optimal transportation maps when we restrict the sign of MTW curvature to be nonnegative. Some of the topics in these lecture notes overlap with other expository articles and books such as [17, 28] (focusing more on geometric implications of conditions on the MTW curvature), [54, 60, 62] (more broader introduction to optimal transportation), however, in addition to giving different point of view, we focus more on the results in [21, 35–37]. Our aim is to give some highlights in the recent developments, so many results/facts are given without detailed proof. Throughout this article, we assume familiarity with basic Riemannian geometry such as exponential map, sectional curvature, cut locus, etc. (cf. [9]).

2. The Ma, Trudinger and Wang curvature condition

2.1. Optimal transportation. Throughout the article we let M be a Riemannian manifold, and let Ω, $\bar{\Omega}$ be bounded open subsets in M. We consider two

2010 *Mathematics Subject Classification.* Primary 35J96; Secondary 53C21.
Key words and phrases. Optimal transportation, regularity, curvature.
The author is supported partly by Canada NSERC discovery grant 371642-09.
This is the final form of the paper.

probability distributions $\rho = \rho(x)\,\mathrm{d}x$, $\bar\rho = \bar\rho(\bar x)\,\mathrm{d}\bar x$ with $\operatorname{supp}\rho \subset \Omega$, $\operatorname{supp}\bar\rho \subset \overline{\Omega}$. Here, we assume $\log\rho, \log\bar\rho \in L^\infty$, namely, the density functions $\rho(x), \bar\rho(\bar x)$ satisfy

$$0 < \lambda \le \rho(x), \quad \bar\rho(\bar x) \le \Lambda \tag{2.1}$$

for some positive constant λ, Λ and for $x \in \Omega$ and $\bar x \in \overline{\Omega}$. Consider the transportation cost function

$$c(x, \bar x) = \operatorname{dist}^2(x, \bar x)/2,$$

where dist denotes the Riemannian distance function. This function c can be regarded a canonical cost function associated to a Riemannian manifold since $-D_x c(x, \bar x) = \exp_x^{-1} \bar x$, where D_x denotes the gradient in the x variable[1] and exp denotes the exponential map. Even though more general cost functions can be considered, *in this lecture we restrict ourselves to this Riemannian distance squared cost function for simplicity of discussion.* In general, due to the cut locus, the distance squared function is not smooth. *To simplify our discussion, we assume throughout these lectures that $c \in C^\infty(\Omega, \overline{\Omega})$: in other words, $\Omega \times \overline{\Omega} \subset M \times M \setminus \mathrm{Cut}$, where* Cut *denotes the cut locus of M.*

By the fundamental result of Brenier [2] and later by McCann [53], the following holds: There exists a unique optimal map T, namely, a Borel measurable map that minimizes the transportation cost

$$\int_\Omega c\big(x, F(x)\big)\,\mathrm{d}\rho(x)$$

among all measurable maps F pushing ρ forward to $\bar\rho$, i.e., $F_\# \rho = \bar\rho$: here, $F_\#\rho(B) = \rho(F^{-1}B)$ for any Boral set B. We can simply denote this as

$$T = \operatorname*{arg\,min}_{F_\#\rho = \bar\rho} \int_\Omega c\big(x, F(x)\big)\,\mathrm{d}\rho(x).$$

This *optimal map* T is almost everywhere differentiable and it satisfies nice analytical properties we list below:

(1) $|\det D_x T(x)| \overset{\text{a.e.}}{=} \rho(x)/\bar\rho\big(T(x)\big)$.

(2) There exists a function $\phi\colon \Omega \to \mathbb{R}$ called *c-potential* such that $T(x) \overset{\text{a.e.}}{=} \exp_x\big(D_x\phi(x)\big)$, i.e., $-D_x c\big(x, T(x)\big) \overset{\text{a.e.}}{=} D_x\phi(x)$. Such a function ϕ is *c-convex*, namely, it has a dual function $\phi^c\colon \overline{\Omega} \to \mathbb{R}$ and

$$\begin{aligned} \phi(x) &= \sup_{\bar x \in \overline{\Omega}} -c(x, \bar x) - \phi^c(\bar x); \\ \phi^c(\bar x) &= \sup_{x \in \Omega} -c(x, \bar x) - \phi(x). \end{aligned} \tag{2.2}$$

Due to boundedness of $\Omega, \overline{\Omega}$ and local Lipschitzness and semiconvexity of $-\operatorname{dist}^2$, one can verify that ϕ and ϕ^c are Lipschitz and semiconvex: see Section 2.2.

(3) The condition $T_\#\rho = \bar\rho$ (see (1)) forces the c-potential ϕ satisfy the *c-Monge – Ampère equation*:

$$\det\big(D^2_{xx}\phi(x) + D^2_{xx}c\big(x, T(x)\big) \overset{\text{a.e.}}{=} \big|\det D_x D_{\bar x} c\big(x, T(x)\big)\big| \frac{\rho(x)}{\bar\rho\big(T(x)\big)}.$$

[1]One may consider D_x as the differential, producing covectors, but we can use the Riemannian metric to identify those with tangent vectors and this is the convention we take in these lectures.

Example 2.1. For $M = \mathbb{R}^n$, $c(x, \bar{x}) = |x - \bar{x}|^2/2$, the function $\phi(x) + |x|^2/2$ is convex and $T(x) = x + \nabla\phi(x)$, with

$$\det(D^2_{xx}\phi(x) + I) = \frac{\rho(x)}{\bar{\rho}(\nabla\phi(x) + x)}.$$

The paper [11] by Cordero-Erausquin, McCann and Schmuckenschlaeger contains many useful results for optimal transport maps on Riemannian manifolds. See also the book [62] by Villani.

From the c-Monge – Ampère equation, the question of regularity of its solution ϕ naturally arises:

Question 2.2. For $\log\rho, \log\bar{\rho} \in L^\infty/C^\infty$, is the optimal map $T \in C^0/C^\infty(\Omega)$ (i.e., $\phi \in C^1/C^\infty(\Omega)$)?

We discuss below key notions related to this question.

2.2. The subdifferential $\partial\phi$ and the c-subdifferential $\partial^c\phi$. Under our assumption that Ω and $\overline{\Omega}$ are bounded, one can show the following fact:

If ϕ is the c-potential defined in (2.2), then ϕ is Lipschitz and semiconvex.

A function is called (locally) semiconvex if it becomes convex by adding a quadratic function (in a local coordinate system). Note that semiconvexity allows the function not to be differentiable at a point. But, also note that if a semiconvex function is differentiable at every point, then it is C^1, i.e., the derivatives are continuous. Moreover, it is a well-known fact (due to A. Alexandrov, see, e.g., [62]) that for a (locally) semiconvex function the set of nondifferentiable points has Hausdorfff dimension less than or equal to $\dim M - 1$. In particular, such set has zero measure. Example of Lipschitz and semiconvex functions include $-dist^2$ for compact Riemannian manifolds. Note that dist^2 is on the other hand is semiconcave but not semiconvex (see, e.g., [11]).

For semiconvex functions we can define the subdifferential.

Definition 2.3 (subdifferential). The subdifferntial $\partial\phi(x)$ at $x \in \Omega$ is defined as the set in the tangent space T_xM given by

$$\partial\phi(x) = \{p \in T_xM \mid \phi(\exp_x v) - \phi(x) \geq \langle p, v\rangle + o(|v|), \forall v \in T_xM \ \& \ |v| \ll 1\}$$

Here, $\langle,\rangle$ and $|\cdot|$ are the Riemannian metric and norm, respectively, and $o(|v|)$ denotes the usual small 'o' error, i.e., $\lim_{|v|\to 0} o(|v|)/|v| = 0$.

Notice that $\partial\phi(x) = \{\nabla\phi(x)\}$ (here $\nabla\phi$ is the gradient) if and only if ϕ is differentiable at x. Moreover, one can check that $\partial\phi(x)$ is a convex set in the affine space T_xM.

While the subdifferential at a point assigns a function a set of tangent vectors, the c-subdifferential gives a set of 'target' points:

Definition 2.4 (c-subdifferential).

$$\partial^c\phi(x) = \{\bar{x} \in \overline{\Omega} \mid \phi(\cdot) \geq \phi(x) - c(\cdot, \bar{x}) + c(x, \bar{x}) \text{ on } \Omega\}$$

The expression $\partial^c\phi$ denotes the graph of this valued map, i.e.

$$\partial^c\phi = \{(x, \bar{x}) \in \Omega \times \overline{\Omega} \mid \bar{x} \in \partial^c\phi(x)\}$$

To be more precise, one may add subscripts as $\partial^c_{\Omega,\overline{\Omega}}\phi$ since the c-subdifferential depends on the source and target domains $\Omega, \overline{\Omega}$. Here, the functions of the form $-c(\cdot,\bar{x}) + c(x,\bar{x}) + const$ are called c-supporting functions.

Regarding an optimal map T and its c-potential ϕ, we have inclusions between the graphs of T, $\partial^c\phi$ and the valued map $\exp\partial\phi$ given by composing the subdifferential $\partial\phi$ with the exponential map, whose graph is defined (by abusing the notation) as

$$\exp\partial\phi = \{(x,\bar{x}) \in \Omega \times \overline{\Omega} \mid \bar{x} \in \exp_x \partial\phi(x)\}.$$

Fact 2.5. $\operatorname{graph} T \subset \partial^c\phi \subset \exp\partial\phi$.

Since T is defined a.e., the first inclusion should be understood in the a.e. sense.

One observes that if $\phi \in C^1$, then $\partial^c\phi = \exp\nabla\phi = \exp\partial\phi$. Regarding the equality between $\partial^c\phi$ and $\exp\partial\phi$ we define *Loeper's maximum principle* [49], a principal notion in these lectures:

Definition 2.6 (Loeper's maximum principle). We say *Loeper's maximum principle* (LMP) holds if for any c-convex function ϕ,

$$\partial^c\phi = \exp\partial\phi.$$

The reason why this is called a maximum principle will be obvious from its another formulation (2.3) given in a later section.

We now state the first main theorem of these lectures, which is due to Loeper [49] (such a result was also hinted by Ma, Trudinger and Wang [52, [Sections 7.3 and 7.5]:

Theorem 2.7 (continuity of OT $\Rightarrow$ LMP). *Suppose that for* each $\log\rho$, $\log\bar{\rho} \in L^\infty$, *the corresponding optimal map T is continuous. Then, Loeper's maximum principle holds.*

In the following section we explain the reason why this theorem should hold.

2.3. Loeper's maximum principle (LMP). In this section, we explain (not a proof) why Loeper's maximum principle (LMP) is a necessary condition to ensure continuity of optimal maps for each $\log\rho, \log\bar{\rho} \in L^\infty$. This will be done by considering the following important example, which shows that if for each $\log\rho, \log\bar{\rho} \in L^\infty$ the corresponding optimal map is continuous, then Loeper's maximum principle (LMP) should hold.

Example 2.8 (Heuristic explanation why (continuity of OT $\Rightarrow$ LMP); see [49, Proposition 4.4] **and also** [52, Section 7.3] **for a similar example).** Fix a point $x \in \Omega$ and two points $\bar{x}_0, \bar{x}_1 \in \overline{\Omega}$. Let

$$\begin{aligned} m_i(\cdot) &= -c(\cdot,\bar{x}_i) + c(x,\bar{x}_i), \quad i = 0,1, \\ \phi(\cdot) &= \max[m_0, m_1] \end{aligned}$$

as functions on Ω. Notice that the c-subdifferential $\partial^c\phi$ pushes forward the uniform measure 1_Ω to the sum of two Dirac measures, $\bar{\rho}_0 := c_0\delta_{\bar{x}_0} + c_1\delta_{\bar{x}_1}$, where $c_0, c_1 \in \mathbb{R}$ are some appropriate constants. We write this as

$$(\partial^c\phi)_{\#}1_\Omega = \bar{\rho}_0.$$

We consider a smooth target probability density $\bar{\rho}_\varepsilon \in C^\infty(\overline{\Omega})$ which is positive on $\overline{\Omega}$ and converges (weakly as measure) to the measure $\bar{\rho}_0$ as $\varepsilon \to 0$. Between 1_Ω and $\bar{\rho}_\varepsilon$, consider the corresponding optimal map with potential ϕ_ε. Namely,

$$(\partial^c \phi_\varepsilon)_\# 1_\Omega = \bar{\rho}_\varepsilon.$$

Suppose "continuity of optimal transport", namely, for every $\log \rho, \log \bar{\rho} \in L^\infty$ *the corresponding optimal map is continuous.* We will see that LMP is a consequence of this assumption, thus a necessary condition. From this assumption we see that $\phi_\varepsilon \in C^1$, since the ensities 1_Ω and $\bar{\rho}_\varepsilon$ are bounded from above and below. Now, because $\phi_\varepsilon \in C^1$, it immediately holds

$$\partial^c \phi_\varepsilon = \exp \nabla \phi_\varepsilon = \exp \partial \phi_\varepsilon.$$

Then, one can show (see Loeper [49]) by taking limit $\varepsilon \to 0$, that

$$\partial^c \phi = \exp \partial \phi$$

The last equality is what is required by Loeper's maximum principle (LMP) for the c-convex function ϕ. One can in fact show that this special case implies LMP for general c-convex functions (see Loeper [49]).

2.4. Geometric interpretation of Loeper's maximum principle. To understand a more geometric meaning of LMP, we first need a notion called c-segment that extends the notion of geodesic. (This definition is due to Ma, Trudinger and Wang [52].)

Definition 2.9 (c-segment). Fix $x \in \Omega$. Let $p\colon t \in [0,1] \to p(t) \in T_x M$ be a line segment, i.e., $p''(t) = 0$. Then, the curve $\bar{x}(t) = \exp_x p(t)$ is called a *c-segment* with respect to x. As a special case, if the line segment $p(t)$ passes through the origin in $T_x M$, the corresponding c-segment is a geodesic passing through x.

One can define similarly a c-segment $x(t)$ with respect to $\bar{x} \in \overline{\Omega}$.

Definition 2.10 (sliding mountain). Let $\bar{x}(t)$ be a c-segment. Let $\bar{x}_i = \bar{x}(i)$, $i = 0, 1$. Define the function

$$m_t(\cdot) = -c\big(\cdot, \bar{x}(t)\big) + c\big(x, \bar{x}(t)\big)$$

We can call this type of functions a *sliding mountain.*

One sees that

$$\{\nabla m_t(x)\}_{0 \le t \le 1} = (\partial \max[m_0, m_1])(x).$$

Now, Loeper's maximum principle (LMP) can be stated as the following: For all $x \in \Omega$ and for any c-segment $\{\bar{x}(t)\}_{0 \le t \le 1}$ with respect to x,

$$\begin{aligned} \text{LMP}: &\quad m_t \le \max[m_0, m_1] \quad \forall\, 0 \le t \le 1 \text{ on } \Omega; \\ \text{local LMP}: &\quad m_t \le \max[m_0, m_1] \quad \forall\, 0 \le t \le 1 \text{ on a neighborhood of } x. \end{aligned} \tag{2.3}$$

Thus, LMP prevents the function $m_t(z)\colon [0,1] \to \mathbb{R}$ (for fixed $z \in \Omega$) from having (in fact, local) maximum in the interior of the interval $[0,1]$.

2.5. Ma – Trudinger – Wang curvature condition and examples. There is an infinitesimal version of Loeper's maximum principle, called the Ma, Trudinger, and Wang's curvature condition [52]. As we will see in Section 3.1, this condition strengthens the sectional curvature nonnegativity condition. Recall $c(x,\bar{x}) = \mathrm{dist}^2(x,\bar{x})/2$.

Consider a pair $(x,\bar{x}) \notin \mathrm{Cut}$, i.e., c is C^∞ near $(x,\bar{x})$. Consider two curves $\{x(s)\}_{s\in[-1,1]} \in \Omega$, $\{\bar{x}(t)\}_{t\in[-1,1]} \in \overline{\Omega}$ where either of them is a c-segment with respect to $\bar{x}$, x, respectively. Let $x(0) = x$ and $\bar{x}(0) = \bar{x}$. Let us use the terminology MTW-*curvature* to describe the tensor quantity of Ma, Trudinger and Wang [52] called c-curvature by Loeper [49], or called cross-curvature in [36, (2.2) and Lemma 4.5; 37, Definition 1.1].

Definition 2.11 (MTW-curvature). Let $(p,\bar{p}) \in T_xM \oplus T_{\bar{x}}M$.

$$\mathrm{MTW}_{(x,\bar{x})}(p,\bar{p}) = -\frac{\mathrm{d}^4}{\mathrm{d}s^2\,\mathrm{d}t^2}\bigg|_{(s,t)=(0,0)} c\big(x(s),\bar{x}(t)\big).$$

Remark 2.12. The MTW-curvature is indeed a curvature since it is induced by the Riemannian curvature tensor of a pseudo-metric defined on the product space $\Omega\times\overline{\Omega}$ as found by McCann and the author [36]. Furthermore, with Warren [39] they extended this result to define another pseudo-metric (a conformal perturbation of the one in [36]) and showed that the graph of the optimal map T in the product space $\Omega\times\overline{\Omega}$ gives a volume maximizing special Lagrangian submanifold, thus finding a connection to symplectic geometry.

Definition 2.13 (MTW condition). We say that MTW condition is satisfied if for all $(x,\bar{x}) \notin \mathrm{Cut}$,

$$\mathrm{MTW}_{(x,\bar{x})}(p,\bar{p}) \geq 0, \quad \forall\, \langle p, (D\exp_x)^{-1}\bar{p}\rangle = 0.$$

Here, the last inner product is with respect to the Riemannian metric and can also be written as $p^i D_{x^i} D_{\bar{x}^j} c\, \bar{p}^j = 0$.

We say that MTW_+ is satisfied if MTW is satisfied and if $p = 0$ or $\bar{p} = 0$ in case of equality in the inequality in Definition 2.13.

These MTW, MTW_+ conditions were originally called A3w, A3, respectively by Ma, Trudinger and Wang [52, 57].

We say that NNCC (nonnegative cross curvature) is satisfied if for all $(x,\bar{x}) \notin$ Cut,

$$\mathrm{MTW}_{(x,\bar{x})}(p,\bar{p}) \geq 0.$$

Notice that $\mathrm{MTW}_+ \Rightarrow$ MTW and NNCC $\Rightarrow$ MTW, but neither MTW_+ or NNCC implies the other.

Remark 2.14.
- The MTW_+ condition was introduced in [52] to get a priori estimates for Monge – Ampère type equations for optimal transport problems, and the method goes back to the work of Wang on reflector antenna problems [63, 64]. Loeper [50] verified MTW_+ for the cost function arising in the far-field reflector antenna problem and then showed regularity of the solution (see also a previous work of Caffarelli, Gutierrez and Huang [8] for a different approach). More recently, there is a work by Karakhanyan and Wang [34] that uses a variant of MTW_+ condition to give a rather complete solution to the regularity of the general (near-field) reflector antenna problem.

• It is a folklore conjecture among experts that for $c = \mathrm{dist}^2/2$ on Riemannian manifold, MTW implies NNCC.

• The condition NNCC has an unexpected application to principal-agent problem in microeconomics theory [19] (see also [54] for an exposition). On the other hand, Sei [56] found applications to statistical problems. In both applications, it was used that under NNCC, the set of c-convex functions is convex, that is, if ϕ_0 and ϕ_1 are c-convex then $(1-t)\phi_0 + t\phi_1$ is c-convex, too [19, 56].

It is immediate to see that for the Euclidean space $(\mathbb{R}^n, g_0)$, the MTW-curvature completely vanishes: MTW $\equiv 0$.

Loeper [49] found a connection of MTW-curvature to the Riemannian curvature.

Theorem 2.15 (Loeper [49, 50]). (1) *For* $x = \bar{x}$, $\mathrm{MTW}_{(x,x)}(p, \bar{p}) = \frac{4}{3}K(p \wedge \bar{p})$, *where* K *denotes the Riemannian sectional curvature. In particular,* MTW *implies nonnegative sectional curvature.*

(2) MTW $\Leftrightarrow$ local LMP (*see* [36] *for an elementary geometric proof for* $\Rightarrow$).

(3) *If the sectional curvature* K *is negative somewhere on* Ω, *then there exists a discontinuous optimal map* $T\colon \Omega \to \overline{\Omega}$, *with smooth densities* $\rho, \bar{\rho}$. (*Here, the domains* $\Omega, \overline{\Omega}$ *can have any nice properties* (*smoothness, convexity, etc.*)).

(4) *The round sphere* (S^n, g_0) *satisfies* MTW_+ *and* LMP (*see also* [36, 38]).

Remark 2.16. Villani [61] showed that the MTW condition is stable under Gromo – Hausdorff convergence, under suitable additional assumptions that give an equivalence between MTW and LMP. Note that LMP is more suitable for synthetic formulation than the MTW involving fourth order derivatives.

The statement (3) in this theorem is basically due to (1) and (2) and the fact that LMP is a necessary condition for regularity of optimal maps. In fact, one can also show that even positive curvature restriction is not enough for MTW, so for the regularity of optimal maps.

Theorem 2.17 ([35]). *There are positively curved manifolds that do not satisfy* MTW.

This is the first result showing the MTW condition is stronger than the positive curvature. Since then, there appeared other examples in this spirit [25; 51, Appendix D]. Especially, in [25], it is shown that two-dimensional ellipsoid surfaces in $\mathbb{R}^3$, when they are thin enough in one direction, do not satisfy MTW condition.

Regarding (4) in Theorem 2.15, in fact, a stronger result holds, from which we can produce a lot of NNCC (thus MTW) examples of Riemannian manifolds:

Theorem 2.18 (see [37]). (1) *The round sphere* (S^n, g_0) *satisfies the condition* NNCC.

(2) *For the Riemannian submersion* $\pi\colon M \to B$, *if* M *satisfies* MTW, MTW_+, NNCC, LMP, *respectively, then* B *satisfies the corresponding conditions, respectively. In particular, the complex projective space* $\mathbf{CP}^n$ *with Fubini – Study metric satisfies all these conditions* (*because* (S^{2n+1}, g_0) *does*).

(3) *Let* M_i, $i = 1, 2$ *satisfy* NNCC *then the Riemannian product* $M_1 \times M_2$ *satisfies* NNCC.

Remark 2.19. • For example, $S^{n_1}(r_1)\times\cdots\times S^{n_j}(r_j)\times\mathbf{CP}^{l_1}\times\cdots\times\mathbf{CP}^{l_k}\times\mathbb{R}^m$ satisfies NNCC. It is shown that [36, 37] this example satisfies LMP. (See also [27; 51; 62, Chapter 12] where they extended the method in [36] for deriving LMP to handle more general cases.)

• In [37], an O'Neill type inequality, which says Riemannian submersions increase curvature (see [9]), is obtained regarding the MTW-curvature (see Definition 2.11), and this yields the statement (2) for MTW, MTW_+ and NNCC in Theorem 2.18.

• Regarding the statement (2) in Theorem 2.18, there are examples of MTW_+ obtained by perturbing the round sphere as shown by Delanoë – Ge [13], Figalli, Rifford and Villani [24, 26, 27]. In particular, Delanoë and Ge showed that small perturbations of the two-dimensional round sphere are NNCC [14]: this is not known in higher dimensions.

• Regarding the statement (3) in Theorem 2.18, the same result does not hold for MTW: see [37].

• There are examples of costs satisfying MTW condition, originated from mechanical action, found by Lee and McCann [45], and also from a modified distance function on the *hyperbolic* space found by Lee and Li [44].

3. Geometry of MTW curvature condition

The goal of this section is twofold: first, we will explain the result in Theorem 2.17, second, we will explain a quantitative version of LMP for MTW_+ (which is due to Loeper [49]), and as an application will show continuity of optimal maps on the round sphere. The latter result was actually shown with Hölder continuity [49].

3.1. MTW is finer than positive sectional curvature. In this subsection, we explain why Theorem 2.17 holds: see [35] for more details.

Consider a surface M such that in a small neighbourhood, say $\mathcal{N}$, of a point $x_0 \in M$, the sectional curvature K is positive and outside the neighbourhood, $K = 0$. Here the curvature is bounded to be sufficiently small. One can construct such a surface by rounding about the vertex of a sufficiently thin cone. We will show that such surface does not satisfy local LMP, thus not MTW (see Theorem 2.15). Of course, this surface is not positively curved, but one can slightly perturb the surface to get a positively curved example that does not satisfy MTW.

Fix a point $x \in M$ outside $\mathcal{N}$. Consider the exponential map $\exp_x : T_xM \to M$. For simplicity we assume that $\exp_x$ is injective everywhere, but in general, under the sufficiently small bound on the curvature, one can find a neighbourhood such that all relevant points in the following discussion are within the injectivity radius (see [35]). Find a line segment $\{\bar p_t\}_{0\le t\le 1}$in T_xM such that (i) the curve $\bar x(t) = \exp_x \bar p_t$ contains x_0, i.e., there is $t_0 \in (0,1)$ such that $x_0 = \bar x(t_0)$, (ii) moreover, its two end points $\bar x(0)$ an $\bar x(1)$ are outside $\mathcal{N}$, and (iii) there exists a geodesic segment γ outside $\mathcal{N}$ passing through the points x and $\bar x(0)$ such that it is orthogonal to the segment $\{\bar x(t)\}_{0\le t\le 1}$ at the point $\bar x(0)$. Let y denote a point in γ close to x but farther from $\bar x(0)$ than x. Let $p \in T_xM$ such that $\exp_x p = y$. We are now ready for the following argument: Define $f(t), \tilde f(t)$ as

$$f(t) = -\operatorname{dist}^2\big(y, \bar x(t)\big) + \operatorname{dist}^2\big(x, \bar x(t)\big)$$
$$\tilde f(t) = -|p - \bar p_t|^2 + |\bar p_t|^2$$

We will violate local LMP (see (2.3)) by showing that $f(t_0) > \max[f(0), f(1)]$. This will be done by comparing the two functions $f(t)$ and $\tilde{f}(t)$. Notice that $f(0) = \tilde{f}(0)$ and $f(1) = \tilde{f}(1)$. Moreover, $\tilde{f}(t) \equiv \tilde{f}(0) = \tilde{f}(1)$ for all $t \in [0,1]$. (This is due to the orthogonality of γ and $\{\bar{x}(t)\}$ at $\bar{x}(0)$.) Now, use the well-known Toponogov theorem (see Cheeger and Ebin [9]),

Theorem 3.1 (Toponogov's comparison theorem). *Let M be a complete Riemannian manifold with sectional curvature $K_M \geq H$, and let M^H be the simply connected 2-dimensional space of constant curvature H. Let $\gamma_i\colon [0,1] \to M$ and $\bar{\gamma}_i\colon [0,1] \to M^H$, $i = 1,2$, be minimal geodesic segments, i.e., they are unique geodesic segments connecting their end points. Suppose that $\gamma_1(0) = \gamma_2(0)$, $\bar{\gamma}_1(0) = \bar{\gamma}_2(0)$; $\measuredangle(\dot{\gamma}_1(0), \dot{\gamma}_2(0)) = \measuredangle(\dot{\bar{\gamma}}_1(0), \dot{\bar{\gamma}}_2(0)) < \pi$, where $\measuredangle$ denotes the angle between tangent vectors. Assume $L[\gamma_i] = L[\bar{\gamma}_i]$, $i = 1,2$, where L denotes arc-length. Then*

$$\operatorname{dist}\big(\gamma_1(1), \gamma_2(1)\big) \leq \operatorname{dist}\big(\bar{\gamma}_1(1), \bar{\gamma}_2(1)\big), \tag{3.1}$$

where dist *denotes the Riemannian distance. Moreover, if there exists a point z on $\gamma_1 \cup \gamma_2 \subset M$ such that $K_M(z) > H$, then the inequality* (3.1) *is strict.*

Applying this theorem, we see that

$$|\bar{p}_{t_0} - p| > \operatorname{dist}\big(y, \bar{x}(t_0)\big)$$

because the sectional curvature of M, $K \geq 0$. Here, the strict inequality is due to the condition $K\big(\bar{x}(t_0)\big) > 0$. Therefore, we see $f(t_0) > \tilde{f}(t_0) = f(0) = f(1)$, and this violates local LMP, thus MTW. Thus, the surface M does not satisfy MTW. This finishes the discussion of this subsection.

3.2. An open question. The result in the previous subsection motivates us to consider finer relation between the sectional curvature and the MTW curvature. Regarding this, the following question is raised by Trudinger:

Question 3.2. Do there exist appropriate norms $\|\cdot\|_1$, $\|\cdot\|_2$ and an $\epsilon > 0$ such that $\|\nabla R\|_1 \leq \epsilon \|R\|_2$ implies MTW?

Here, R denotes the Riemannian curvature tensor, and ∇R is its covariant derivative (thus a tensor itself). See [13,24] for partial results in this direction, in the case of perturbations of the round sphere. Answers to Question 3.2 are useful to obtain a robust method of finding Riemannian manifolds that satisfy MTW, especially since it is very hard to verify the MTW condition in general. Note that there are a few works for finding effective criteria of MTW condition (see [25,43]). Also, there is a relation between the MTW condition and the convexity of tangential injectivity domains [24,26,27,51].

3.3. Loeper's quantitative maximum principle (LQMP) under strong MTW condition (MTW$_+$). We now state a quantitative version of Loeper's maximum principle and its applications to the continuity of optimal maps. The original results in this subsection are due to Loeper [49].

Theorem 3.3 (LQMP). *Assume that the cost c satisfies* MTW$_+$. *Namely,* $\mathrm{MTW}(p,\bar{p}) \geq K_0|p|^2|\bar{p}|^2$ *for $\langle p, D\exp^{-1}\bar{p}\rangle = 0$. Let $\bar{x}(t)$ be a c-segment with respect to x. Define*

$$m_t(\cdot) = -c\big(\cdot, \bar{x}(t)\big) + c\big(x, \bar{x}(t)\big).$$

Then, there exist r_0 and K_1 (both depending on the cost function c, especially on K_0) such that $\forall 0 < r \leq r_0$ and $\forall z \in B_r(x)$, $\forall 0 \leq t \leq 1$,

$$m_t(z) \leq \max[m_0(z), m_1(z)] - K_1 t(1-t)\,\mathrm{dist}^2(z,x)\,\mathrm{dist}^2\big(\bar{x}(0), \bar{x}(1)\big) + \|c\|_{C^3}\,\mathrm{dist}^3(z,x).$$

This theorem was originally proved in [49]: see [37, Appendix] for a different proof. Also, see [27,51] for an improved version of this result.

Loeper's quantitative maximum principle (LQMP) shows that the gap between the double mountain like function $\max[m_0, m_1]$ and the sliding mountain like function m_t, is quadratic in the distance near the point x along where they coincide. To have this estimate turns out to be very useful for showing regularity (continuity) of optimal maps as we see below.

3.3.1. *Application of Loeper's quantitative maximum principle* (LQMP). The quantitative maximum principle of Loeper is a powerful tool for proving regularity of optimal maps on positively curved domains with MTW_+. In particular, one can show Hölder continuity of optimal maps on S^n [49, 50], $\mathbf{CP}^n$ [37], and perturbations of S^n [24, 27, 51] for source and target measures bounded from above and below, i.e., $\log\rho, \log\bar{\rho} \in L^\infty$.

We show the following, originally due to Loeper, as an example.

Theorem 3.4 ([49, 50]; **see also** [38]). *Let $\Omega = \bar{\Omega} = S^n$, the round sphere. Assume that the source and target measures satisfy $0 < \lambda \leq \rho, \bar{\rho} \leq \Lambda$ where λ, Λ are constants. Let T be the optimal map $T_{\#}\rho = \bar{\rho}$. Then, $T \in C^0(\Omega)$ (in fact $T \in C^\alpha(\Omega)$ for some $0 < \alpha < 1$, depending on the dimension n).*

We note here that a sharp Hölder exponent $\alpha = (n+1)/(2n^2+n-1)$ is obtained by Liu [46]. Also, if further $\rho, \bar{\rho} \in C^\infty$, then by applying [52], $T \in C^\infty$ as in [50]. The continuity method of [52] also applies to get smooth optimal maps on perturbations of the round sphere [13].

Proof of Theorem 3.4. In this proof, we ignore the technical problem that the cost fucntion dist^2 is not smooth for antipodal pairs in S^n: these points where dist^2 fails to be smooth is called *cut-locus*: we discuss this issue in Section 7.

Recall that the round sphere satisfies MTW_+, thus both LMP and LQMP. The following argument can be called "Sausage-Meat Ball Argument."

Let ϕ be the c-potential function for the optimal map T, i.e., $\mathrm{graph}\, T \subset \partial^c\phi$. Suppose by contradiction that $T \notin C^0$. Then there exists a point x such that the c-subdifferential $\partial^c\phi(x)$ has two distinct points $\bar{x}_0, \bar{x}_1$ ($\bar{x}_0 \neq \bar{x}_1$). Let $\bar{x}(t)$ be a c-segment between $\bar{x}_0 = \bar{x}(0)$ and $\bar{x}_1 = \bar{x}(1)$ with respect to x. Define for each $\delta > 0$, a tubular neighborhood

$$N_\delta = \{\bar{z} \mid \mathrm{dist}\big(z, \bar{x}(t)\big) \leq \delta, \tfrac{1}{3} \leq t \leq \tfrac{2}{3}\}.$$

We will use the following result which we will show later:

Claim 3.5. *Recall r_0 from* LQMP (*Theorem* 3.3). *There exists $C_1 > 0$ (depending only on the cost c and $\mathrm{dist}\big(\bar{x}(0), \bar{x}(1)\big)$) small enough such that for all $r < r_0$, if $\delta = C_1 r$, then $N_\delta \subset \partial\phi\big(B_r(x)\big)$.*

Choose r and δ as in this claim. Now, observe that

$$\bar{\rho}\Big(\partial^c\phi(B_r(x))\Big) = \bar{\rho}\big(T(B_r(x)\big) = \rho\big(B_r(x)\big)$$

Use Claim 3.5 and $\lambda \leq \rho, \bar{\rho} \leq \Lambda$, to see that

$$\bar{\rho}\Big(\partial^c\phi\big(B_r(x)\big)\Big) \geq \bar{\rho}(N_\delta) \gtrsim \delta^{n-1} \sim r^{n-1}.$$

But, on the other hand, $\rho\big(B_r(x)\big) \sim r^n$. Thus, comparing these, we have $r^n \gtrsim r^{n-1}$. Letting $r \to 0$, we get a contradiction. This shows Theorem 3.4 □

PROOF OF CLAIM 3.5. Recall $m_t(\cdot) = -c\big(\cdot, \bar{x}(t)\big) + c\big(x, \bar{x}(t)\big)$. Define

$$m_{\bar{z}}(\cdot) = -c(\cdot, \bar{z}) + c(x, \bar{z}).$$

The following simple estimates will be useful:

$$m_{\bar{z}}(z) - m_t(z) \leq \|D_x D_{\bar{x}} c\| \operatorname{dist}(z, x) \operatorname{dist}\big(\bar{z}, \bar{x}(t)\big) + \text{higher order terms of } \operatorname{dist}^2(z, x), \text{ etc.} \tag{3.2}$$

We will show that there exists $C_1 > 0$ that for each $r \leq r_0$, $\delta = C_1 r$, $\bar{z} \in N_\delta$,

$$m_{\bar{z}}(z) \leq \phi(z) \quad \forall z \in \partial B_r(x). \tag{3.3}$$

This then will imply, by comparison principle and the definition of $\partial^c\phi$, that $\bar{z} \in \partial^c\phi(B_r)$, completing the proof. Now it remains to show (3.3). Notice that $\phi(\cdot) \geq \max[m_0(\cdot), m_1(\cdot)]$. Let $z \in B_r(x)$, $r \leq r_0$. Then, by (3.2) and LQMP,

$$\begin{aligned} m_{\bar{z}}(z) &\leq m_t(z) + \|D_x D_{\bar{x}} c\| \operatorname{dist}(z, x) \operatorname{dist}\big(\bar{z}, \bar{x}(t)\big) \\ &\leq \max[m_0(z), m_1(z)] - K_1 t(1-t) r^2 \operatorname{dist}^2\big(\bar{x}(0), \bar{x}(1)\big) \\ &\qquad + \|c\|_{C^2} C_1 r^2 + \text{higher order in } r. \end{aligned}$$

By choosing C_1 small enough, we see that for $\frac{1}{3} \leq t \leq \frac{2}{3}$, the last line is bounded above by $\max[m_0(z), m_1(z)]$, thus by $\phi(z)$. We showed (3.3). □

4. Hölder continuity of optimal transport maps under MTW without LQMP

Under only MTW (without MTW$_+$), the Loeper's quantitative maximum principle (LQMP) is not available anymore, making the analysis of optimal maps more difficult. Nevertheless, we can show (interior) regularity of optimal maps in this case [21]. Notice that the special case $c(x, \bar{x}) = -x \cdot \bar{x}$ (equivalently $c(x, \bar{x}) = \frac{1}{2}|x - \bar{x}|^2$) in $\mathbb{R}^n$, is addressed in the pioneering work of Delanoë [12], Urbas [59] and Caffarelli [3–7] on the regularity of Monge–Ampère equation. Especially, in this case, Caffarelli has obtained Hölder continuity of optimal maps with measurable data, namely, assuming only L^∞ (upper and lower) bounds (2.1) on the source and target densities ρ, $\bar{\rho}$. One can view the results below (Theorem 4.3) as an extension of Caffarelli's methods and results to more general cost functions. One of the novelties here is that one now can handle domains in Riemannian manifolds: e.g., products of round spheres [20]. We remark that Liu, Trudinger and Wang have obtained higher regularity results with more regular data, using continuity methods [48, 57] (see also [47]). Such continuity method is not available for merely measurable data, and we need more geometric arguments for the analysis.

4.1. (Interior) Hölder continuity of optimal transport maps. Throughout this section we consider domains Ω', $\overline{\Omega}$ in an n-dimensional Riemannian manifold M with $(\Omega' \times \overline{\Omega}) \cap \mathrm{Cut} = \varnothing$ so that the cost function $c(x,y) = \mathrm{dist}^2(x,y)/2$ is smooth on $\Omega' \times \overline{\Omega}$. Here, the domain Ω' is an open set containing the source domain Ω of the optimal transportation.

Definition 4.1 ((strong) c-convexity of $\overline{\Omega}$ with respect to Ω' [52]**).** We say that $\overline{\Omega}$ is (strongly) c-convex with respect to Ω', if for all $x \in \Omega'$, the inverse image $\exp_x^{-1} \overline{\Omega}$ is (strongly) convex as a subset in the tangent space $T_x\Omega$: use the Riemannian metric in $T_x\Omega$ to measure how strong the set is convex. Recall that a set in $\mathbb{R}^n$ is strongly convex if the set is an intersection of balls of uniformly upper bounded radius. The smaller this bound, the stronger the convexity.

Similary, we define (strong) c-convexity of Ω' with respect to $\overline{\Omega}$.

Remark 4.2. The c-convexity on the domain $\overline{\Omega}$ is a necessary condition for regularity theory of optimal maps. For instance, for the case $c(x,\bar{x}) = -x \cdot \bar{x}$, Caffarelli showed a counterexample [5] to regularity (in fact, continuity) where the target domain is not convex (thus not c-convex). Ma, Trudinger and Wang [52, Section 7.3] showed similar example for more general cost functions.

We now state the main theorem of this section:

Theorem 4.3 ([21, Theorem 2.1])**.** *Assume that*

- $\Omega', \overline{\Omega}$ *are strongly c-convex with respect to each other*;
- $\log \bar{\rho} \in L^\infty(\overline{\Omega})$, $\log \rho \in L^\infty(\Omega)$, $\Omega \subset \Omega'$ *is an open set. The set Ω is not necessarily c-convex*;
- T *is an optimal map with* $T_\# \rho = \bar{\rho}$;
- *the cost* $c = \mathrm{dist}^2/2$ *satisfies* MTW.

Then,

(1) $T \in C^\alpha_{\mathrm{loc}}(\Omega)$.
(2) *The restriction $T|_\Omega$ of T to Ω, is one-to-one.*

Remark 4.4. • In two dimensions, Figalli and Loeper [23] obtained continuity of optimal maps without assuming that the source measure ρ is bounded from below. Their method goes back to the classical work of Alexandrov [1].

• The Hölder exponent α depends only on the dimension n and the upper and lower bounds of ρ and $\bar{\rho}$, in particular not on the specific cost c: see [21, Section 9].

item When $\rho, \bar{\rho} \in C^\infty$, one can apply the above injectivity (Theorem 4.3(2)) to the result of [48] (see also [47]) to get $T \in C^\infty(\Omega)$.

4.2. Tools for regularity of optimal transport maps. In this subsection, we explain a few tools for the regularity of optimal maps as in Theorem 4.3, which are available without the MTW assumption.

4.2.1. (*weak*) *c-Monge–Ampère equation.* Let $T_\# \rho = \bar{\rho}$ be the optimal map push-forwarding ρ onto $\bar{\rho}$ with the corresponding c-potential ϕ. Assume that ρ, $\bar{\rho}$ are bounded away from zero and infinity, namely, $\log \rho \in L^\infty(\Omega)$ and $\log \bar{\rho} \in L^\infty(\overline{\Omega})$. Then, it is well-known that ϕ satisfies the following weak form of the c-Monge–Ampère equation: see for example, [21, Lemma 3.1(e)]. Namely, there exists a constant $\lambda > 0$ such that

$$(\mathrm{MA}_\lambda) \qquad \lambda |B| \le |\partial^c \phi(B)| \le \frac{1}{\lambda}|B| \quad \forall \text{ Borel susbset } B \subset \Omega$$

Here, $\partial^c\phi(B) = \bigcup_{x\in B} \partial^c\phi(x)$ and $|B| = \int_B \mathrm{d\,vol}$.

The above condition MA_λ can be denoted simply as $|\partial^c\phi| \sim 1_\Omega$, since it says the c-Monge–Ampère measure $|\partial^c\phi|$, defined as $|\partial^c\phi|(B) = |\partial^c\phi(B)|$, is equivalent to the uniform measure on Ω.

Example 4.5. Let ϕ be the c-cone on $\mathbb{R}^n$ (with $c(x,y) = |x-y|^2/2$),

$$\phi(x) = \sup_{y\in B_1(0)} -|x-y|^2 + |y|^2$$

Then, one can see that $|\partial^c\phi| \sim \delta_0$, the Dirac-delta measure at 0, because $\partial^c\phi(B) = B_1(0)$ for any B containing 0. Thus ϕ does not satisfy the above c-Monge–Ampère equation MA_λ.

4.2.2. *Interior-not-to-boundary result for optimal maps.* Another important tool is a lemma that assures that the optimal transport map does not mix the interior points with the boundary, at least if the domains satisfy appropriate convexity conditions. More precisely,

Theorem 4.6 (Interior-not-to-boundary; see [21, Theorem 5.1]). *Assume that*

- $(\Omega' \times \overline{\Omega}) \cap \mathrm{Cut} = \varnothing$;
- $\Omega', \overline{\Omega}$ *strongly c-convex with respect to each other*;
- $\Omega \subset \Omega'$;
- $\partial^c\phi(\Omega) \subset \overline{\Omega}$;
- $|\partial^c\phi| \sim 1_\Omega$.

Then,

(1) $\partial^c\phi(\mathrm{int}\,\Omega) \cap \partial\overline{\Omega} = \varnothing$;
(2) $\partial^c\phi(\partial\Omega') \cap \mathrm{int}\,\overline{\Omega} = \varnothing$

Remark 4.7.
- Notice that the MTW condition is not assumed in this theorem.
- This theorem was a necessary ingredient in the paper of Figall and Loeper [23] where they showed the same result in the two dimensions without the lower bound of the source measure ρ.

For the proof of the above theorem, we recall the c-monotonicity of $\partial^c\phi$. When $c(x,y) = -x\cdot y$ on $\mathbb{R}^n$ where $\partial^c\phi = \partial\phi$ (the subdifferential) and c-convex functions are nothing but convex functions, it reads as

$$(x,\bar{x}),(z,\bar{z}) \in \partial\phi \implies \langle z-x, \bar{z}-\bar{x}\rangle \geq 0.$$

For more general case, it reads as

$$(x,\bar{x}),(z,\bar{z}) \in \partial^c\phi \implies -c(z,\bar{z}) + c(x,\bar{z}) + c(z,\bar{x}) - c(x,\bar{x}) \geq 0.$$

The c-monotonicity roughly says that infinitesimally, the (valued) map $\partial^c\phi$ is irrotational.

Idea of proof of Theorem 4.6. We will only show the assertion (1). This is enough for showing the idea. For simplicity, we present only the case when $M = \mathbb{R}^2$ and $c(x,y) = -x\cdot y$. Then, the function ϕ is convex (thus $\partial^c\phi = \partial\phi$) and $\overline{\Omega}$ is a strongly convex set.

Suppose by contradiction that there is a pair $(x,\bar{x})$ with $x \in \mathrm{int}\,\Omega$ and $\bar{x} \in \partial\phi(x)\cap\partial\overline{\Omega}$. Find a vector v such that the normal plane $N_v(\bar{x}) = \{z \mid \langle z-\bar{x}, v\rangle = 0\}$

at $\bar{x}$ has the unique intersection $\bar{x}$ with $\overline{\Omega}$. Moreover, by the strong convexity of $\overline{\Omega}$, the boundary $\partial\overline{\Omega}$ looks like the graph of a quadradic function over $N_v(\bar{x})$: we can give coordinates $(\bar{x}^1, \bar{x}^2) \in N_v(\bar{x}) \times \mathbb{R}$, such that $\overline{\Omega}$ is above the graph of the function $\bar{x}^2 = C|\bar{x}^1|^2$, and that $v = (0, -1)$. Now, consider for $\theta, \epsilon > 0$ small enough, the conical set $E_{\theta,\epsilon} \subset \Omega$ defined as

$$E_{\theta,\epsilon} = \{z \mid \langle z - x, v\rangle \geq \cos\theta |z - x||v| \ \& \ |z - x| \leq \epsilon\}$$

Note that $|E_{\theta,\epsilon}| \sim \theta\epsilon^2$. Define

$$\overline{E}_\theta = \left\{\bar{z} \,\middle|\, \langle \bar{z} - \bar{x}, v\rangle \geq \cos\left(\frac{\pi}{2} + \theta\right)\right\} \cap \overline{\Omega}\Big\}.$$

One can compute using the quadratic function $\bar{x}^2 = C|\bar{x}^1|^2$ as above, that $|\overline{E}_\theta| \lesssim \theta^3$.

But, by monotonicity, $\partial\phi(E_{\theta,\epsilon}) \subset \overline{E}_\theta$. Thus, the desired contradiction follows by comparing volumes:

$$\begin{aligned}\theta\epsilon^2 \sim |E_{\theta,\epsilon}| &\sim |\partial\phi(E_{\theta,\epsilon})| \quad (\text{by } |\partial^c\phi| \sim 1_\Omega)\\ &\leq |\overline{E}_\theta| \lesssim \theta^3.\end{aligned}$$

Let $\theta \to 0$ while fixing ϵ, then we get a contradiction. This finishes the proof. □

4.3. Why is MTW good for regularity of optimal transport maps? Appeared convexity! To see how the MTW condition affects the geometry of c-convex functions, we consider the following transformation of the coordinates and the functions.

We first assume that all relevant points are outside the cut-locus so that the cost function is smooth, and moreover the exponential map is invertible on such points.

Fix $\bar{x}_0 \in M$. Give correspondence between $q \in T_{\bar{x}_0}M$ and $x \in M$ as

$$x(q) = \exp_{\bar{x}_0} q, \quad q(x) = \exp_{\bar{x}_0}^{-1} x.$$

Now modify the cost function and a c-convex function ϕ in the q variable as

$$\begin{aligned}\tilde{c}(q, \bar{x}) &:= c(x(q), \bar{x}) - c(x(q), \bar{x}_0),\\ \tilde{\phi}(q) &:= \phi\big(x(q)\big) + c(x(q), \bar{x}_0)\end{aligned}$$

Notice that $\tilde{\phi}$ is $\tilde{c}$-convex and if $B = \exp_{\bar{x}_0} \widetilde{B}$ then $\partial^c\phi(B) = \partial^{\tilde{c}}\tilde{\phi}(\widetilde{B})$, namely the image of c-subdifferential is not changed under this transformation of c and ϕ.

This transformation is very useful under the MTW condition because of the following result [21, 46]:

Theorem 4.8 (appeared convexity). *Assume $\Omega, \overline{\Omega}$ be c-convex with respect to each other. Let ϕ be c-convex in the variable $x \in \Omega$.*

1. *If* MTW *holds for c, then $\tilde{\phi}$ is level set convex in the variable $q \in \exp_{\bar{x}_0}^{-1} \Omega$.*
2. *If* NNCC *holds for c, then $\tilde{\phi}$ is convex in the variable $q \in \exp_{\bar{x}_0}^{-1} \Omega$.*

Here, level-set convexity means that each sub-level set $\{\tilde{\phi} \leq k\}$, $k \in \mathbb{R}$, is a convex set.

Remark 4.9. Notice that the above statement (1) is a direct consequence of Loeper's maximum principle LMP (2.3).

Example 4.10. Recall that in $\mathbb{R}^n$ with the quadratic cost $\tilde{c}(x,\bar{x}) = |x-\bar{x}|^2/2$, the cross curvature in Definition 2.11 vanishes identically, thus NNCC holds. In this special case, the exponential map $x(q) = \exp_{\bar{x}_0} q = q + \bar{x}_0$, thus, the above transform becomes

$$\begin{aligned}\tilde{c}(q,\bar{x}) &= \tfrac{1}{2}|q+\bar{x}_0-\bar{x}|^2 - \tfrac{1}{2}|q|^2\\ &= -q\cdot(\bar{x}-\bar{x}_0) + \tfrac{1}{2}|\bar{x}-\bar{x}_0|^2,\end{aligned}$$

which is linear (thus convex) in q. Note that in this case c-convex functions are not convex; for example, consider the c-convex function

$$\phi(x) = \max[-\frac{1}{2}|x-\bar{x}_1|^2, -\frac{1}{2}|x-\bar{x}_2|^2]$$

for two fixed $\bar{x}_0 \neq \bar{x}_1 \in \mathbb{R}^n$. However, the transformed function $\tilde{\phi}(q) = \phi(x(q)) + \frac{1}{2}|x(q), \bar{x}_0|^2$ is convex, for example,

$$\begin{aligned}&\max\left[-\tfrac{1}{2}|x-\bar{x}_1|^2, -\tfrac{1}{2}|x-\bar{x}_2|^2\right] + \tfrac{1}{2}|x(q)-\bar{x}_0|^2\\ &\quad= \max\left[-q\cdot(\bar{x}_1-\bar{x}_0) - \tfrac{1}{2}|\bar{x}_1-\bar{x}_0|^2, q\cdot(\bar{x}_2-\bar{x}_0) - \tfrac{1}{2}|\bar{x}_1-\bar{x}_0|^2\right].\end{aligned}$$

5. Alexandrov type estimates

The appeared convexity (see Theorem 4.8) tells us that under the MTW condition c-convex functions can be transformed to level-set convex functions in appropriate exponential coordinates. For this observation to be useful in applications, we extend the Alexandrov estimates well-known for convex functions, to level-set convex c-convex functions.

First consider a basic tool in convex analysis, the so-called Fritz John's ellipsoid lemma:

Theorem 5.1 (John's ellipsoid lemma [33]). *Let $Z \subset \mathbb{R}^n$ be an open bounded convex set. Then, there exists an ellipsoid E, such that*

$$E \subset Z \subset nE \tag{5.1}$$

where nE is the dilation of E with respect to its centre. Equivalently, there exists an invertible affine map $L\colon \mathbb{R}^n \to \mathbb{R}^n$ such that

$$B_1(0) \subset L^{-1}(Z) \subset B_n(0),$$

for the balls $B_1(0), B_n(0)$ centred at the origin with radius 1 and n, respectively.

The following theorem in [21] extends the classical Alexandrov estimates for convex functions to the c-convex functions under the MTW condition.

Theorem 5.2 (Alexandrov upper and lower bound [21]). *Use the notation in the previous section and Theorem 4.8. Assume that*

- *the function $-\tilde{c}$ is level set convex (which holds under* MTW*);*
- *$\tilde{\phi}$ is $\tilde{c}$-convex (thus it is also level-set convex since $-\tilde{c}$ is level set convex);*
- $1/\lambda \geq |\partial^{\tilde{c}}\tilde{\phi}| \geq \lambda > 0$;
- *the set $Z := \{z \in \exp_{\bar{x}_0}^{-1}\Omega \mid \tilde{\phi} < 0\} \subset\subset \exp_{\bar{x}_0}^{-1}\Omega$.*

Then, we have the following:

(1) *(Alexandrov lower bound) There exists a constant $C(n,\lambda) > 0$ such that*

$$|Z|^2 \leq C(n,\lambda)\left(\sup_Z|\tilde{\phi}|\right)^n. \tag{5.2}$$

(2) (*Alexandrov upper bound*) *If* $\operatorname{diam} Z \ll 1$ *and* Z *is sufficiently far from* $\partial \exp_{\bar{x}_0}^{-1} \Omega$, *then*

$$|\tilde{\phi}(q_t)|^n \lesssim (1-t)^{1/2^{n-1}} |Z|^2 \tag{5.3}$$

where $q_t \in t\partial Z$, *for* $0 < t < 1$. *Here,* $t\partial Z$ *is the dilation of* ∂Z *by the factor* t *with respect to the centre of the ellipsoid for the convex set* Z *as in the John's lemma* (*Theorem* 5.1).

An important idea behind Alexandrov type estimates is that convex functions behaves like quadratic functions, and for quadratic functions such estimates are straightforward:

Example 5.3. For $x = (x_1, x_2) \in \mathbb{R}^2$, let $\tilde{\phi}(x) = a_1 x^2 + a_2 x_2^2 - b$ for $0 < a, b \in \mathbb{R}$. Assume $|\det D^2 \tilde{\phi}| \sim 1$, i.e., $a_1 a_2 \sim 1$.

Now, for $Z = \{x \mid \tilde{\phi}(x) \le 0\}$,

$$\begin{aligned} |Z|^2 &\sim \frac{b^2}{a_1 a_2} \sim b^2 \quad (\text{since } |\det D^2 \tilde{\phi}| \sim 1) \\ &= \sup_Z |\tilde{\phi}|^2. \end{aligned}$$

This gives the Alexandrov lower bound.

On the other hand we see for $q_t \in t\partial Z$,

$$|\tilde{\phi}(q_t)|^2 = |t^2 b - b|^2 = (1-t)^2 (1+t)^2 |b|^2 \sim (1-t)^2 |Z|^2$$

which gives the Alexandrov upper bound.

The point is that the estimates (5.2) (5.3) hold regardless the shape of the convex set Z, i.e., it can be very thin. This latter case is unavoidable, since if $\tilde{\phi}$ is merely $C^{1,\alpha}$ (which is the optimal regularity for potentials of optimal maps with merely measurable source and target densities), there is no control on how thin the section can be. Of course, for C^2 (uniformly) convex functions, there is a uniform control on the shape of the sections.

5.1. Alexandrov lower bound. Let us first discuss the proof of the Alexandrov lower bound (5.2). Instead of giving the full proof of (5.2), we show the well-known special case $\tilde{c}(x,y) = -x \cdot y$ and for convex $\tilde{\phi}$ so that $\partial^{\tilde{c}} \tilde{\phi} = \partial \tilde{\phi}$. We will then discuss the more general case.

5.1.1. *Proof of the Alexandrov lower bound* (5.2) *for the case* $\tilde{c}(x,y) = -x \cdot y$ *and* $\tilde{\phi}$ *is convex.* This is a standard proof and one can find it elsewhere (for example, in the book of Gutierrez [31]). We give the proof here for user's convenience and also for discussion of the more general case.

We first renormalize the set Z. Find an affine map $L : \mathbb{R}^n \to \mathbb{R}^n$ such that the set $Z^* := L^{-1} Z$ is comparable to $B_1(0)$ i.e., $B_1(0) \subset Z^* \subset B_n(0)$ and $|Z| \sim |\det L|$. Let

$$\tilde{\phi}^*(z^*) := \frac{1}{(\det L)^{2/n}} \tilde{\phi}(L z^*).$$

Then, for all Borel set B,

$$|\partial \tilde{\phi}^*(L^{-1}(B)| = \frac{1}{|\det L|} |\partial \phi(B)|.$$

We also see that

$$|\partial \tilde{\phi}| \sim 1 \iff |\partial \tilde{\phi}^*| \sim 1.$$

Pick any $p \in \partial\tilde{\phi}^*(\frac{1}{2}Z^*)$, where the set $\frac{1}{2}Z^*$ is the dilation of the set Z^* by factor $\frac{1}{2}$ with respect to the origin 0. Then, we see by convexity of $\tilde{\phi}^*$ and the fact that Z^* is in shape comparable to the unit ball,

$$|p| \lesssim h^*$$

where $h^* := \sup_{Z^*}|\tilde{\phi}^*|$. This shows that

$$\partial\tilde{\phi}^*(\tfrac{1}{2}Z^*) \subset B_{Ch^*}(0) \quad \text{for some constant } C > 0,$$

and thus,

$$|\partial\tilde{\phi}^*(\tfrac{1}{2}Z^*)| \lesssim |h^*|^n.$$

Now the left-hand side is the same as

$$\begin{aligned}\frac{1}{|\det L|}\left|\partial\tilde{\phi}\left(\frac{1}{2}Z\right)\right| &\gtrsim \frac{1}{|\det L|}|Z| \quad (\text{since } |\partial\tilde{\phi}| \gtrsim \lambda > 0)\\ &\sim \frac{1}{|Z|}|Z| = 1 \quad (\text{since } |Z| \sim |\det L|).\end{aligned}$$

On the other hand, the right-hand side is

$$\frac{1}{(\det L)^2}\left(\sup_Z|\tilde{\phi}|\right)^n \sim \frac{(\sup_Z|\tilde{\phi}|)^n}{|Z|^2}$$

Comparing these, we see

$$|Z|^2 \leq \left(\sup_Z|\tilde{\phi}|\right)^n$$

as desired.

5.1.2. *Discussion for more general case.* Let us discuss the more general case.

Under NNCC condition, the corresponding inequality reduces to the case $\tilde{c}(x,y) = -x \cdot y$, mainly because of the two reasons below:

- $\tilde{\phi}$ is convex (due to Theorem 4.8(2));
- $|\partial^c\tilde{\phi}| \lesssim |\partial\tilde{\phi}|$.

The second assertion holds since (ignoring differentiability),

$$\begin{aligned}\left|\det \mathrm{Jac}\, \partial^{\tilde{c}}\tilde{\phi}\right| &= |\det(D_x D_{\bar{x}}\tilde{c})|^{-1} \det\left(D^2_{xx}\tilde{\phi} + D^2_{xx}\tilde{c}\right)\\ &\lesssim \left|\det D^2_{xx}\phi\right| = |\det \mathrm{Jac}\, \partial\phi|\end{aligned}$$

where the inequality holds because $\det(D_x D_{\bar{x}}\tilde{c})$ is bounded and NNCC condition implies $D^2_{xx}\tilde{c} \leq 0$ thus $D^2_{xx}\tilde{\phi} + D^2_{xx}\tilde{c} \leq D^2_{xx}\tilde{\phi}$ as matrices.

On the other hand, the MTW case is much more difficult, because both of above key properties under NNCC do not hold anymore.

One may argue that a renormalization method as in the proof for the classical affine cost case $\tilde{c}(x,y) = -x \cdot y$, would work, especially letting the relevant set smaller and smaller so to make the cost function more close to the affine cost (when x and y are close any smooth cost looks like an affine cost asymptotically). But, there is a serious difficulty in this argument. Namely, for more general cost function $\tilde{c}$, the term $D_{\bar{x}}D^2_{xx}\tilde{c}$ (which measures how much the local behavior of the cost function differs from that of the affine cost) may blow-up under the renormalization, if the set Z before normalization is very thin.

Thus, it does not seem reasonable to use the renormalization method to treat general MTW case. But, by the appeared convexity (Theorem 4.8(1)) we can still use John's ellipsoid lemma (Theorem 5.1) to treat a convex body geometrically as an ellipsoid. The actual proof is not so simple, and we refer the reader to the paper [21].

5.2. Alexandrov upper bound. We now discuss the proof of the Alexandrov upper bound (5.3). As an auxiliary result, the following lemma is proved in [21, Lemma 6.10], whose proof manipulates the fact that the cost $\tilde{c}$ is close to the linear cost in small scale.

Lemma 5.4 (see [21]). *Use the same assumption and notation as in Theorem 5.2. Let Π^+, Π^- be two parallel hyperplanes contained in $T_{\bar{x}_0}M \setminus Z$ and touching ∂Z from two opposite sides. If $\operatorname{diam} Z \ll 1$ and Z is sufficiently far from $\partial \exp_{\bar{x}_0}^{-1} \Omega$, then*

$$|\tilde{\phi}(\tilde{q})|^n \lesssim \frac{\min\{\operatorname{dist}(\tilde{q}, \Pi^+), \operatorname{dist}(\tilde{q}, \Pi^-)\}}{\ell_{\Pi^+}} |\partial^{\tilde{c}} \tilde{\phi}|(Z) \mathscr{L}^n(Z),$$

where ℓ_{Π^+} denotes the maximal length among all the segments obtained by intersecting Z with a line orthogonal to Π^+.

This lemma is enough to show the injectivity and continuity of T [21, Section 7 & 8]. But, for this to be applicable to the Hölder continuity of T [21, Section 9] (this section uses the method of Gutierrez and Huang [32] and Forzani and Maldonado [29]), it is important to know that for any Z as in the Lemma 5.4, one can choose parallel hyperplanes Π^+, Π^- in such way that the ratio

$$\min\{\operatorname{dist}(\tilde{q}, \Pi^+), \operatorname{dist}(\tilde{q}, \Pi^-)\}/\ell_{\Pi^+} \to 0$$

gets close to zero as q is close to the boundary ∂Z, in a 'uniform' way independent of the particular shape of Z. In other words, we need an estimate which quantifies the dimensional dependence of the claim that corresponding to any (interior) point near the boundary of a convex set, is a supporting hyperplane much closer than the thickness of the set in the orthogonal direction. Such estimate is obtained in the following new result in convex geometry, whose proof is elementary but quite nontrivial:

Theorem 5.5 (Convex bodies and supporting hyperplanes [22]). *Let $\widetilde{Q} \subset \mathbb{R}^n$ be a convex body (with nonempty interior) such that (5.1) holds for some ellipsoid E centered at the origin. Fix $0 \le s \le 1/(2n)$. To each $y \in (1-s)\partial\widetilde{Q}$ corresponds at least one line ℓ through the origin and hyperplane Π supporting $\widetilde{Q}$ such that: Π is orthogonal to ℓ and*

$$\operatorname{dist}(y, \Pi) \le c(n) s^{1/2^{n-1}} \operatorname{diam}(\ell \cap \widetilde{Q}). \tag{5.4}$$

Here, $c(n)$ is a constant depending only on n.

We refer the reader to the paper [22] for more discussions about this estimate and its proof.

Lemma 5.4 and Theorem 5.5 implies (5.3).

6. How to prove injectivity of optimal transport maps under MTW

To illustrate how the previous results are used, we explain as an example, how to show injectivity of the optimal map T under the MTW condition and conditions on the source and the target domains. (A similar method can be used to show the continuity of T.) Here, we use the same assumptions as given in Theorem 4.3.

Definition 6.1 (Contact set). For each $\bar{x} \in \overline{\Omega} = \operatorname{int}\overline{\Omega}$, the contact set $S(\bar{x})$ for $\bar{x}$ is the set

$$S(\bar{x}) = \{x \in \Omega \mid \partial^c \phi(x) = \bar{x}\}.$$

The injectivity of T is equivalent to that $S(\bar{x})$ is singleton for all $\bar{x} \in \overline{\Omega}$.

Let us briefly explain the idea how to show $S(\bar{x})$ is singleton under some technical conditions. For simplicity of exposition, let us consider the case $c(x, \bar{x}) = -x \cdot \bar{x}$, which is due to Cafferelli: [3,5,7]. The more general case as in Theorem 4.3 is a bit more complicated: see [21]. We emphasize here that the reason why we can carry out Caffarelli's idea is because we now have

- MTW, in particular, appeared convexity (Theorem 4.8),
- Interior-not-to-boundary (Theorem 4.6),
- Alexandrov-type estimates (Theorem 5.2).

Now, let us explain Caffarelli's localization argument in [3]: the expository article [7] is very useful. We include it here to demonstrate how the previous results in Sections 4 and 5 are used. In the case $c(x, \bar{x}) = -x \cdot \bar{x}$, the c-potential function ϕ is convex on $\mathbb{R}^n$, the c-convex domains Ω, $\overline{\Omega}$ are convex in $\mathbb{R}^n$, and the strong c-convexity of $\overline{\Omega}$ implies strong convexity. We assume the Monge–Ampère equation $|\partial \phi| \sim 1$. (Of course, for more general cost functions, this is replaced by $|\partial^c \phi| \sim 1$.)

Now, let us show the injectivity of T, i.e., $\#S(\bar{x}_0) = 1$ for each $\bar{x} \in \overline{\Omega}$. Suppose $\#S(\bar{x}_0) > 1$ for some $\bar{x}_0 \in \overline{\Omega}$. The goal is to contradict this.

Step 1. By Theorem 4.6 (Interior-not-to-boundary), we see

$$S(\bar{x}_0) \subset \operatorname{int}\Omega, \qquad \text{since } \partial^c \phi(\partial\Omega) \cap \operatorname{int}\overline{\Omega} = \varnothing.$$

Remark 6.2. In fact, for $c(x, \bar{x}) = -x \cdot \bar{x}$, this interior-not-to-boundary result is not necessary, since in this case one can extend the convex function ϕ to the whole $\mathbb{R}^n$. Here, the case to exclude is when the contact set contains an infinite line. If this occurs, then as pointed out in [5] the convex function has zero Monge–Ampère measure, i.e., $|\partial \phi(B)| = 0$ for any Borel set B, thus this case is excluded by our assumption $|\partial \phi| \sim 1$.

Step 2. Now, by appeared convexity (Theorem 4.8), if we let $S = \exp_{\bar{x}_0}^{-1}(S(\bar{x}_0)) \subset T_{\bar{x}_0}\overline{\Omega}$, then

$$S \text{ is convex and bounded.}$$

Of course, in the current special case, this convexity immediately follows from the convexity of ϕ.

Step 3. We now find an exposed point, say x_e of S. Exposed point is by definition, such a point where a hyperplane touches the convex set only at that point.

Step 4. For a family $\bar{x}_\theta \in \overline{\Omega}$, $0 \leq \theta \leq 1$, (thus $\bar{x}_\theta = \bar{x}_0$ for $\theta = 0$) and a point x_0, define.

$$m_\theta(\cdot) := -c(\cdot, \bar{x}_\theta) + c(x_0, \bar{x}_\theta)$$

Let $\phi_\theta := \phi - m_\theta$ and let $Z_\theta := \{z \mid \phi_\theta < 0\}$. Since x_e is an exposed point of $S = \{z \mid \partial^c \phi(z) = \bar{x}_0\}$, we can choose the point $x_0 \in S$, nearby x_e, and the family $\bar{x}_\theta \in \overline{\Omega}$, so that

$$Z_\theta \to S \text{ and } \operatorname{dist}(x_e, \partial Z_\theta) \to 0, \text{ as } \theta \to 0$$

Step 5. One can also show that for $\theta \ll 1$,

$$\inf_{Z_\theta} |\phi_\theta| \sim \phi_\theta(x_e).$$

Step 6. From the assumption $|\partial\phi| \sim 1$, we can apply the Alexandrov estimates (see Theorem 5.2).

$$|Z_\theta|^2 \lesssim (\inf_{Z_\theta} |\phi_\theta|)^n$$

$$|\phi_\theta(x_e)|^n \lesssim \eta(\theta)|Z_\theta|^2 \qquad \text{(some function } \eta(\theta) \text{ such that } \lim_{\theta\to 0+} \eta(\theta) = 0\text{)}$$

Step 7. Now apply $\theta \to 0$ in Step 6 and Step 5, and we get a contradiction $1 \lesssim 0$. This shows that $\#S(\bar{x}_0) = 1$ for each $\bar{x} \in \overline{\Omega}$, thus, the injectivity of T.

7. Regularity of optimal maps on global domains

We now give a few remarks on regularity of optimal transportation on global domains. Here, by a global domain, we mean a closed manifold M, $\Omega = \overline{\Omega} = M$, with $\log\rho, \log\bar{\rho} \in L^\infty(M)$. Loeper gave the first such regularity result ($T \in C^\alpha/C^\infty$ for $\log\rho, \log\bar{\rho} \in L^\infty(M)/C^\infty(M)$) by showing it on the round sphere S^n [50]. It was then followed by work of many researchers including the author.

First, as a necessary condition for regularity of optimal maps, Loeper's maximum principle LMP needs to be verified. It was first shown by Loeper [49] on domains in $\mathbb{R}^n$ (for cost functions satisfying the MTW condition) using the regularity results of Trudinger and Wang [57]. To treat more global manifold domains (e.g. products of round spheres), an elementary method for deriving LMP from MTW and appropriate geometric conditions, was introduced by McCann and the author [36] (see the work of Trudinger and Wang [58] for other approach, obtained independently from [36]), which later was strengthened by Loeper, Figalli, Rifford and Villani [24, 27, 51, 62]. Up to now, the examples of domains with LMP includes Riemannian distance squared costs on the products $S^{n_1}(r_1)\times\cdots\times S^{n_j}(r_j)\times \mathbf{CP}^{l_1}\times\cdots\times\mathbf{CP}^{l_k}\times\mathbb{R}^m$ of Euclidean spaces, round spheres, complex projective spaces, and their appropriate quotient spaces (this product example trivially includes flat tori) [37], as well as perturbations of the round sphere and its discrete quotients [27, 51]. See, e.g., [26] for more detailed list.

As we mentioned before, for manifold domains, a problem arises due to nonsmoothness of the cost function dist^2 along the cut-locus. So, a key step is to show that the optimal map T stays away from the cut locus, so that one can assume that the cost function is smooth. Namely,

Question 7.1 (Stay away from cut-locus). Fix a Riemannian manifold M. Suppose $\log\rho, \log\bar{\rho} \in L^\infty(M)$. Is

$$T(x) \cap \text{cut}(x) = \varnothing \quad \text{for each } x \in M \text{ and its cut-locus cut}(x)? \tag{7.1}$$

The property (7.1) is a necessary condition for higher regularity (e.g., C^1, C^∞) of optimal maps. Such *stay-away* result was obtained affirmatively for the case of the round sphere S^n by Delanoë and Loeper [15], its perturbation by Delanoë and Ge [13] (but with further restriction on $\rho, \bar{\rho}$ depending on the perturbation), and the product of round spheres $S^{n_1}(r_1)\times\cdots\times S^{n_k}(r_k)$ (of arbitrary dimensions and size) by Figalli, McCann and the author [20]. The latter result [20] gives the first regularity result for optimal transport ($T \in C^\alpha/C^\infty$ for $\log\rho, \log\bar{\rho} \in L^\infty(M)/C^\infty(M)$) on global domains that are not positively curved and not totally flat. This case differs significantly from the known regularity results on positivley curved domains, e.g., $\mathbf{CP}^n$ [37], RP^n and its perturbation [51], perturbations of the round sphere and their discrete quotients [13, 24, 27], where Loeper's quantified maximum principle LQMP (or a strong a priopri estimates of Ma, Trudinger and Wang [52]) can be applied. Note that on the flat tori, Cordero-Erausquiun [10] showed regularity of the optimal map T by lifting the situation to the universal covering space $\mathbb{R}^n$, where Caffarelli's regularity theory [3–7] applies.

This *stay-away-from-cut-locus* problem is not well understood. For example, we do not yet have such result (for $\rho, \bar{\rho}$ independent on the perturbation) for the perturbation of the round sphere. Note that even without such stay-away-from-singularity it is still possible to show that T is continuous on the perturbation of the sphere as in [24, 26]. However, the stay-away result will, if it holds, show Hölder continuity and higher regularity.[2]

8. Additional remarks on the literature

We close these lectures with a few remarks on the literature about some directions involving the Ma-Trudinger-Wang curvature which are not mentioned in the above discussion.

8.1. Parabolic optimal transport. The parabolic problem of optimal transportation theory is considered by Street, Warren and the author [40] under MTW_+ condition on manifold domains (e.g., $S^n, \mathbf{CP}^n$) and Kitagawa [41] under MTW condition on domains in $\mathbb{R}^n$ with appropriate geometric assumptions. They considered the parabolic Monge – Ampère-type equation

$$\frac{\partial u}{\partial t} = \ln\det\Big(D^2_{xx}u + D^2_{xx}c\big(x, T(x)\big)\Big) - \ln\rho(x) + \ln\bar{\rho}\big(T(x)\big) - \ln\det\big|D^2_{x\bar{x}}c\big(x, T(x)\big)\big|,$$

and have obtained long-time existence results and convergence (exponential convergence under MTW_+ [40]) to the solution to optimal transportation problem. This parabolic approach, in particular, gives a natural algorithm for finding optimal maps.

8.2. Multivalued optimal maps. In general, optimal transportation between two measures gives not a single-valued map but a multivalued map. The analysis of such multivalued maps is not well understood, though there are results by Gangbo and McCann [30] and McCann and Sosio [55] who considered the multivalued (bivalent) optimal maps for the cost given by the Euclidian distance squared restricted to the round sphere $S^{n-1} \subset \mathbb{R}^n$. McCann and Sosio used techniques involving Loeper's quantitative maximum principle LQMP (see Section 3.3) to study Hölder continuity of such bivalent maps. Note that however, even in this case, if the source and target measure $\rho, \bar{\rho}$ are sufficiently close to constant densities, Kitagawa and Warren [42] showed that the optimal transportation is given by a single-valued map and smooth.

[2]Notice that Delanoë and Ge [13] showed smoothness of optimal maps on perturbations of the round sphere, however, their perturbation of the domain is restricted by the source and target measures $\rho, \bar{\rho}$, or in other words, ρ and $\bar{\rho}$ has to be chosen appropriately depending on the perturbation. Higher regularity of optimal maps on a fixed small perturbation of the sphere, but with arbitrary smooth $\rho, \bar{\rho}$, is still an open problem.

8.3. Regularity/partial regularity without MTW or convexity assumptions on the domain. As we have discussed in these lecture notes, we now have counterexamples to continuity of optimal maps when either MTW [49] (see Theorem 2.7) or appropriate convexity assumptions [5, 52] are not satisfied. Notice that the known necessary (sometimes sufficient) conditions do not involve other key players in the transportation problem, namely the source and target distributions ρ and $\bar{\rho}$. In particular, there still is a possibility to have regularity of the optimal map T without MTW condition by imposing further restrictions on $\rho, \bar{\rho}$. For example, it is obvious that if $\rho = \bar{\rho}$ or if $\bar{\rho}$ is a Dirac-delta measure, then, T is the identity map or a trivial map, respectively, thus C^∞ on the support of ρ. In this spirit, Warren [65] obtained regularity of optimal maps between *narrow enough* Gaussian measures, regardless of MTW condition. It is an important wide open question to find a necessary and sufficient condition for regularity of optimal transport, which contains all the relevant data ρ,$\bar{\rho}$, the cost function c, and the geometry of source and target domains. A guess is that one may try to find some curvature condition for the pseudometric in [39].

A related outstanding open problem is to get partial regularity of optimal maps without MTW or appropriate convexity assumptions on the domains. It is well-known that the singular set (the set of discontinuity) of the optimal map (for cost $=$ dist$^2/2$) has Hausdorff dimension less than or equal to $n-1$ in n-dimensional domain. So, the point is to get a *sharper* description of the singular set. No such result has been known regarding the violation of MTW condition, however, for the Euclidean distance squared (cross $\equiv 0$, thus MTW), a partial regularity is known by Figalli and the author [18] when the convexity assumption on the domains is violated, extending the two-dimensional results of Yu [66] and Figalli [16]. In [18], the singular set is shown to be contained in a measure zero *closed* set, thus for example, excluding the case it to be dense. However, this result lacks of the more precise description on the structure of the singular set as given in two dimensions in [16, 66].

Acknolwegement. The author thanks the SMS school and the organizers, Galia Dafni and Alina Stancu, for their hospitality, Robert McCann for introducing him to the world of optimal transportation, and also Tongseok Lim (the author's PhD student) and the referee for useful comments about the manuscript. The present lecture notes and the author's work presented here, were not possible without the sincere support, encouragement and love of his wife, Dong Soon Shim, and the two daughters, Joyce and Ashley.

References

1. A. Alexandroff, *Smoothness of the convex surface of bounded Gaussian curvature*, C. R. (Doklady) Acad. Sci. URSS (N.S.) **36** (1942), 195–199.
2. Y. Brenier, *Polar factorization and monotone rearrangement of vector-valued functions*, Comm. Pure Appl. Math. **44** (1991), no. 4, 375–417.
3. L. A. Caffarelli, *A localization property of viscosity solutions to the Monge – Ampère equation and their strict convexity*, Ann. of Math. (2) **131** (1990), no. 1, 129–134.
4. ______, *Some regularity properties of solutions of Monge Ampère equation*, Comm. Pure Appl. Math. **44** (1991), no. 8-9, 965–969.
5. ______, *The regularity of mappings with a convex potential*, J. Amer. Math. Soc. **5** (1992), no. 1, 99–104.
6. ______, *Boundary regularity of maps with convex potentials.* II, Ann. of Math. (2) **144** (1996), no. 3, 453–496.

7. ______, *A priori estimates and the geometry of the Monge Ampère equation*, Nonlinear Partial Differential Equations in Differential Geometry (Park City, UT, 1992) (R. Hardt and M. Wolf, eds.), IAS/Park City Math. Ser., vol. 2, Amer. Math. Soc., Providence, RI, 1996, pp. 5–63.
8. L. A. Caffarelli, C. E. Gutiérrez, and Q. Huang, *On the regularity of reflector antennas*, Ann. of Math. (2) **167** (2008), no. 1, 299–323.
9. J. Cheeger and D. G. Ebin, *Comparison theorems in Riemannian geometry*, Vol. 9, North-Holland, Amsterdam, 1975. North-Holland Math. Library.
10. D. Cordero-Erausquin, *Sur le transport de mesures périodiques*, C. R. Acad. Sci. Paris Sér. I Math. **329** (1999), no. 3, 199–202.
11. D. Cordero-Erausquin, R. J. McCann, and M. Schmuckenschläger, *A Riemannian interpolation inequality à la Borell, Brascamp and Lieb*, Invent. Math. **146** (2001), no. 2, 219–257.
12. P. Delanoë, *Classical solvability in dimension two of the second boundary-value problem associated with the Monge – Ampère operator*, Ann. Inst. H. Poincaré Anal. Non Linéaire **8** (1991), no. 5, 443–457.
13. P. Delanoë and Y. Ge, *Regularity of optimal transport on compact, locally nearly spherical, manifolds*, J. Reine Angew. Math. **646** (2010), 65–115.
14. ______, *Locally nearly spherical surfaces are almost-positively curved*, Methods Appl. Anal., to appear.
15. P. Delanoë and G. Loeper, *Gradient estimates for potentials of invertible gradient-mappings on the sphere*, Calc. Var. Partial Differential Equations **26** (2006), no. 3, 297–311.
16. A. Figalli, *Regularity properties of optimal maps between nonconvex domains in the plane*, Comm. Partial Differential Equations **35** (2010), no. 3, 465–479.
17. ______, *Regularity of optimal transport maps [after Ma – Trudinger – Wang and Loeper]*, Astérisque **332** (2010), Exp. No. 1009, ix, 341–368.
18. A. Figalli and Y.-H. Kim, *Partial regularity of Brenier solutions of the Monge – Ampère equation*, Discrete Contin. Dyn. Syst. **28** (2010), no. 2, 559–565.
19. A. Figalli, Y.-H. Kim, and R. J. McCann, *When is multidimensional screening a convex program?*, J. Econom. Theory **146** (2011), 454-478.
20. ______, *Regularity of optimal transport maps on multiple products of spheres*, J. Eur. Math. Soc. (JEMS), to appear, available at `arXiv:1006.1957`.
21. ______, *Hölder continuity and injectivity of optimal maps*, available at `arXiv:1107.1014`.
22. ______, *On supporting hyperplanes to convex bodies*, available at `arXiv:1107.1016`.
23. A. Figalli and G. Loeper, *C^1 regularity of solutions of the Monge – Ampère equation for optimal transport in dimension two*, Calc. Var. Partial Differential Equations **35** (2009), no. 4, 537–550.
24. A. Figalli and L. Rifford, *Continuity of optimal transport maps and convexity of injectivity domains on small deformations of* $\mathbb{S}^2$, Comm. Pure Appl. Math. **62** (2009), no. 12, 1670–1706.
25. A. Figalli, L. Rifford, and C. Villani, *On the Ma – Trudinger – Wang curvature on surfaces*, Calc. Var. Partial Differential Equations **39** (2010), no. 3-4, 307–332.
26. ______, *Necessary and sufficient conditions for continuity of optimal transport maps on Riemannian manifolds*, Tohoku Math. J. (2) **63** (2011), no. 4, 855–876.
27. ______, *Nearly round spheres look convex*, Amer. J. Math. **134** (2012), no. 1, 109–139.
28. A. Figalli and C. Villani, *Optimal transport and curvature*, Nonlinear PDE's and Applications (Cetraro, 2008) (L. Ambrosio and Savaré. Giuseppe, eds.), Lecture Notes in Math., vol. 2028, Springer, Heidelberg, 2011, pp. 171–217.
29. L. Forzani and D. Maldonado, *Properties of the solutions to the Monge – Ampère equation*, Nonlinear Anal. **57** (2004), no. 5-6, 815–829.
30. W. Gangbo and R. J. McCann, *Shape recognition via Wasserstein distance*, Quart. Appl. Math. **58** (2000), no. 4, 705–737.
31. C. E. Gutiérrez, *The Monge – Ampère equation*, Progr. Nonlinear Differential Equations Appl., vol. 44, Birkhäuser, Boston, MA, 2001.
32. C. E. Gutiérrez and Q. Huang, *Geometric properties of the sections of solutions to the Monge – Ampère equation*, Trans. Amer. Math. Soc. **352** (2000), no. 9, 4381–4396.
33. F. John, *Extremum problems with inequalities as subsidiary conditions*, Studies and Essays Presented to R. Courant on his 60th Birthday, January 8, 1948, Interscience Publ., New York, 1948, pp. 187–204.

34. A. Karakhanyan and X.-J. Wang, *On the reflector shape design*, J. Differential Geom. **84** (2010), no. 3, 561–610.
35. Y.-H. Kim, *Counterexamples to continuity of optimal transport maps on positively curved Riemannian manifolds*, Int. Math. Res. Not. IMRN (2008), Art. ID rnn120.
36. Y.-H. Kim and R. J. McCann, *Continuity, curvature, and the general covariance of optimal transportation*, J. Eur. Math. Soc. (JEMS) **12** (2010), no. 4, 1009–1040.
37. ———, *Towards the smoothness of optimal maps on Riemannian submersions and Riemannian products (of round spheres in particular)*, J. Reine Angew. Math., to appear, available at `arXiv:math/0806.0351`.
38. ———, Appendices to original version of *Continuity, curvature, and the general covariance of optimal transportation*, available at `arXiv:math/0712.3077`.
39. Y.-H. Kim, R. J. McCann, and M. Warren, *Pseudo-Riemannian geometry calibrates optimal transportation*, Math. Res. Lett. **17** (2010), no. 6, 1183–1197.
40. Y.-H. Kim, J. Streets, and M. Warren, *Parabolic optimal transport equations on manifolds*, Int. Math. Res. Not. IMRN, to appear, available at `arXiv:1008.3824`.
41. J. Kitagawa, *A parabolic flow toward solutions of the optimal transportation problem on domains with boundary*, J. Reine Angew. Math., to appear.
42. J. Kitagawa and M. Warren, *Regularity of optimal transport with Euclidean distance squared cost on the embedded sphere*, available at `arXiv:1101.5146`.
43. P. W. Y. Lee, *New computable necessary conditions for the regularity theory of optimal transportation*, SIAM J. Math. Anal. **42** (2010), no. 6, 3054–3075.
44. P. W. Y. Lee and J. Li, *New examples satisfying Ma – Trudinger – Wang conditions*, SIAM J. Math. Anal. **44** (2012), no. 1, 61–73.
45. P. W. Y. Lee and R. J. McCann, *The Ma – Trudinger – Wang curvature for natural mechanical actions*, Calc. Var. Partial Differential Equations **41** (2011), no. 1-2, 285–299.
46. J. Liu, *Hölder regularity of optimal mappings in optimal transportation*, Calc. Var. Partial Differential Equations **34** (2009), no. 4, 435–451.
47. J. Liu and N. S. Trudinger, *On Pogorelov estimates for Monge – Ampère type equations*, Discrete Contin. Dyn. Syst. **28** (2010), no. 3, 1121–1135.
48. J. Liu, N. S. Trudinger, and X.-J. Wang, *Interior $C^{2,\alpha}$ regularity for potential functions in optimal transportation*, Comm. Partial Differential Equations **35** (2010), no. 1, 165–184.
49. G. Loeper, *On the regularity of solutions of optimal transportation problems*, Acta Math. **202** (2009), no. 2, 241–283.
50. ———, *Regularity of optimal maps on the sphere: the quadratic cost and the reflector antenna*, Arch. Ration. Mech. Anal. **199** (2011), no. 1, 269–289.
51. G. Loeper and C. Villani, *Regularity of optimal transport in curved geometry: the nonfocal case*, Duke Math. J. **151** (2010), no. 3, 431–485.
52. X.-N. Ma, N. S. Trudinger, and X.-J. Wang, *Regularity of potential functions of the optimal transportation problem*, Arch. Ration. Mech. Anal. **177** (2005), no. 2, 151–183.
53. R. J. McCann, *Polar factorization of maps on Riemannian manifolds*, Geom. Funct. Anal. **11** (2001), no. 3, 589–608.
54. R. J. McCann and N. Guillen, *Five lectures on optimal transportation: geometry, regularity and applications*. in this book.
55. R. J. McCann and M. Sosio, *Hölder continuity for optimal multivalued mappings*, SIAM J. Math. Anal. **43** (2011), no. 4, 1855–1871.
56. T. Sei, *A Jacobian inequality for gradient maps on the sphere and its application to directional statistics*, Comm. Statist. Theory Methods, to appear, available at `arXiv:0906.0874`.
57. N. S. Trudinger and X.-J. Wang, *On the second boundary value problem for Monge – Ampère type equations and optimal transportation*, Ann. Sc. Norm. Super. Pisa Cl. Sci. (5) **8** (2009), no. 1, 143–174.
58. ———, *On strict convexity and continuous differentiability of potential functions in optimal transportation*, Arch. Ration. Mech. Anal. **192** (2009), no. 3, 403–418.
59. J. Urbas, *On the second boundary value problem for equations of Monge – Ampère type*, J. Reine Angew. Math. **487** (1997), 115–124.
60. C. Villani, *Topics in optimal transportation*, Grad. Stud. Math., vol. 58, Amer. Math. Soc., Providence, RI, 2003.
61. ———, *Stability of a 4th-order curvature condition arising in optimal transport theory*, J. Funct. Anal. **255** (2008), no. 9, 2683–2708.

62. ______, *Optimal transport: Old and new*, Grundlehren Math. Wiss., vol. 338, Springer, Berlin, 2009.
63. X.-J. Wang, *On the design of a reflector antenna*, Inverse Problems **12** (1996), no. 3, 351–375.
64. ______, *On the design of a reflector antenna. II*, Calc. Var. Partial Differential Equations **20** (2004), no. 3, 329–341.
65. M. Warren, *Regularity for a log-concave to log-concave mass transfer problem with near Euclidean cost*, Comm. Anal. Geom. **19** (2011), no. 1, 191–208.
66. Y. Yu, *Singular set of a convex potential in two dimensions*, Comm. Partial Differential Equations **32** (2007), no. 10-12, 1883–1894.

Department of Mathematics, University of British Columbia, Vancouver, BC V6T 1Z2, Canada
E-mail address: yhkim@math.ubc.ca

Centre de Recherches Mathématiques
CRM Proceedings and Lecture Notes
Volume **56**, 2013

Five Lectures on Optimal Transportation: Geometry, Regularity and Applications

Robert J. McCann and Nestor Guillen

Abstract. In this series of lectures we introduce the Monge–Kantorovich problem of optimally transporting one distribution of mass onto another, where optimality is measured against a cost function $c(x, y)$. Connections to geometry, inequalities, and partial differential equations will be discussed, focusing in particular on recent developments in the regularity theory for Monge–Ampère type equations. An application to microeconomics will also be described, which amounts to finding the equilibrium price distribution for a monopolist marketing a multidimensional line of products to a population of anonymous agents whose preferences are known only statistically.

Preamble

This survey is based on a series of five lectures by Robert McCann, delivered at a summer school on "New Vistas in Image Processing and Partial Differential Equations" organized 7–12 June 2010 by Irene Fonseca, Giovanni Leoni, and Dejan Slepcev of Carnegie Mellon University on behalf of the Center for Nonlinear Analysis there. The starting point for the manuscript which emerged was a detailed set of notes taken during those lectures by Nestor Guillen. Since that time the notes have been used as the basis for minicourses on a number of occasions, including the 50th anniversary edition of the Séminaire de mathématiques supérieures (2011) at Montréal's Centre de recherches mathématiques, organized by Galia Dafni and Alina Stancu of Concordia University along with [RJM].

These notes are intended to convey a flavor for the subject, without getting bogged down in too many technical details. Part of the discussion is therefore impressionistic, and some of the results are stated under the tacit requirement that the supports of the measures $\mu^{\pm}$ be compact, with the understanding that they extend to noncompactly supported measures under appropriate hypotheses [53, 54, 67, 101] concerning the behavior near infinity of the measures and the costs. The choice of topics to be covered in a series of lectures is necessarily idiosyncratic. General references for these and other topics include papers of the first author posted on the website `www.math.toronto.edu/mccann`, the User's Guide of Ambrosio and

2010 *Mathematics Subject Classification.* Primary 49N15; Secondary 35J70, 53C50, 58J05, 90B06, 91B24.

[RJM]'s research was supported in part by grant 217006-08 of the Natural Sciences and Engineering Research Council of Canada.

This is the final form of the paper.

Gigli [7], and the two books by Villani [144, 145]. Earlier surveys include the ones by Ambrosio [6], Evans [51], Urbas [142] and Rachev and Rüschendorf [121]. Many detailed references to the literature may be found there, to augment the bibliography of selected works included below.

1. An introduction to optimal transportation

1.1. Monge – Kantorovich problem: transporting ore from mines to factories. The problem to be discussed can be caricatured as follows: imagine we have a distribution of iron mines across the countryside, producing a total of 1000 tonnes of iron ore weekly, and a distribution of factories across the countryside that consume a total of 1000 tonnes of iron ore weekly. Knowing the *cost* $c(x, y)$ per ton of ore transported from a mine at x to a factory at y, the problem is to decide which mines should be supplying which factories so as to minimize the total transportation costs.

To model this problem mathematically, let the triples $(M^{\pm}, d^{\pm}, \omega^{\pm})$ denote two complete separable metric spaces $M^{\pm}$ — also called *Polish* spaces — equipped with distance functions $d^{\pm}$ and Borel reference measures $\omega^{\pm}$. These two metric spaces will represent the landscapes containing the mines and the factories. They will often be assumed to be geodesic spaces, and/or to coincide. Here a *geodesic space* $M(= M^{\pm})$ refers to a metric space in which every pair of points $x_0, x_1 \in M$ is connected by a curve $s \in [0, 1] \to x_s \in M$ satisfying

$$d(x_0, x_s) = sd(x_0, x_1) \qquad \text{and} \qquad d(x_s, x_1) = (1 - s)d(x_0, x_1) \quad \forall\, s \in [0, 1]. \tag{1.1}$$

Such a curve is called a *geodesic segment.*

E.g. (1) Euclidean space $M = \mathbf{R}^n$, $d(x, y) = |x - y|$, $\omega = \mathrm{Vol} = \mathcal{H}^n =$ Hausdorff n-dimensional measure, geodesic segments take the form $x_s = (1-s)x_0 + sx_1$.

(2) Complete connected Riemannian manifold $(M = M^{\pm}, g_{ij})$, with or without boundary: $\mathrm{d}\omega = \mathrm{dVol} = \mathrm{d}\mathcal{H}^n = (\det g_{ij})^{1/2}\, \mathrm{d}^n x$,

$$\frac{1}{2} d^2(x_0, x_1) := \inf_{\{x_s | x(0) = x_0, x(1) = x_1\}} \frac{1}{2} \int_0^1 \langle \dot{x}_s \dot{x}_s \rangle_{g(x_s)}\, \mathrm{d}s.$$

A minimizing curve $s \in [0, 1] \mapsto x_s \in M$ exists by the Hopf – Rinow theorem; it satisfies (1.1), and is called a Riemannian *geodesic.*

The distributions of mines and factories will be modeled by Borel probability measures μ^+ on M^+ and μ^- on M^-, respectively. Any Borel map $G\colon M^+ \to M^-$ defines an image or *pushed-forward* measure $\nu = G_{\#}\mu^+$ on M^- by

$$(G_{\#}\mu^+)[V] := \mu^+[G^{-1}(V)] \quad \forall\, V \subset M^-. \tag{1.2}$$

A central problem in optimal transportation is to find, among all maps $G : M^+ \to M^-$ pushing μ^+ forward to μ^-, one which minimizes the total cost

$$\mathrm{cost}(G) = \int_{M^+} c\big(x, G(x)\big)\, \mathrm{d}\mu^+(x). \tag{1.3}$$

This problem was first proposed by Monge in 1781, taking the Euclidean distance $c(x, y) = |x - y|$ as his cost function [112]. For more generic costs, some basic mathematical issues such as the existence, uniqueness, and mathematical structure of the optimizers are addressed in the second lecture below. However nonlinearity of the objective functional and a lack of compactness or convexity for its domain make Monge's formulation of the problem awkward to work with. One and a half centuries later, Kantorovich's relaxation of the problem to an (infinite-dimesional) linear program provided a revolutionary tool [75, 76].

1.2. Wasserstein distance and geometric applications. The minimal cost of transport between μ^+ and μ^- associated to $c(\cdot,\cdot)$ will be provisionally denoted by

$$W_c(\mu^+,\mu^-) = \inf_{G_\#\mu^+=\mu^-} \mathrm{cost}(G). \tag{1.4}$$

It can be thought of as quantifying the discrepancy between μ^+ and μ^-, and is more properly defined using Kantorovich's formulation (2.1), though we shall eventually show the two definitions coincide in many cases of interest. When $M = M^\pm$, the costs $c(x,y) = d^p(x,y)$ with $0 < p < 1$ occur naturally in economics and operations research, where it is often the case that there is an economy of scale for long trips [104]. In this case, the quantity $W_c(\mu^+,\mu^-)$ defines a metric on the space $\mathcal{P}(M)$ of Borel probability measures on M. For $p \geq 1$ on the other hand, it is necessary to extract a *p-th* root to obtain a metric

$$d_p(\mu^+,\mu^-) := W_c(\mu^+,u^-)^{1/p} \tag{1.5}$$

on $\mathcal{P}(M)$ which satisfies the triangle inequality.

Though the initial references dealt specifically with the case $p = 1$ [77, 151], the whole family of distances are now called *Kantorovich – Rubinstein* or *Wasserstein* metrics [49]. Apart from the interesting exception of the limiting case $d_\infty = \lim_{p\to\infty} d_p$ [31, 106], on a compact metric space M all these metrics give rise to the same topology, namely weak-$*$ convergence. For noncompact M, the d_p topologies differ from each other only in the number of moments of a sequence of measures which are required to converge. Moreover, $(\mathcal{P}(M), d_p)$ inherits geometric properties from (M,d), such as being a geodesic space. Notions such as Ricci curvature in the underlying space (M,d) can be characterized by the geodesic convexity first explored by McCann [102] of certain functionals on the larger space $(\mathcal{P}(M), d_2)$— such as Boltzmann's entropy. One direction of this equivalence was proved by Cordero-Erausquin, McCann, and Schmuckenschäger [36] and Otto and Villani [116] in projects which were initially independent (see also [35, 37]), while the converse was established by von Renesse and Sturm [148] confirming the formal arguments of [116]. This equivalence forms the basis of Lott – Villani and Sturm's definition of lower bounds for Ricci curvature in the metric measure space setting, without reference to any underlying Riemannian structure [96, 133]. McCann and Topping used a similar idea to characterize the Ricci flow [107], which led Lott [95] and Topping [136] to simpler proofs of Perelman's celebrated monotonicity results [118]. Despite the interest of these recent developments, we shall not pursue them farther in these lectures, apart from sketching a transportation-based proof of the isoperimetric theorem whose ideas underpin many such geometric connections.

1.3. Brenier's theorem and convex gradients. It turns out that Monge's cost $c(x,y) = |x-y|$ is among the hardest to deal with, due to its lack of strict convexity. For this cost, the minimizer of (1.4) is not generally unique, even on the line $M^\pm = \mathbf{R}$. Existence of solutions is tricky to establish: the first 'proof', due to Sudakov [134], relied on an unsubstantiated claim which turned out to be correct only in the plane $M^{\pm} = \mathbf{R}^2$ [15]; higher dimensional arguments were given increasing generality by Evans – Gangbo [52], and then Ambrosio [6], Caffarelli – Feldman – McCann [24], and Trudinger – Wang [138] independently. Simpler approaches were proposed by Champion – DePascale [30] and Bianchini – Cavalletti [15] more recently.

The situation for the quadratic cost $c(x,y) = |x-y|^2$ is much simpler, mirroring the relative simplicity of the Hilbert geometry of L^2 among Banach spaces L^p with $p \geq 1$. Brenier [18,19] (and others around the same time [1,39,40,120,128,130]) showed that there is a unique [1,40] solution [39] , and characterized it as a convex gradient [120,128,130].

Theorem 1.1 (A version of Brenier's theorem). *If $\mu^+ \ll dVol$ and μ^- are Borel probability measures on $M^{\pm} = \mathbf{R}^n$, then there exists a convex function $u \colon \mathbf{R}^n \to \mathbf{R} \cup \{+\infty\}$ whose gradient $G = Du \colon \mathbf{R}^n \to \mathbf{R}^n$ pushes μ^+ forward to μ^-. Apart from changes on a set of measure zero, G is the only map to arise in this way. Moreover, G uniquely minimizes Monge's problem* (1.4) *for the cost $c(x,y) = |x-y|^2$.*

Remark. In this generality the theorem was established by McCann [101], where the assumption $\mu^+ \ll$ dVol was also relaxed. A further relaxation by Gangbo – McCann [67] is shown to be sharp in Gigli [69].

1.4. Fully-nonlinear degenerate-elliptic Monge – Ampère type PDE. How do partial differential equations (more specifically, fully nonlinear degenerate elliptic PDE) enter the picture? Let's consider the constraint $G_{\#}\mu^+ = \mu^-$, assuming moreover that $\mu^{\pm} = f^{\pm} dVol^{\pm}$ on $\mathbf{R}^n$ or on Riemannian manifolds $M^{\pm}$. Then if $\phi \in C(M^-)$ is a test function, it follows that

$$\int_{M^+} \phi(G(x)) f^+(x)\, \mathrm{dVol}^+(x) = \int_{M^-} \phi(y) f^-(y)\, \mathrm{dVol}^-(y).$$

In the case where G is a diffeomorphism, we could combine the Jacobian factor $\mathrm{d}^n y = |\det DG(x)|\, \mathrm{d}^n x$ from the change of variables $y = G(x)$ with arbitrariness of $\phi \circ G$ to conclude $f^+(x) = |\det DG(x)| f^-\big(G(x)\big)$ for all x. We will actually see that this nonlinear equation holds f^+-a.e. as a consequence of Theorem 3.2.

In the case of Brenier's map $G(x) = Du(x)$, convexity of u implies nonnegativity of the Jacobian $DG(x) = D^2u(x) \geq 0$. It also guarantees almost everywhere differentiability of G by Alexandrov's theorem (or by Lebesgue's theorem in one dimension); see Theorem 3.2 for the sketch of a proof. Thus u solves the Monge – Ampère equation [19]

$$f^+(x) = \det\big(D^2u(x)\big) f^-\big(Du(x)\big) \tag{1.6}$$

a.e. [102] subject to the condition $Du(x) \in M^-$ for $x \in M^+$. This is known as the 2*nd boundary value problem* in the partial differential equations literature. We shall see that linearization of this equation around a *convex* solution leads to a (degenerate) elliptic operator (3.8), whose ellipticity becomes uniform if the solution u is smooth and *strongly* convex, meaning positivite-definiteness of its Hessian is strict: $D^2u(x) > 0$.

1.5. Applications. The Monge – Kantorovich theory has found a wide variety of applications in pure and applied mathematics. On the pure side, these include connections to inequalities [2, 32, 38, 60, 98, 100, 102, 137], geometry (including sectional [80, 92], Ricci [36, 95, 96, 107, 133] and mean [83] curvature), nonlinear partial differential equations [18, 21, 23, 97, 141], and dynamical systems (weak KAM theory [13]; nonlinear diffusions [115]; gradient flows [8]). On the applied side these include applications to vision (image registration and morphing [72]), economics (equilibration of supply with demand [33, 50], structure of cities [27], maximization of profits [26, 55, 123] or social welfare [55]), physics [48, 59, 103, 135], engineering (optimal shape/material design [16, 17], reflector antenna design [71, 149, 150], aerodynamic resistance [119]), atmosphere and ocean dynamics (the semigeostrophic theory [41–44]), biology (irrigation [14], leaf growth [152]), and statistics [121, 129]. See [144, 145] for further directions, references, and discussion.

1.6. Euclidean isoperimetric inequality. It was observed (independently by McCann [100, 102] and Trudinger [137]) that a solution to the second boundary value problem for the Monge – Ampère equation (1.6) yields a simple proof of the isoperimetric inequality (with its sharp constant): for $M^+ \subset \mathbf{R}^n$

$$\mathrm{Vol}(M^+) = \mathrm{Vol}(B_1) \implies \mathcal{H}^{n-1}(\partial M^+) \geq \mathcal{H}^{n-1}(\partial B_1). \tag{1.7}$$

The following streamlined argument was perfected later; it combines optimal maps with an earlier approach from Gromov's appendix to [109].

PROOF. Take $f^+ = \chi_{M^+}$ and $f^- = \chi_{B_1}$ to be uniformly distributed. Brenier's theorem then gives a volume-preserving map $G = Du$ between M^+ and B_1:

$$1 = \det{}^{1/n}\big(D^2u(x)\big).$$

The expression on the right is the geometric mean of the eigenvalues of $D^2u(x)$, which are nonnegative by convexity of u, so the arithmetic-geometric mean inequality yields

$$1 \leq (\text{arithmetic mean of eigenvalues}) = \frac{1}{n}\Delta u \tag{1.8}$$

almost everywhere in M^+. (The right-hand side is the absolutely continuous part of the distributional Laplacian; convexity of u allows it to be replaced by the full distributional Laplacian of u without spoiling the inequality.) Integrating inequality (1.8) on M^+ yields

$$\mathrm{Vol}(B_1) = \mathrm{Vol}(M^+) \leq \frac{1}{n}\int_{M^+} \Delta u \, \mathrm{d}^n x = \frac{1}{n}\int_{\partial M^+} Du(x)\cdot \hat{n}_{M^+}(x)\, \mathrm{d}\mathcal{H}^{n-1}(x). \tag{1.9}$$

Now, since $G = Du \in B_1$ whenever $x \in M^+$, we have $|Du| \leq 1$, thus

$$\mathrm{Vol}(B_1) \leq \frac{1}{n}\int_{\partial M^+} 1 \, \mathrm{d}\mathcal{H}^{n-1} = \frac{1}{n}\mathcal{H}^{n-1}(\partial M^+). \tag{1.10}$$

In the case special $M^+ = B_1$, Brenier's map coincides with the identity map so equalities hold throughout (1.8) – (1.10), yielding the desired conclusion (1.7)! □

As the preceding proof shows, one of the important uses of optimal transportation in analysis and geometry is to encode nonlocal 'shape' information into a map which can be localized, reducing global geometric inequalities to algebraic inequalities under an integral. For subsequent developments in this direction, see works of Alesker, Ambrosio, Barthe, Cordero-Erausquin, Carrillo, Castillon, Dar, Figalli, Gigli, Lott, Maggi, McCann, Milman, Nazaret, Otto, Pratelli, von Renesse,

Schmuckenschläger, Sturm, Topping and Villani in [5, 8, 11, 28, 29, 36–38, 60, 95, 96, 103, 107, 116, 133, 148].

1.7. Kantorovich's reformulation of Monge's problem. Now let us turn to the proof of Brenier's theorem and the ideas it involves. A significant breakthrough was made by Kantorovich [75, 76], who *relaxed* our optimization problem (the Monge problem), by dropping the requirement that all the ore from a given mine goes to a single factory. In other words:

Replace $G\colon M^+ \to M^-$ by a measure $0 \leq \gamma$ on $M^+ \times M^-$ whose marginals are μ^+ and μ^-, respectively, and among such measures choose γ to minimize the functional

$$\mathrm{cost}(\gamma) = \int_{M^+\times M^-} c(x,y)\,\mathrm{d}\gamma(x,y).$$

Such a joint measure γ is sometimes known as a "transport plan" (in analogy with "transport map"). This is better than Monge's original formulation for at least two reasons:

(1) The functional to be minimized now depends linearly on γ.

(2) The set $\Gamma(\mu^+, \mu^-)$ of admissible competitors γ is a nonempty convex subset of a suitable Banach space: namely, the dual space to continuous functions $(C(M^+ \times M^-), \|\cdot\|_\infty)$ (which decay to zero at infinity in case the compactness of $M^\pm$ is merely local).

In this context, well-known results in functional analysis guarantee existence of a minimizer γ for general $\mu^\pm$ and merely bounded and continuous c. Our primary task will be to understand when the solution will be unique, and to characterize it. At least one minimizer will be an extreme point of the convex set $\Gamma(\mu^+, \mu^-)$, but its uniqueness remains an issue. Necessary and sufficient conditions will come from the duality theory of (infinite-dimensional) linear programming [9, 78].

2. Existence, uniqueness, and characterization of optimal maps

Let's get back to the Kantorovich problem:

$$W_c(\mu^+,\mu^-) := \min_{\gamma\in\Gamma(\mu^+,\mu^-)} \int_{M^+\times M^-} c(x,y)\,\mathrm{d}\gamma(x,y) = \min_{\gamma\in\Gamma(\mu^+,\mu^-)} \mathrm{cost}(\gamma) \tag{2.1}$$

The basic geometric object of interest to us will be the support $\operatorname{spt}\gamma := S$ of a competitor γ, namely the smallest closed subset $S \subset M^+ \times M^-$ carrying the full mass of γ.

What are some of the competing candidates for the minimizer?

E.g. (1) Product measure: $\mu^+ \otimes \mu^- \in \Gamma(\mu^+, \mu^-)$, for which $\operatorname{spt}\gamma = \operatorname{spt}\mu^+ \times \operatorname{spt}\mu^-$.

(2) Monge measure: if $G\colon M^+ \to M^-$ with $G_\#\mu^+ = \mu^-$ then $\mathrm{id} \times G\colon M^+ \to M^+ \times M^-$ and $\gamma = (\mathrm{id} \times G)_\#\mu^+ \in \Gamma(\mu^+, \mu^-)$ has $\mathrm{cost}(\gamma) = \mathrm{cost}(G)$.

The second example shows in what sense Kantorovich's formulation is a relaxation of Monge's problem, and why (1.4) must be at least as big as (2.1). In this example $\operatorname{spt}\gamma$ will be the (closure of the) graph of $G\colon M^+ \to M^-$, which suggests how Monge's map G might in principle be reconstructed from a minimizing Kantorovich measure γ. Before attempting this, let us recall a notion which characterizes optimality in the Kantorovich problem.

Definition 2.1 (c-cyclically monotone sets). $S \subset M^+ \times M^-$ is *c-cyclically monotone* if and only if all $k \in \mathbf{N}$ and $(x_1, y_1), \ldots, (x_k, y_k) \in S$ satisfy

$$\sum_{i=1}^{k} c(x_i, y_i) \leq \sum_{i=1}^{k} c(x_i, y_{\sigma(i)}) \tag{2.2}$$

for each permutation σ of k letters.

The following result was deduced by Smith and Knott [131] from a theorem of Rüschendorf [125]. A more direct proof was given by Gangbo and McCann [67]; its converse is true as well.

Theorem 2.2 (Smith and Knott '92). *If* $c \in C(M^+ \times M^-)$*, then optimality of* $\gamma \in \Gamma(\mu^+, \mu^-)$ *implies* $\operatorname{spt} \gamma$ *is a c-cyclically monotone set.*

The idea of the proof in [67] is that if $\operatorname{spt} \gamma$ is not cyclically monotone, then setting $o_i = (x_i, y_i)$ and $z_i = (x_i, y_{\sigma(i)})$ we could with some care define a perturbation

$$\gamma_\epsilon = \gamma + \epsilon(\text{near the } z\text{'s}) - \epsilon(\text{near the } o\text{'s})$$

in $\Gamma(\mu^+, \mu^-)$ of γ for which $\operatorname{cost}(\gamma_\epsilon) < \operatorname{cost}(\gamma_0)$, thus precluding the optimality of γ.

E.g. If $c(x, y) = -x \cdot y$ or equivalently if $c(x, y) = \frac{1}{2}|x - y|^2$ then (2.2) becomes $\sum_{i=1}^{k} \langle y_i, x_i - x_{i-1} \rangle \geq 0$ with the convention $x_0 := x_k$. This is simply called *cyclical monotonicity*, and can be viewed as a discretization of

$$\oint \bar{y}(x) \cdot \mathrm{d}x \geq 0,$$

a necessary and sufficient condition for the vector field $\bar{y}(x)$ to be conservative, meaning $y = Du(x)$. This heuristic underlies a theorem of Rockafellar [124]:

Theorem 2.3 (Rockafellar 66). *The set* $S \subset \mathbf{R}^n \times \mathbf{R}^n$ *is cyclically monotone if and only if there exists a convex function* $u \colon \mathbf{R}^n \to \mathbf{R} \cup \{\infty\}$ *such that* $S \subset \partial u$ *where*

$$\partial u := \{(x, y) \in \mathbf{R}^n \times \mathbf{R}^n \mid u(z) \geq u(x) + \langle z - x, y \rangle + o(|z - x|)\ \forall z \in \mathbf{R}^n\}. \tag{2.3}$$

The *subdifferential* ∂u defined by (2.3) consists of the set of (point, slope) pairs for which y is the slope of a hyperplane supporting the graph of u at $\big(x, u(x)\big)$.

Remark 2.4 (Special case (*monotonicity*)). Note that when $c(x, y) = -x \cdot y$ and $k = 2$, (2.2) implies for all $(x_1, y_1), (x_2, y_2) \in S$ that

$$\langle \Delta x, \Delta y \rangle := \langle x_2 - x_1, y_2 - y_1 \rangle \geq 0. \tag{2.4}$$

This condition implies that y_2 is constrained to lie in a halfspace with y_1 on its boundary and Δx as its inward normal. Should $y = Du(x)$ already be known to be conservative, the monotonicity inequalities (2.4) alone become equivalent to convexity of u.

2.1. Linear programming duality. An even more useful perspective on these linear programming problems is given by the the duality theorem discovered by Kantorovich [75] and Koopmans [84] — for which they later shared the Nobel

Memorial Prize in economics. It states that our minimization problem is equivalent to a maximization problem

$$\begin{aligned}(2.5)\qquad \min_{\gamma\in\Gamma(\mu^+,\mu^-)} \int_{M^+\times M^-} c\,\mathrm{d}\gamma \\ = \sup_{(-u,-v)\in \mathrm{Lip}_c} -\int_{M^+} u(x)\,\mathrm{d}\mu^+(x) - \int_{M^-} v(y)\,\mathrm{d}\mu^-(y).\end{aligned}$$

Here

$$\mathrm{Lip}_c = \{(u^+,u^-) \text{ with } u^\pm \in L^1(M^\pm, d\mu^\pm) \mid \\ c(x,y) \geq u^+(x)+u^-(y)\ \forall (x,y)\in M^+\times M^-\}.$$

One of the two inequalities ($\geq$) in (2.5) follows at once from the definition of $-(u,v)\in \mathrm{Lip}_c$ by integrating

$$c(x,y)\geq -u(x)-v(y)$$

against $\gamma\in\Gamma(\mu^+,\mu^-)$. The magic of duality is that *equality* holds in (2.5).

2.2. Game theory. Some intuition for why this magic works can be gleaned from the the theory of (two-player, zero-sum) games. In that context, Player 1 chooses strategy $x\in X$, Player 2 chooses strategy $y\in Y$, and the outcome is that Player 1 pays $P(x,y)$ to Player 2. The payoff function $P\in C(X\times Y)$ is predetermined and known in advance to both players; P1 wants to minimize the resulting payment and P2 wants to maximize it.

Now, what if one the players declares his or her strategy (x or y) to the other player in advance? If P1 declares first, the outcome is better for P2, who has a chance to optimize his response to the announced strategy x, and conversely. This implies that

$$(2.6)\qquad \inf_{x\in X}\sup_{y\in Y} P(x,y) \geq \sup_{y\in Y}\inf_{x\in X} P(x,y);$$

(Player 1 declares first vs. Player 2 declares first.) von Neumann [146] identified structural conditions on the payoff function to have a saddle point (2.7), in which case equality holds in (2.5); see also Kakutani's reference to [147] in [74].

Theorem 2.5 (convex/concave min-max). *If $X\subset \mathbf{R}^m$ and $Y\subset\mathbf{R}^n$ are compact and convex, then equality holds in* (2.6) *provided for each $(x_0,y_0)\in X\times Y$ both functions $x\in X\mapsto P(x,y_0)$ and $y\in Y\mapsto -P(x_0,y)$ are convex. (In fact, convexity of all sublevel sets of both functions is enough.)*

PROOF. Let

$$x_b(y)\in \arg\min_{x\in X} P(x,y) \quad y_b(x)\in\arg\max_{y\in Y} P(x,y)$$

denote the best responses of P1 and P2 to each other's strategies y and x. Note x_b and y_b are continuous if the convexity and concavity assumed of the payoff function are both strict. In that case, Brouwer's theorem asserts the the function $y_b\circ x_b\colon Y\to Y$ has a fixed point y_0. Setting $x_0=x_b(y_0)$, since $y_0=y_b(x_0)$ we have found a saddle point

$$\begin{aligned}(2.7)\qquad \inf_{x\in X}\max_{y\in Y} P(x,y) \leq \max_{y\in Y} P(x_0,y) = P(x_0,y_0) \\ = \min_{x\in X} P(x,y_0) \leq \sup_{y\in Y}\min_{x\in X} P(x,y)\end{aligned}$$

of the payoff function, which proves equality holds in (2.6). If the convexity and concavity of the payoff function are not strict, apply the theorem just proved to the perturbed payoff $P_\epsilon(x,y) = P(x,y) + \epsilon(|x|^2 - |y|^2)$ and take the limit $\epsilon \to 0$. $\square$

E.g. Expected payoff [146]: von Neumann's original example of a function to which the theorem and its conclusion applies is the expected payoff $P(x,y) = \sum_{i=1}^m \sum_{j=1}^n p_{ij} x_i y_j$ of *mixed* or *randomized* strategies x and y for a game in which P1 and P2 each have only finitely many pure strategies, and the payoff corresponding to strategy $1 \le i \le m$ and $1 \le j \le n$ is p_{ij}. In this case $X = \{x \in [0,1]^m \mid \sum x_i = 1\}$ and $Y = \{y \in [0,1]^n \mid \sum y_j = 1\}$ are standard simplices of the appropriate dimension, whose vertices correspond to the pure strategies.

2.3. Relevance to optimal transport: Kantorovich – Koopmans duality. Infinite-dimensional versions of von Neumann's theorem can also be formulated where X and Y lie in Banach spaces; they are proved using Schauder's fixed point theorem instead of Brouwer's. A payoff function germane to optimal transportation is defined on the strategy spaces $X = \{0 \le \gamma \text{ on } M^+ \times M^-\}$ and $Y = \{(u,v) \in L^1(M^+,\mu^+) \oplus L^1(M^-,\mu^-)\}$ by

$$P(\gamma,(u,v)) = \int_{M^+\times M^-} (c(x,y) + u(x) + v(y))\, d\gamma(x,y) - \int_{M^+} u\, d\mu^+ - \int_{M^-} v\, d\mu^-.$$

Note the bilinearity of P on $X \times Y$. Since

$$\inf_{\gamma\in X} P(\gamma,(u,v)) = \begin{cases} -\infty & \text{unless } (-u,-v) \in \mathrm{Lip}_c, \\ 0 - \int u\, d\mu^+ - \int v\, d\mu^- & \text{otherwise,} \end{cases}$$

the Kantorovich – Koopmans dual problem is recovered from the version of the game in which P2 is compelled to declare his strategy first:

$$\sup_{(u,v)\in Y} \inf_{\gamma\in X} P(\gamma,(u,v)) = \sup_{(-u,-v)\in \mathrm{Lip}_c} \int (-u)\, d\mu^+ + \int (-v)\, d\mu^-. \tag{2.8}$$

On the other hand, rewriting

$$P(\gamma,(u,v)) = \int_{M^+\times M^-} c\, d\gamma + \int u(d\gamma - d\mu^+) + \int v(y)\big(d\gamma(x,y) - d\mu^-(y)\big)$$

we see

$$\sup_{(u,v)\in Y} P(\gamma,(u,v)) = \begin{cases} +\infty & \text{unless } \gamma \in \Gamma(\mu^+,\mu^-) \\ \int_{M^+\times M^-} c\, d\gamma & \text{if } \gamma \in \Gamma(\mu^+,\mu^-). \end{cases} \tag{2.9}$$

Thus the primal transportation problem of Kantorovich and Koopmans

$$\inf_{\gamma\in X} \sup_{(u,v)\in Y} P(\gamma,(u,v)) = \inf_{\gamma\in\Gamma(\mu^+,\mu^-)} \int_{M^+\times M^-} c\, d\gamma$$

corresponds to the version of the game in which P1 declares his strategy first. The equality between (2.8) and (2.9) asserted by an appropriate generalization of von Neumann's theorem implies the duality (2.5):

$$\min_{\gamma\in\Gamma(\mu^+,\mu^-)} \int_{M^+\times M^-} c\, d\gamma = \sup_{(-u,-v)\in \mathrm{Lip}_c} \int_{M^+} (-u)\, d\mu^+ + \int_{M^-} (-v)\, d\mu^-.$$

2.4. Characterizing optimality by duality. The following theorem can be deduced as an immediate corollary of this duality. We may think of the potentials u and v as being Lagrange multipliers enforcing the constraints on the marginals of γ; in the economics literature they are interpreted as shadow prices which reflect the geographic variation in scarcity or abundance of supply and demand. The geography is encoded in the choice of cost.

Theorem 2.6 (Necessary and sufficient conditions for optimality). *The existence of $-(u,v) \in \mathrm{Lip}_c$ such that γ vanishes outside the zero set of the nonnegative function $k(x,y) = c(x,y) + u(x) + v(y) \geq 0$ on $M^+ \times M^-$ is necessary and sufficient for the optimality of $\gamma \in \Gamma(\mu^+, \nu^-)$ with respect to $c \in C(M^+ \times M^-)$.*

Corollary 2.7 (First- and second-order conditions on potentials). *Optimality of γ and (u,v) implies $Dk = 0$ and $D^2k \geq 0$ at any point $(x,y) \in \operatorname{spt}\gamma$ where these derivatives exist. In particular, $D_x[c(x,y) + u(x) + v(y)] = 0$ and $D_x^2[c(x,y) + u(x) + v(y)] \geq 0$ holds γ-a.e., and likewise for y-derivatives.*

E.g. Consider the special case of the bilinear cost: $c(x,y) = -x \cdot y$. Here the first and second order conditions of the corollary become

$$y = Du(x) \quad \text{and} \quad D^2u(x) \geq 0,$$

suggesting y is the graph of the gradient of a convex function. In this case, convexity of u guarantees D^2u is defined a.e. with respect to Lebesgue measure, by Alexandrov's theorem.

2.5. Existence of optimal maps and uniqueness of optimal measures. More generally, we claim u inherits Lipschitz and semiconvexity bounds (3.2) – (3.3) from $c(x,y)$, which guarantee the existence of x-derivatives in the preceding corollary — at least Lebesgue almost everywhere. This motivates the following theorem of Gangbo [65] and Levin [88]; variations appeared independently in Caffarelli [22], Gangbo and McCann [66, 67], and Rüschendorf [126, 127] at around the same time, and subsequently in [97].

Definition 2.8 (Twist conditions). A function $c \in C(M^+ \times M^-)$ differentiable with respect to $x \in M^+$ is said to be *twisted* if

(2.10) $\forall\, x_0 \in M^+$, the map $y \in M^- \mapsto D_x c(x_0, y) \in T_x^* M^+$ is one-to-one. $\quad$ (A1)$^+$

For $(x,p) \in T^*M^+$ denote the unique $y \in M^-$ solving $D_x c(x,y) + p = 0$ by $y = Y(x,p)$ when it exists. When the same condition holds for the cost $\tilde{c}(y,x) := c(x,y)$, we denote it by (A1)$^-$. When both c and $\tilde{c}$ satisfy (A1)$^+$, we say the cost is *bi-twisted*, and denote this by (A1).

Theorem 2.9 (Existence of Monge solutions; uniqueness of Kantorovich solutions). *Fix Polish probability spaces $(M^\pm, \mu^\pm)$ and assume M^+ is a n-dimensional manifold and $\mathrm{d}\mu^+ \ll \mathrm{d}^n x$ is absolutely continuous (in coordinates). Let $c \in C(M^+ \times M^-)$ differentiable with respect to $x \in M^+$ satisfy the twist condition* (2.10) *and assume $D_x c(x,y)$ is bounded locally in $x \in M^+$ uniformly in $y \in M^-$. Then, there exists a locally Lipschitz (moreover, c-convex, as in Definition* 2.10*) function $u \colon M^+ \to \mathbf{R}$ such that*

(a) *$G(x) := Y\big(x, Du(x)\big)$ pushes μ^+ forward to μ^-;*
(b) *this map is unique, and uniquely solves Monge's minimization* (1.4);
(c) *Kantorovich's minimization* (2.1) *has a unique solution γ;*

(d) $\gamma = (\mathrm{id} \times G)_{\#}\mu^+$.

Definition 2.10 (c-convex). A function $u\colon M^+ \to \mathbf{R}\cup\{+\infty\}$ (not identically infinite) is *c-convex* if and only if $u = (u^{\tilde{c}})^c$, where

$$u^{\tilde{c}}(y) = \sup_{x\in M^+} -c(x,y) - u(x) \quad \text{and} \quad v^c(x) = \sup_{y\in M^-} -c(x,y) - v(y). \tag{2.11}$$

Remark 2.11 (Legendre – Fenchel transform and convex dual functions). When $c(x,y) = -\langle x, y\rangle$, then $u^{\tilde{c}}(y)$ is manifestly convex: it is the Legendre – Fenchel transform or convex dual function of $u(x)$. In this case, $(u^{\tilde{c}})^c$ is well-known to yield the lower semicontinuous convex hull of the graph of u, so that $u = (u^{\tilde{c}})^c$ holds if and only if u is already lower semicontinuous and convex. More generally, we interpret the condition $u = u^{\tilde{c}c}$ as being the correct adaptation of the notion of convexity to the geometry of the cost function c.

SKETCH OF PROOF OF THEOREM 2.9. The key idea of the proof is to establish existence of a maximizer $-(u,v) \in \mathrm{Lip}_c$ of (2.5) with the additional property that $(u,v) = (v^c, u^{\tilde{c}})$. Differentiability of $u = u^{\tilde{c}c}$ on a set $\mathrm{dom}\, Du$ of full $\mathrm{d}\mu^+ \ll \mathrm{d}^n x$ measure then follows from Rademacher's theorem and Lemma 3.1. The map $G(x) := Y\big(x, Du(x)\big)$ is well-defined on $\mathrm{dom}\, Du$ by the twist condition (assuming the supremum (2.11) defining $(u^{\tilde{c}})^c(x)$ is attained). Corollary 2.7 shows any minimizer γ vanishes outside the graph of this map, and it then follows easily that $\gamma = (\mathrm{id} \times G)_{\#}\mu^+$ and hence γ is uniquely determined by u [3]. Conversely any other c-convex $\tilde{u}$ for which $\widetilde{G}(x) = Y\big(x, D\tilde{u}(x)\big)$ pushes μ^+ forward to μ^- can be shown to maximize the dual problem by checking that $\tilde{\gamma} = (\mathrm{id} \times \widetilde{G})_{\#}\mu^+$ vanishes outside the support of $\widetilde{G}$. Thus $\tilde{\gamma} = \gamma$ and $\widetilde{G} = G$ holds μ^+-a.e..

To extract the desired $-(u,v) \in \mathrm{Lip}_c$ from a maximizing sequence $-(u_k, v_k)$ requires some compactness. (This would come from the convexity of u_k and v_k in case $c(x,y) = -x \cdot y$ via the Blaschke selection theorem.) Observe $-(u,v) \in \mathrm{Lip}_c$ implies

$$u(x) \geq \sup_{y\in M^-} -c(x,y) - v(y) =: v^c(x).$$

Moreover, $-(v^c, v) \in \mathrm{Lip}_c$ and $-(v^c) \geq -u$ can only increase the value of the objective functional relative to $-(u,v)$. Thus $-(v^c, v)$ is a better candidate for a maximizer than $-(u,v)$. Repeating the process shows $-(v^c, v^{c\tilde{c}})$ and $-(v^{c\tilde{c}c}, v^{c\tilde{c}}) \in Lip_c$ are better still, since $(v^c)^{\tilde{c}} \leq v$ and $(v^{c\tilde{c}})^c \leq v^c$ by the same logic. On the other hand, starting from $v^{c\tilde{c}} \leq v$, the negative coefficient in definition (2.11) implies the opposite inequality $(v^{c\tilde{c}})^c \geq v^c$. Thus $v^{c\tilde{c}c} = v^c$ quite generally. (This is precisely analogous to the fact that the second Legendre transform u^{**} does not change a function $u = v^*$ which is already convex and lower semicontinuous; see Remark 2.11.)

Replacing a maximizing sequence $-(u_k, v_k)$ with $-(v_k^c, v_k^{c\tilde{c}})$ therefore yields a new maximizing sequence at least as good which moreover consists of c-convex functions. Lemma 3.1 shows this new family is locally equi-Lipschitz, hence we only need local boundedness for the Arzelà – Ascoli theorem to yield a limiting maximizer $-(u,v)$, which will in fact be c-convex, though we can also replace it by $-(v^c, v^{c\tilde{c}})$ just to be sure. Local boundedness also follows from Lemma 3.1, after fixing $x_0 \in \mathrm{spt}\,\mu^+ \subset M^+$ and replacing $-(u_k, v_k)$ by $-(u_k - \lambda_k, v_k + \lambda_k)$ with $\lambda_k = u_k(x_0)$. This replacement does not change the value of the objective functional (2.8), yet ensures that $u(x_0) = 0 \neq \pm\infty$. □

3. Methods for obtaining regularity of optimal mappings

Given mines and factories $(M^{\pm}, \mu^{\pm})$ and a cost function $c \in C(M^+ \times M^-)$, in the preceding section we found conditions which guarantee the existence and uniqueness of a map $G(x) = Y\big(x, Du(x)\big)$ such that $G_{\#}\mu^+ = \mu^-$ with $u = u^{\tilde{c}c}$, i.e., c-convex. Under the same conditions, the map G is the unique minimizer of Monge's problem (1.4). The space M^+ was assumed to be an n-dimensional manifold, and the following twist hypothesis $(\mathrm{A1})^+$, equivalent to (2.10), was crucial to specifying $Y(x, \cdot)$:

(3.1) $\forall y_1 \neq y_2 \in M^-$assume $x \in M^+ \mapsto c(x, y_1) - c(x, y_2)$ has no critical points.

Notice, however, that (3.1) cannot be satisfied by any cost function which is *differentiable* throughout a compact manifold M^+ (without boundary). In case $M^+ = \mathbf{S}^n$, Monge solutions do not generally exist [68], but criteria are given in [3, 33] which guarantee uniqueness of the Kantorovich minimizer. On the other hand, it is an interesting open problem to find a criterion on $c \in C^1(M^+ \times M^-)$ which guarantees uniqueness of Kantorovich solutions for all $\mu^{\pm} \in L^1(M^{\pm})$ in more complicated topologies, such as the torus $M^{\pm} = \mathbf{T}^2$ for example. Here differentiability of the cost function is crucial; for costs such as Riemannian distance squared, the desired uniqueness is known [34, 105], but the cost fails to be differentiable at the cut locus.

3.1. Rectifiability: differentiability almost everywhere. The current section is devoted to reviewing methods for exploring the smoothness properties of the optimal map G found above, or equivalently of its c-convex potential u. The following lemma shows that all c-convex functions inherit Lipschitz and semiconvexity properties directly from the cost function c; it has already been exploited to prove Theorem 2.9.

Lemma 3.1 (Inherent regularity of c-convex functions). *If $u = u^{\tilde{c}c}$ and $c(\cdot, y) \in C^k_{\mathrm{loc}}(M^+)$ for each $y \in M^-$, then $k = 1$ implies* (3.2) *and $k = 2$ implies* (3.3):

$$|Du(x)| \leq \sup_{y \in M^-} |D_x c(x, y)| \qquad \text{(local Lipschitz regularity)}; \tag{3.2}$$

$$D^2 u(x) \geq \inf_{y \in M^-} -D^2_{xx} c(x, y) \qquad \text{(semiconvexity)}. \tag{3.3}$$

(*Inequalities* (3.2) – (3.3) *may be interpreted in the weak or the viscosity senses.*)

Similarly, c-cyclically monotone sets $S \subset M^+ \times M^-$ turn out to be contained in Lipschitz submanifolds of dimension $n = \dim M^{\pm}$ when the cost function is nondegenerate (3.4). The following recent theorem of McCann, Pass and Warren [108] combines with Rademacher's theorem — which asserts the differentiability Lebesgue a.e. of Lipschitz functions — to give a simple tool for establishing that $f^+(x) = \big|\det\big(DG(x)\big)\big|^- \big(G(x)\big)$ holds f^+-a.e..

Theorem 3.2 (Rectifiability of optimal transport [108]**).** *Assume $M^{\pm}$ are n-dimensional manifolds, at least in a neighbourhood U of $(x_0, y_0) \in M^+ \times M^-$, where $c \in C^2(U)$ and*

$$\det D^2_{x^i y^j} c(x_0, y_0) \neq 0. \tag{3.4}$$
(A2)

If $S \subset M^+ \times M^-$ is c-cyclically monotone, then $S \cap V$ lies in an n-dimensional Lipschitz submanifold, for some neighbourhood $V \subset U$ of (x_0, y_0).

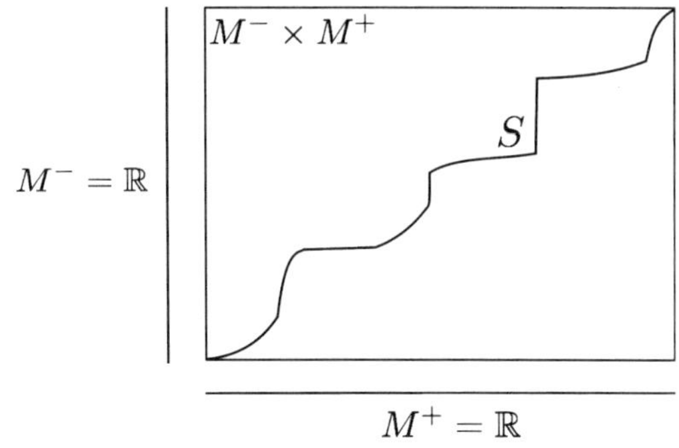

FIGURE 1. Optimizers have locally monotone support S in the plane.

In view of Theorem 2.2, this conclusion applies either to the graph $S = \mathrm{Graph}(G)$ of any optimal map (1.4) or the support $S = \mathrm{spt}(\gamma)$ of any optimal measure (2.1) in the transportation problem.

Motivation. On the line $M^{\pm} = \mathbf{R}$, without further assumptions on c or $\mu^{\pm}$, a transport map may not exist nor be monotone, yet the theorem above says that even so all pieces of $spt(\gamma)$ lie along along Lipschitz arcs in the plane. These curves will actually be locally monotone — nondecreasing or nonincreasing depending on the sign of $D^2_{xy}c(x_0, y_0)$; see Figure 1.

IDEA OF PROOF. Introduce the notation $b = -c$. In case $c(x,y) = -x \cdot y$ on $M^{\pm} = \mathbf{R}^n$, monotonicity asserts for all $(x_0, y_0), (x_1, y_1) \in S$ that $\Delta x = x_1 - x_0$ and $\Delta y = y_1 - y_0$ satisfy

$$0 \le \langle \Delta x, \Delta y \rangle = \left\langle \frac{\Delta z - \Delta w}{\sqrt{2}}, \frac{\Delta z + \Delta w}{\sqrt{2}} \right\rangle$$

where

$$(z, w) := \left(\frac{x+y}{\sqrt{2}}, \frac{x-y}{\sqrt{2}} \right). \tag{3.5}$$

This implies that $|\Delta w|^2 \le |\Delta z|^2$ meaning $w = w(z)$ has Lipschitz constant 1 as a graph over $z \in \mathbf{R}^n$. Equivalently, S has Lipschitz constant 1 as a graph over the diagonal in $M^+ \times M^-$. This special case was established by Alberti and Ambrosio [4], using an argument of Minty [110].

For more general costs $b = -c$ and any $\epsilon > 0$, the nondegeneracy (3.4) implies the existence of new coordinates $\tilde{y} = \tilde{y}(y)$ on M^- in a neighbourhood of y_0 such that $\tilde{b}\big(x, \tilde{y}(y)\big) = b(x,y)$ satisfies $|D^2_{x\tilde{y}}\tilde{b}(x,\tilde{y}) - I| < \epsilon$ in a neighborhood $\widetilde{V}$ of $(x_0, \tilde{y}_0)$ which is convex in coordinates.

Now, $\langle \Delta x, \Delta \tilde{y} \rangle \ge -\epsilon |\Delta x||\Delta \tilde{y}|$ follows from

$$0 \le \tilde{b}(x_0, \tilde{y}_0) + \tilde{b}(x_1, \tilde{y}_1) - \tilde{b}(x_0, \tilde{y}_1) - \tilde{b}(x_1, \tilde{y}_0) = D^2_{x\tilde{y}}\tilde{b}(x^*, \tilde{y}^*)(x_1 - x_0)(\tilde{y}_1 - \tilde{y}_0) \tag{3.6}$$

and the change of variables analogous to (3.5) yields

$$|\Delta z|^2 - |\Delta w|^2 \ge -\epsilon |\Delta w - \Delta z||\Delta w + \Delta z| \ge -\epsilon\big(|\Delta w|^2 + |\Delta z|^2\big).$$

Thus $(1+\epsilon)|\Delta z|^2 \ge (1-\epsilon)|\Delta w|^2$, which shows $w = w(z)$ is again a Lipschitz function of $z \in \mathbf{R}^n$ in the chosen coordinates. □

3.2. From regularity a.e. to regularity everywhere. The regularity results discussed so far — Lipschitz continuity of the potential u, and of $\mathrm{Graph}(G) \subset M^+ \times M^-$ rather than of the map $G(x) = Y\big(x, Du(x)\big)$ itself — required no hypotheses on the probability measures $\mu^- = G_\# \mu^+$. To address the continuity, differentiability, and higher regularity everywhere for the map $G \colon M^+ \to M^-$ is a much more delicate issue which certainly requires further hypotheses on the data $\mu^\pm$ and c. For example, if $\operatorname{spt} \mu^-$ is connected but $\operatorname{spt} \mu^+$ is not, then G cannot be continuous. The same reasoning makes it clear that ellipticity of the Monge – Ampère equation (1.6) cannot be nondegenerate for all convex solutions; regularity must propagate from boundary conditions since the *purely local* effect of the equation is insufficent to conclude $u \in C^1_{\mathrm{loc}}$. It is often easier to work with the scalar potential u rather than the mapping G; we shall see this reduces the problem to a question in the theory of second-order, fully-nonlinear, degenerate-elliptic partial differential equations (4.2) generalizing the Monge – Ampère equation. However, this question was answered first in the special case $c(x, y) = -x \cdot y$ corresponding to the case (1.6) by Delanoë in the plane $n = 2$ [45], and by Caffarelli and Urbas in higher dimensions $M^\pm = \mathbf{R}^n$ [20, 21, 23, 141].

Remark 3.3. Note for $c \in C^{k+1}(M^+ \times M^-)$ that $u \in C^{k+1}$ implies $G \in C^k$ by the following remark. Whereas the twist condition $(\mathrm{A1})^+$ asserts that the definition $Y(x, p)$ by $D_x c\big(x, Y(x, p)\big) + p = 0$ is unambiguous, nondegeneracy (A2) allows the implicit function theorem to be applied to conclude C^k smoothness of $Y(x, p)$ (where defined).

3.3. Regularity methods for the Monge – Ampère equation; renormalization. There are several methods for obtaining regularity results for convex solutions of the Monge – Ampère equation. The first to be discussed here is the *continuity method*, used for example by Delanoë '91 and Urbas '97. This approach requires relatively strong assumptions on the smoothness of the measures $d\mu^\pm = f^\pm \,\mathrm{dVol}$ and on the convexity and smoothness of their domains $M^\pm \subset \mathbf{R}^n$. When it applies, it yields global regularity of the resulting potential up to the boundary of ∂M^+ from the same fixed point argument which shows a solution exists. The second method is the *renormalization method* pioneered by Caffarelli '92 – '96, which starts from the unique (weak) solution to the 2nd boundary value problem and uses affine invariance of the equation to blow-up the solution near a putative singularity and derive a contradiction in the limit. This method is quite flexible: it has the advantage of yielding certain conclusions under weaker assumptions on the data, and has therefore proven useful for addressing such phenomena as the free boundary which arises in partial transport problems, where the densities $f^\pm$ need not be continuous and are constrained but not specified a priori [25]. Using this method, Caffarelli '92 was able to prove the following regularity result on the *interior* M^+_{int} of $M^+ \subset \mathbf{R}^n$:

Theorem 3.4 (Local regularity [21]). *Fix $c = -x \cdot y$ and let $M^- \subset \mathbf{R}^n$ be convex. Assume $\mu^\pm = f^\pm \, dx$ with $\log f^\pm \in L^\infty(M^\pm)$. If $\log f^\pm \in W^{k,\infty}_{\mathrm{loc}}(M^\pm_{\mathrm{int}})$, there exists $\alpha \in (0, 1)$(depending only on n, k and the bounds on $\log f^\pm$) such that $u \in C^{k+1,\alpha}_{\mathrm{loc}}(M^+_{\mathrm{int}})$, where u is the convex function with $Du(M^+) \subset M^-$ such that $Du_\# \mu^+ = \mu^-$.*

Remark 3.5 (Degenerate ellipticity). As shown also in [21], when one drops the convexity assumption on M^- the gradient map may be discontinuous at interior points. This goes in hand with the claim made in the previous subsection that regularity (even in the interior) must propagate from the boundary.

Remark 3.6 (Local versus global regularity). This is a local regularity result. At present, global regularity (up to ∂M^+) requires both domains $\partial M^\pm$ to be strongly convex and smooth (Caffarelli '96) [23]. Here *strong* convexity means the principal curvatures (or second fundamental form) of the domain boundaries should be positive-definite.

Remark 3.7 (Higher regularity via uniformly elliptic linearization). The cases of primary relevance are $\log f^\pm$ merely bounded and measurable ($k = 0$), and $\log f^\pm$ also locally Lipschitz ($k = 1$). Once $u \in C^{2,\alpha}_{\mathrm{loc}}$ has been deduced from these assumptions, higher regularity in the interior of $M^\pm$ follows from uniform ellipticity of the Monge–Ampère equation (1.6):

$$0 = \log\Big(\det\big(D^2u(x) + \tau D^2w(x)\big)\Big) + \log\Big(f^-\big(Du(x) + \tau Dw(x)\big)\Big) - \log\big(f^+(x)\big). \tag{3.7}$$

For example, linearizing this equation at $\tau = 0$ yields the equation

$$0 = \mathrm{Tr}\big(D^2u(x)^{-1}D^2w(x)\big) + D\log f^-|_{Du(x)} \cdot Dw(x), \tag{3.8}$$

which must be satisfied by spatial derivatives $w = D_{x^i}u$ of u. Convexity combines with $u \in C^{2,\alpha}_{\mathrm{loc}}$ and the equation (1.6) itself to bound $\|\Lambda^{n-1}f^-/f^+\|^{-1}_{L^\infty} \le D^2u(x) \le \Lambda$ on compact subsets of M^+_{int}. The derivatives of u thus satisfy a uniformly elliptic linear equation (3.8) with Hölder continuous coefficients, so Schauder estimates [70] and bootstrapping yield as much as regularity as can be expected when $k \ge 2$.

3.4. The continuity method (schematic) (cf. Delanoë '91, Urbas '97). To apply the continuity method, we assume $M^\pm \subset \mathbf{R}^n$ are smooth and strongly convex, and $\log f^\pm \in C^{2,\alpha}(M^\pm) \cap L^\infty$.

Choose a dilation by $\epsilon > 0$ sufficiently small and translation $M_0 = G_0(M^+)$ of M^+ by $x_0 \in \mathbf{R}^n$ such that $M_0 \subset\subset M^-$. Let $\mu_0 := (G_0)_\# \mu^+$ be the push-forward of μ^+ through the corresponding dilation and translation $G_0(x) = \epsilon x - x_0$. Notice that G_0 is the gradient of the smooth convex function $u_0(x) = \epsilon|x|^2/2 - x_0 \cdot x$, and as such gives the optimal map between μ^+ and μ_0. The idea behind the continuity method is to construct a family of target measures $d\mu_t = f_t\, d\mathrm{Vol}$ interpolating between $d\mu_0$ and $d\mu_1 := d\mu^-$, and to study the set T of $t \in [0,1]$ for which the optimal transportation problem of Brenier admits a solution with a convex potential $u_t \in C^{2,\alpha}(M^+)$. The interpolating measures must be constructed so that the $C^{2,\alpha}(M^\pm) \cap L^\infty$ norms of $\log f_t$, and the strong convexity and smoothness of $M_t = \mathrm{spt}\, f_t$, can be quantified independently of $t \in [0,1]$. We then hope to show $T \subset [0,1]$ is both open and closed. If so, it must exhaust the entire interval (since $0 \in T$), therefore $1 \in T$ as desired.

Closed. To show closedness of this set requires an a priori estimate of the form $\|u_t\|_{C^{2,\alpha}(M^+)} \le C(\|\log f^\pm\|_{C^{2,\alpha}(M^\pm)}, \|\partial M^\pm\|_{C^{2,\alpha},\text{strong convexity}})$ for any smooth solution $u_t \in C^4(M^+)$ of the 2nd boundary value problem

$$\det D^2u_t(x) = \frac{f_0(x)}{f_t(Du_t(x))} \quad \text{with } Du_t(M^+) \subset M_t. \tag{3.9}$$

Such estimates are delicate, but can be obtained by using the maximum principle after differentiating the equation twice, and constructing barriers. Once obtained, they imply that if $t_k \in T$ and $t_k \to t_\infty$ then $t_\infty \in T$ also; the corresponding solutions u_{t_k} belong to $C^4(M^+)$ as in Remark 3.7.

Open. The fact that T is open is shown using an infinite-dimensional implicit function theorem. This requires knowing that the linearized operator is invertible (i.e., uniformly elliptic), and can be solved for the relevant boundary conditions.

For $u \in C^{2,\alpha}(M^+)$ we have already argued the uniform ellipticity of the linearization (3.8). To linearize the boundary conditions $Du_t(M^+) \subset M_t$, introduce a sufficiently smooth and strongly convex function $h\colon \mathbf{R}^n \to \mathbf{R}$ whose level sets $M_t = \{y \in \mathbf{R}^n \mid h(y) \leq t\}$ give the domains $M_t := \operatorname{spt} f_t$, and rewrite the nonlinear boundary condition in the form $h\big(Du_t(x)\big) \leq t$ with equality on ∂M^+. Linearizing this in u yields the boundary condition of the linear equation for w:

$$Dh\big(Du_t(x)\big) \cdot Dw(x) = 0 \quad \text{on } \partial M^+. \tag{3.10}$$

For the linear problem (3.8) to be well-posed, we need a *uniformly nontangential* prescribed gradient for w on ∂M^+. Since Dh parallels the normal $\hat{n}_{M_t}$ to M_t, this amounts to the uniform obliqueness estimate

$$\hat{n}_{M_t}(Du_t(x)) \cdot \hat{n}_{M^+}(x) \geq \delta > 0 \tag{obliqueness}$$

provided by Urbas [141], with δ depending only on coarse bounds for the data. This concludes the sketch that $T \subset [0,1]$ is open: well-posedness of the linear problem (3.8)–(3.10) when $t = t_0 \in T$ implies the existence of solutions $u_t \in C^{2,\alpha}(M^+)$ to the nonlinear problem (3.9) for any t close enough to t_0.

Both approaches (renormalization and continuity method) have been extended in recent years to more general costs, and this will be the topic of the next lectures.

4. Regularity and counterexamples for general costs

4.1. Examples. The development of a regularity theory for general cost functions satisfying appropriate hypotheses on compact domains $M^\pm \subset \mathbf{R}^n$ began with the work of Ma, Trudinger and Wang [97]. Prior to that there were regularity results only for a few special costs, such as:

Example 4.1 (Bilinear cost). $c(x,y) = -x \cdot y$ or equivalently $c(x,y) = |x-y|^2/2$ [21,23,45,141], and its restriction to boundaries of strongly convex sets such as $M^\pm = \partial B_1(0)$ in $\mathbf{R}^n$ (Gangbo and McCann [68]);

Example 4.2 (Logarithmic cost; conformal geometry and reflector antenna design). $c(x,y) = -\log|x-y|$ appearing in conformal geometry (cf. Viaclovsky's review [143]), and its restriction to the Euclidean unit sphere, which is relevant to reflector antenna design (Glimm and Oliker [71], X.-J. Wang [149, 150]) and helped to inspire Wang's subsequent collaborations [91,97,139,140] with Trudinger, Ma, and Liu.

In the wake of Ma, Trudinger and Wang's [97] results, many new examples have emerged of cost functions which satisfy [46,47,61–63,79–81,85–87,89,93,94, 97] — or which violate [92,97] — their sufficient conditions (A0) – (A4) and $(\mathrm{A3})_\mathrm{s}$ for regularity from §4.4 below — not to mention the subsequent variants (A3) and (B3) introduced by Trudinger and Wang [139] and Kim and McCann [80] respectively, on the crucial condition $(\mathrm{A3})_\mathrm{s}$. Among the most interesting of these are the geometrical examples and counterexamples of Loeper:

Example 4.3 (Sphere). $c(x,y) = \frac{1}{2}d^2_{\mathbf{S}^n}(x,y)$ on the round sphere satisfies $(A3)_s$ [93] (and (B3) [81]);

Example 4.4 (Saddle). $c(x,y) = \frac{1}{2}d^2_M(x,y)$ on hyperbolic space $M = \mathbf{H}^n$ violates $(A3)_s$ (and (A3)) — as does the Riemannian distance squared cost on any Riemannian manifold $M^\pm = M$ which has (at least one) negative sectional curvature at some point $x \in M$ [92].

4.2. Counterexamples to the continuity of optimal maps. For any cost function which violates (A3), Loeper went further to show there are probability measures $d\mu^\pm = f^\pm d^n x$ with smooth positive densities bounded above and below — so that $\log f^\pm \in C^\infty(M^\pm)$ — for which the unique optimal map $G\colon M^+ \to M^-$ is discontinuous [92]. Let's see why this is so for the quadratic cost given on either the hyperbolic plane or a saddle surface as in Example 4.4.

Consider transportation from the uniform measure μ^+ on a sufficiently small ball to a target measure consisting of three point masses $\mu^- = \frac{1}{3}\sum_{i\leq 3}\delta_{y_i}$ near the center of the ball, choosing y_2 to be the midpoint of y_1 and y_3. In this case Theorem 2.9 provides constants $v_1, \ldots, v_3$ and a c-convex function

$$u(x) = \max\{u_i(x) \mid i = 1,2,3\} \quad \text{where } u_i(x) = -c(x,y_i) - v_i, \tag{4.1}$$

such that the optimal map $G_\#\mu^+ = \mu^-$ satisfies $G^{-1}(y_i) = \{x \in M^+ \mid u(x) = u_i(x)\}$. We interpret $-v_i$ to be the value of the good at the potential destination $y_i \in \operatorname{spt}\mu^-$; the producer at $x \in M^+$ will ship his good to whichever target point y_i provides the greatest value after transportation costs are deducted (4.1); here the values $v_1, \ldots, v_3$ are adjusted to balance supply with demand, so that each of the three regions $G^{-1}(y_i)$ contains $\frac{1}{3}$ of the mass of μ^+. For the Euclidean distance-squared cost these three regions are easily seen to be convex sets, while for the spherical distance-squared they remain connected. For the hyperbolic distance-squared, however, the 'middle' region $G^{-1}(y_2)$ consists of two disconnected components, near opposite sides of the ball $\operatorname{spt}\mu^+$ (see figure below, or for instance [80, Figure 1]). This disconnectedness is the hallmark of costs for which (A3) fails, and allowed Loeper to construct counterexamples to the continuity of optimal mappings as follows.

In the preceding discussion, μ^- was not given by a smooth positive density; still it can be approximated by a sequence of measures $\mu^-_\epsilon := \mu^- * \eta_\epsilon$ which are. Now consider the reverse problem of transporting μ^-_ϵ to μ^+. Call the optimal map for this new problem $x = G^-_\epsilon(y)$. For $\delta > 0$, taking $\epsilon > 0$ sufficiently small ensures for each $1 \leq i \leq 3$ that nearly $\frac{1}{3}$ of the mass of μ^-_ϵ concentrates near y_i and is mapped into a δ-neighbourhood of $G^{-1}(y_i)$. Intuitively, for δ sufficiently small, this forces a discontinuity of G^-_ϵ which tears the region near y_2 into at least two disconnected components: nearly half of the mass near this point must map to each disconnected component of $G^{-1}(y_2)$; see Figure 2. This construction shows why the distance-squared cost on a hyperbolic or saddle surface cannot generally produce smooth optimal transport maps.

There is another of obstruction to the continuity of G, namely the convexity (at least when $M^\pm = \mathbb{R}^n$) of the support of μ^-. This was shown by Caffarelli [21] with the following elementary example: consider $u\colon \mathbb{R}^2 \to \mathbb{R}$ given by

$$u(x) = |x_1| + \tfrac{1}{2}|x|^2, \quad x = (x_1, x_2).$$

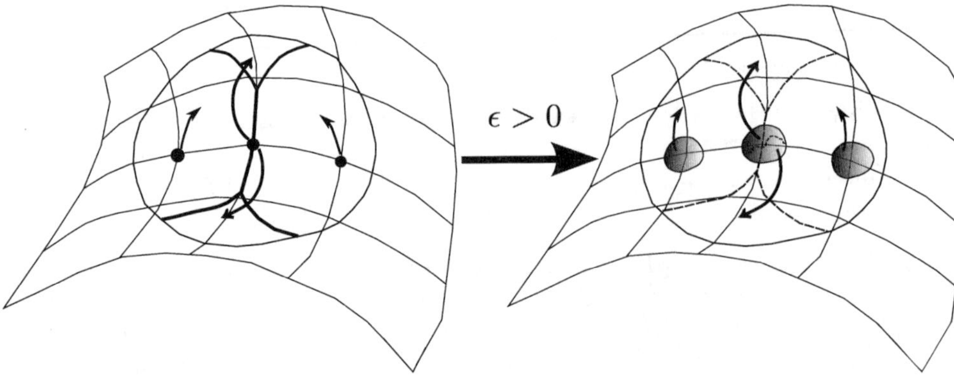

FIGURE 2. A tear occurs when spreading a triply peaked density uniformly over the saddle.

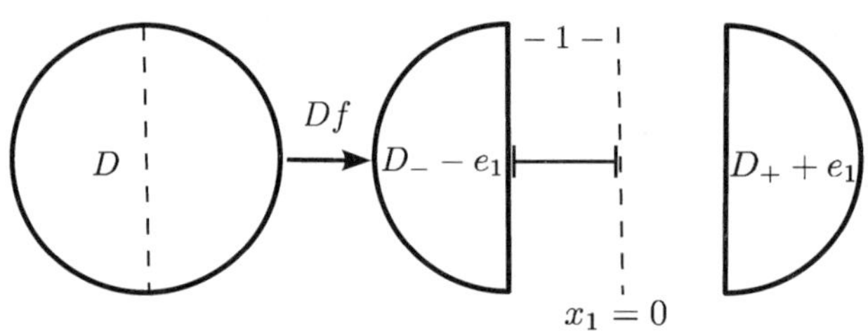

FIGURE 3. Disconnected targets also produce tears, as do nonconvex targets (spt μ^-).

If we consider the cost $c(x, y) = -x \cdot y$, then $y = Du(x)$ gives the optimal transport map between the unit disc (with Lebesgue measure) into two shifted half discs (Figure 3); in particular, the transport map is discontinuous across $\{x_1 = 0\}$. If the two half-disks $D_\pm$ are joined by a bridge of width $\epsilon > 0$, he went on to argue that the optimal map $y = Du_\epsilon$ cannot remain continuous for small ϵ, without contradicting weak convergence $|\partial u_\epsilon| \rightharpoonup |\partial u_0|$ of the Monge–Ampère measures by producing a limiting jump $|\partial u_0(D)| > \limsup_{\epsilon\to 0} |\partial u_\epsilon(D)|$ in total Monge–Ampère mass.

We turn now to conditions which rule out these type of examples, and lead to positive regularity results.

4.3. Monge–Ampère type equations. For the quadratic cost, finding a smooth optimal map was equivalent to solving the 2nd boundary value problem for the Monge–Ampère equation (1.6). Let us now derive the analogous equation for a more general cost, keeping in mind that whatever PDE we end up with cannot generally be better than degenerate-elliptic, since vanishing of $d\mu^- = f^- \,\mathrm{dVol}$ can lead to nonsmooth solutions in the interior of the support of $d\mu^+ = f^+ \,\mathrm{dVol}$.

Let us see what specific PDE emerges from the local expression $|\det DG(x)| = f^+(x)/f^-(G(x))$ for $G_\# \mu^+ = \mu^-$. Recall from Corollary 2.7 that $D^2_{xx} c(x, G(x)) + D^2 u(x) \geq 0$ and $D_x c(x, G(x)) + Du(x) = 0$. Differentiating the latter expression gives a relation

$$D^2_{xx}c\big(x, G(x)\big) + D^2_{xy}c\big(x, G(x)\big)DG(x) + D^2u(x) = 0$$

which can be solved for $DG(x)$ to yield

$$\det\Big(D^2u(x) + D^2_{xx}c\Big(x, Y\big(x, Du(x)\big)\Big)\Big) = \left|\det\big(D^2_{xy}c(x,y)\big)\frac{f^+(x)}{f^-(y)}\right|_{y=Y(x,Du(x))}. \tag{4.2}$$

Here we have assumed (A1)$^+$–(A2), the form $G(x) = Y\big(x, Du(x)\big)$ of the optimal map is from Theorem 2.9, and the boundary condition is $Y\big(x, Du(x)\big) \subset M^-$ for all $x \in M^+$. We have arrived as before at a fully-nonlinear second-order equation, whose linearization around any c-convex solution $u = u^{\tilde{c}c}$ is (at least degenerate-) elliptic.

4.4. Ma – Trudinger – Wang conditions for regularity. Sufficient conditions for the c-optimal map $G\colon M^+ \to M^-$ to be smooth between a pair of smooth bounded probability densities satisfying $\log f^\pm \in C^\infty(M^\pm)$ on compact domains $M^\pm \subset \mathbf{R}^n$ were found by Ma – Trudinger – Wang and Trudinger – Wang [97, 139]. The crucial condition on the cost $c(x,y)$ distinguishing Examples 4.1 – 4.3 from Example 4.4 above involves a quantity they identified, which other authors have variously dubbed the *Ma – Trudinger – Wang tensor* [145], *c-sectional curvature* [92], or *cross-curvature* [80]; cf. §6 below. To define it, let us adopt their convention that subscripts such as $c_{i,j} = \partial^2 c/\partial x^i \partial y^j$ and $c_{ij,kl} = \partial^4 c/\partial x^i \partial x^j \partial y^k \partial y^\ell$ indicate iterated derivatives in coordinates, with commas separating derivatives with respect to $x \in M^+$ from those with respect to variables $y \in M^-$. Let $c^{j,i}(x,y)$ denote the inverse matrix to $c_{i,j}(x,y)$.

Definition 4.5 (Cross-curvature). Given tangent vectors $p \in T_{x_0}M^+$ and $q \in T_{y_0}M^-$, define $\operatorname{cross}(p,q) := \big(-c_{ij,kl} + c_{ij,r}c^{r,m}c_{m,kl}\big)p^i p^j q^k q^l$. Here and subsequently, the Einstein summation convention is in effect.

The conditions assumed by Ma, Trudinger and Wang were the following [97]; our designations (A3)$_\mathrm{s}$ and (A3) correspond to their (A3) and (A3w) from [139]:

(A0) $c \in C^4(M^+ \times M^-)$, and for all (x_0, y_0) in the compact set $M^+ \times M^- \subset \mathbf{R}^n \times \mathbf{R}^n$;
(A1) $y \in M^- \mapsto D_x c(x_0, y)$ and $x \in M^+ \mapsto D_y c(x, y_0)$ are injective;
(A2) $\det D^2_{x^i,y^j} c(x_0, y_0) = \det(c_{i,j}) \neq 0$;
(A3) $\operatorname{cross}(p,q) \geq 0$ for all $(p,q) \in T_{(x_0,y_0)}M^+ \times M^-$ such that $p^i c_{i,j} q^j = 0$;
(A4) $M^-_{x_0} := D_x c(x_0, M^-) \subset \mathbf{R}^n$ and $M^+_{y_0} := D_y c(M^+, y_0) \subset T^*_{y_0}M^-$ are convex.

Among the variants on (A3) subsequently proposed [61, 80, 94], let us recall the *nonnegative cross-curvature* condition [80]:

(B3) $\operatorname{cross}(p,q) \geq 0$ for all $(p,q) \in T_{(x_0,y_0)}M^+ \times M^-$.

The first two conditions above are familiar from Theorems 2.9 and 3.2; (A1) was proposed independently of [97] in [65, 88], while there is an antecedent for (A2) in the economics literature [99]. The last condition (A4) adapts the convexity required by Delanoë, Caffarelli (Theorem 3.4) and Urbas, to the geometry of the cost function $c(x,y)$; when $M^+_{x_0}$ and $M^-_{y_0}$ are smooth and their convexity is strong — meaning the principal curvatures of their boundaries are all strictly positive — we denote it by $(A4)_s$. When inequality (A3) or (B3) holds strictly — and hence uniformly on the compact set $M^+ \times M^-$ — we denote that fact by $(A3)_s$ or $(B3)_s$ respectively.

Remark 4.6. The quadratic cost $c(x, y) = -x \cdot y$ of Brenier satisfies (B3) but not $(A3)_s$. Since we have already seen that (A3) is necessary [92] as well as sufficient for the continuity of optimal maps, the quadratic cost is actually a delicate borderline case. The negative $-c$ of any cost c satisfying $(A3)_s$ — including those of Examples 4.2–4.3 — necessarily violates (A3).

4.5. Regularity results. Assume (A0)–(A4) and $\log f^\pm \in C^\infty(M^\pm)$. Under the stronger condition $(A3)_s$, Ma, Trudinger and Wang [97] proved the interior regularity of the optimal map G and corresponding c-convex potential $u \in C^\infty(M^+_{\mathrm{int}})$; a flaw in their argument was later repaired in [140] (see [80] for another approach). Substituting strong convexity $(A4)_s$ for $(A3)_s$, but retaining (A3), Trudinger and Wang [139] used the continuity method to establish regularity up to the boundary $u \in C^\infty(M^+)$. Relaxing strong convexity to (A4) in that context is an open problem.

For densities merely satisfying $f^+/f^- \in L^\infty(M^+ \times M^-)$, under the strong condition $(A3)_s$, Loeper was able to establish local Hölder continuity of the optimal map — or equivalently $u \in C^{1,\alpha}_{\mathrm{loc}}(M^+_{\mathrm{int}})$ — with explicit Hölder exponent $\alpha = 1/(4n-1)$, using a direct argument [92] that we sketch out below. This exponent was later improved to its sharp value $\alpha = 1/(2n-1)$ by Liu [90]. For the quadratic cost $c(x, y) = -x \cdot y$, the best known estimates [64] for the Hölder exponent α are much worse, and depend on bounds for $\log(f^+/f^-)$. Assuming (A3) and $(A4)_s$ instead of $(A3)_s$, Figalli, Kim and McCann adapted Caffarelli's renormalization techniques [57] to derive continuity and injectivity of optimal maps, with a Hölder exponent on $U \subset\subset M^+$ which depends on the bound $\|\log(f^+/f^-)\|_{L^\infty(U\times M^-)}$ but not on choice of cost; using one of their arguments, a similar conclusion was also obtained by Figalli and Loeper [58] in the special case $n = 2$. Liu, Trudinger and Wang showed that higher regularity then follows from further assumptions on f^+/f^- in any dimension [91].

Using this theory, regularity results have now been obtained in geometries such as the round sphere [93], perturbations [46, 61, 63], submersions [46, 81] and products [56] thereof, and hyperbolic space [86, 89]. Significant cut-locus issues arise in this context. Loeper and Villani [94] conjecture, and in some cases have proved, that condition $(A3)_s$on the quadratic cost $c(x, y) = d^2(x, y)$ actually implies convexity of the domain of injectivity of the Riemannian exponential map $\exp_x : T_xM \to M$, unexpectedly linking the cross-curvature tensor to quite classical questions in Riemannian geometry.

4.6. Ruling out discontinuities: Loeper's maximum principle. Let us discuss how the condition (A3) rules out the tearing phenomenon which we saw on the saddle surface of Example 4.4.

Discontinuities in the optimal map $G(x) = Y\big(x, Du(x)\big)$ correspond to locations $x_0 \in M^+$ where differentiability of the potential function $u = u^{\tilde{c}c}$ fails, such as locations where the supremum

$$u(x_0) = \sup_{y \in M^-} -c(x_0, y) - u^{\tilde{c}}(y) \tag{4.3}$$

is attained by two or more points $y_0 \neq y_1 \in M^-$.

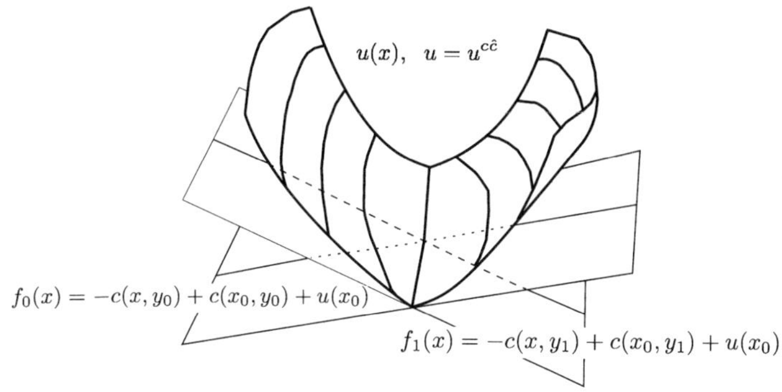

FIGURE 4. Discontinuous optimal maps arise from distinct supporting hyperplanes.

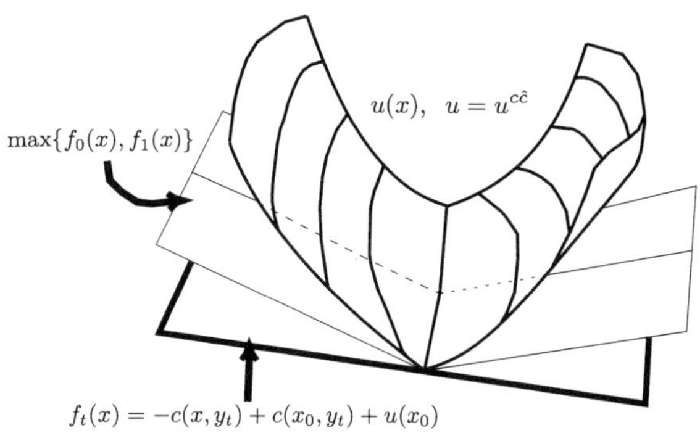

FIGURE 5. Can one c-affine support at x_0 be rotated to another, without exceeding u?

The set of such y is denoted by $\partial^c u(x_0)$, while the set of such pairs is denoted $\partial^c u \subset M^+ \times M^-$. Unless we can find a continuous curve $t \in [0,1] \mapsto y_t \in \partial^c u(x_0)$ which connects y_0 to y_1, it will be possible [92] to construct probability densities satisfying $\log f^\pm \in C^\infty(M^\pm) \cap L^\infty$ with a discontinuous optimal map as in §4.2 above. But there are not many possibilities to have such a curve.

In the classical case $c(x,y) = -x \cdot y$ (see Figures 4 and 5), $\partial^c u = \partial u$ and we have a convex function u with two different supporting planes at x_0.

In particular one may continuously rotate the first plane about the point $\big(x_0, u(x_0)\big)$ without ever crossing the graph of u until it agrees with the second plane giving a one-parameter family of supporting planes to u at the same point. This way one sees that $\partial u(x_0)$ contains a "segment" $\{y_t\}_{t\in(0,1)}$. In this special case $y_t = (1-t)y_0 + ty_1$ where y_0 and y_1 are the slopes of the original supporting hyperplanes. In the general case, the corresponding local picture forces the "c-segment" $\{y_t\}_{t\in(0,1)}$ given by

$$D_x c(x_0, y_t) = (1-t) D_x c(x_0, y_0) + t D_x c(x_0, y_1) \tag{4.4}$$

to be our only hope for a continuous path connecting y_0 to y_1 in $\partial^c u(x_0)$.

Now, were G to exhibit a discontinuity, this construction suggests G has to transport a very small mass around x_0 into a set with very large mass, which would give a contradiction given the constraint $G_\# \mu^+ = \mu^-$ and the assumptions on $\mu^\pm$.

This is indeed the case, at least under the stronger assumption $(A3)_s$, as we shall see below (Proposition 4.12 and Theorem 4.11).

All of this is of course contingent on whether the entire family of functions $\{f_t\}_t$ lie below $u(x)$, which might not be true for an arbitrary cost c; see Figure 5. Indeed, Loeper's key observation is that the Ma – Trudinger – Wang condition (A3) is what guarantees that any family of functions $f_t(y) = -c(x, y_t) + c(x_0, y_t)$ with y_t satisfying (4.4) never goes above $u(x)$. More precisely, it remains below $\max\{f_0(x), f_1(x)\}$.

Theorem 4.7 (Loeper's maximum principle [80, 92]**).** *If* (A0) – (A4) *hold and* $x_0 \in M^+$ *and* $(y_t)_{t\in[0,1]} \subset M^-$ *satisfy* (4.4), *then*

$$f(x,t) := -c(x, y_t) + c(x_0, y_t) \le \max\{f(x,0), f(x,1)\} \quad \forall (x,t) \in M^+ \times [0,1]. \tag{4.5}$$

Remark 4.8 ([81])**.** If in addition, (B3) holds, then $t \in [0,1] \to f(x,t)$ is convex.

Loeper's original proof was quite tortuous, relying on global regularity results for optimal transportation already established by Trudinger and Wang [139]. Here we sketch instead a simple, direct proof due to Kim and McCann [80], who later added Remark 4.8. A preliminary lemma gives some insight into the relevance of the cross-curvature.

Lemma 4.9 (A nontensorial expression for cross-curvature [80]**).** *Assuming* (A0) – (A4), *if* $\big(x(s)\big)_{-1\le s\le 1} \subset M^+$ *and* $\big(y(t)\big)_{-1\le t\le 1} \subset M^-$ *satisfy either*

$$\frac{\mathrm{d}^2}{\mathrm{d}t^2}\bigg|_{t=0} D_x c\big(x(0), y(t)\big) = 0 \quad \text{or} \quad \frac{\mathrm{d}^2}{\mathrm{d}s^2}\bigg|_{s=0} D_y c\big(x(s), y(0)\big) = 0,$$

then $\mathrm{cross}\big(\dot x(0), \dot y(0)\big) = -\big(\partial^4/(\partial^2 s \partial^2 t)\big)\big|_{s=0=t} c\big(x(s), y(t)\big)$.

This lemma is precisely analogous to the formula for the distance between two arclength parameterized geodesics $x(s)$ and $y(t)$ passing through $x(0) = y(0)$ in a Riemannian manifold:

$$d^2\big(x(s), \bar x(t)\big) = s^2 + t^2 - 2st\cos\theta - \frac{k}{3} s^2 t^2 \sin^2\theta + O\big((s^2+t^2)^{5/2}\big)$$

where θ is the angle between $\dot x(0)$ and $\dot y(0)$ and k is the sectional curvature of the plane which they span. Therefore, we will not give its proof.

Proof of Remark 4.8 and sketch of Theorem 4.7. Assume $(A3)_s$ for simplicity. It suffices to prove the following claim.

Claim 1. *if* $(\partial f/\partial t)(x, t_0) = 0$ *then* $(\partial^2 f/\partial t^2)(x, t_0) > 0$.

Proof of Claim 1. Convexity (A4) allows us to define $s \in [0,1] \mapsto x(s)$ by

$$D_y c\big(x(s), y(t_0)\big) = (1-s) D_y c\big(x_0, y(t_0)\big) + s D_y c\big(x, y(t_0)\big) \tag{4.6}$$

and $g(s) = (\partial^2 f/\partial t^2)(x(s), t_0)$. Our claim is that $g(1) > 0$. Since $f(x_0, t) = 0$ and hence $g(0) = 0$, to prove Claim 1 it suffices to establish strict convexity in Claim 2. □

Claim 2. $g\colon [0,1] \to \mathbf{R}$ *is convex, and minimized at* $s = 0$.

PROOF OF CLAIM 2. Once $g(s)$ is known to be convex, we need only observe that

$$g'(0) = -\frac{\partial^3}{\partial s \partial t^2}\Big|_{\substack{s=0\\t=t_0}} c\big(x(s), y(t)\big)$$

vanishes by our choice (4.4) of $y(t) = y_t$, to conclude $g(s)$ is minimized at $s = 0$. □

Why should $g(s)$ be convex? Note that $g''(s) = -\big(\partial^4/(\partial s^2 \partial t^2)\big)\big|_{t=t_0} c\big(x(s), y(t)\big)$ is already nonnegative according to Lemma 4.9 if we assume (B3). Remark 4.8 is thereby established. Under the weaker condition (A3), we need $\dot{x}^i(s) c_{i,j} \dot{y}^j(t_0) = 0$ to conclude $g(s)$ is convex — and strictly convex if $(\mathrm{A3})_{\mathrm{s}}$ holds. But

$$0 = \frac{\partial f}{\partial t}(x, t_0) = \int_0^1 c_{i,j}\big(x(s), y(t_0)\big) \dot{x}^i(s) \dot{y}^j(t_0) ds$$

and the integrand is constant by our construction (4.6) of $x(s)$. □

To deduce the continuity result of the next section, the following corollary is crucial.

Corollary 4.10. *Assume* (A0) – (A4) *and fix* $(x_0, y_0) \in M^+_{\mathrm{int}} \times M^-$. *If* $u = u^{\tilde{c}c}$ *satisfies*

$$u(x) \geq -c(x, y_0) + c(x_0, y_0) + u(x_0) \tag{4.7}$$

in a neighbourhood of x_0, *the same equality holds for all* $x \in M^+$.

PROOF. The local inequality (4.7) implies $p_0 := -D_x c(x_0, y_0) \in \partial u(x_0)$. If $x_0 \in \operatorname{dom} Du$, the conclusion is easy. The global inequality

$$u(x) \geq -c(x, y_1) + c(x_0, y_1) + u(x_0)$$

holds for any $(x_0, y_1) \in \partial^c u$, and for $y_1 \in \arg\max_{y \in M^-} -c(x_0, y) - u^{\tilde{c}}(y)$ in particular. The twist condition (A1) then implies $y_0 = y_1$.

Even if $x_0 \notin \operatorname{dom} Du$, taking, e.g., $p = -D_x c(x_0, y_1)$ yields

$$-c\big(x, Y(x_0, p)\big) + c\big(x_0, Y(x_0, p)\big) \leq u(x) - u(x_0) \quad \forall\, x \in M^+. \tag{4.8}$$

In fact, the set $P = \{p \in M^-_{x_0} \mid (4.8) \text{ holds}\}$ is convex, according to Theorem 4.7. On the other hand, P includes all the extreme points p of $\partial u(x_0)$, since the preceding argument can be applied to a sequence $(x_k, y_k) \in \partial^c u \cap (\operatorname{dom} Du \times M^-)$ with $\big(x_k, Du(x_k)\big) \to (x_0, p)$. Thus $P \supset \partial u(x_0)$, whence $p_0 \in P$ as desired. (In fact, $P = \partial u(x_0)$). □

4.7. Interior Hölder continuity for optimal maps. To conclude our discussion on regularity of optimal mappings, let us sketch Loeper's Hölder continuity result [92].

Theorem 4.11 (Loeper '09). *Assume* (A0) – (A2)*and* (A4).

(i) *If* (A3) *is violated, there exist probability densities with* $\log f^\pm \in C^\infty(M^\pm) \cap L^\infty$ *and a discontinuous optimal map* $G\colon M^+_{\mathrm{int}} \to M^-$ *satisfying* $G_\#(f^+ \,\mathrm{dVol}) = f^- \,\mathrm{dVol}$.

(ii) *Conversely, if* $(\mathrm{A3})_{\mathrm{s}}$ *holds and* $f^+/f^- \in L^\infty(M^+ \times M^-)$, *then* $G \in C^{1/(4n-1)}_{\mathrm{loc}}(M^+_{\mathrm{int}}, M^-)$.

In one dimension $n = 1$, we see G is Lipschitz directly from the equation $G'(x) = f^+(x)/f^-\big(G(x)\big)$. In higher dimensions, this theorem is a direct consequence of the following proposition, whose inequalities $\sim$ hold up to multiplicative constants depending only on the cost c, and in particular on the size of the uniform modulus of positivity in condition $(\mathrm{A3})_{\mathrm{s}}$.

Proposition 4.12 (Sausage into ball [92]). *Assuming the hypotheses and notation of Theorem 4.11(ii), take $x_0, x_1 \in M^+$ and set $\Delta x = x_1 - x_0$ and $\Delta y = y_1 - y_0$ where $y_i = G(x_i)$. If $|\Delta x| \lesssim |\Delta y|^5$, there is a ball $B_\epsilon(x) \supset G^{-1}(S_\delta)$ of radius $\epsilon \sim \sqrt{|\Delta x|/|\Delta y|}$ centered on the line segment joining x_0 to x_1, containing the preimage of the "sausage"*

$$S_\delta = \left\{ y \in M^- \;\Big|\; \inf_{t \in [1/3, 2/3]} |y - y_t| \le \delta \right\}$$

of radius $\delta \sim \epsilon|\Delta y|^2$ around the middle third of the curve $(y_t)_{t\in[0,1]} \subset M^-$ satisfying $0 = (\mathrm{d}^2/\,\mathrm{d}t^2) D_x c(x, y_t)$.

PROOF OF THEOREM 4.11(ii). At pairs of points $y_i = G(x_i)$ where $|\Delta x| \gtrsim |\Delta y|^5$ we already have Hölder exponent $\frac{1}{5}$ — even better than claimed. At other points, using the fact that G is a transport map between $\mu^\pm = f^\pm dx$, the Proposition yields $\mu^-(S_\delta) = \mu^+(G^{-1}(S_\delta)) \lesssim \|f^+\|_\infty \epsilon^n$, but also $\delta^{n-1}|\Delta y| \inf_{M^-} f^- \lesssim \mu^-(S_\delta)$. Combining the squares of these two inequalities, our choices $\delta \sim \epsilon|\Delta y|^2$ and $\epsilon^2 \sim |\Delta x|/|\Delta y|$ yield the desired Hölder estimate:

$$\|(f^-)^{-1}\|_\infty^{-2} \epsilon^{2n-2} |\Delta y|^{4n-2} \lesssim \|f^+\|_\infty^2 \epsilon^{2n-2} \frac{|\Delta x|}{|\Delta y|}.$$

Thus $G \in C^{1/(4n-1)}_{\mathrm{loc}}$. □

The proposition relies delicately on Corollary 4.10 and the correct choice of δ and ϵ:

PROOF SKETCH OF PROPOSITION 4.12 (cf. [82]). According to Theorem 2.9, the optimal map $G(x) = Y\big(x, Du(x)\big)$ is given by a potential $u = u^{\tilde{c}c}$ and $\mathrm{Graph}(G) \subset \partial^c u$. Thus $(x_i, y_i) \in \partial^c u$, meaning $f_i(x) = -c(x, y_i) + c(x_i, y_i) + u(x_i)$ satisfies $u(x) \ge \max\{f_0(x), f_1(x)\}$ with equality at x_0 and x_1. Take x to be the point on the segment joining x_0 to x_1 where $f_0(x) = f_1(x)$ ($= 0$ without loss of generality). The semiconvexity of u shown in Lemma 3.1 then yields the bound $u(x) \lesssim |\Delta x||\Delta y| + |\Delta x|^2$. Assumption $(\mathrm{A3})_{\mathrm{s}}$ allows Theorem 4.7 to be quantified, so that $f_t(\cdot) := -c(\cdot, y_t) + c(x, y_t) + u(x) \le u(\cdot)$ actually satisfies

$$f_t(x') - u(x') \lesssim -t(1-t)|x' - x|^2 |\Delta y|^2$$

for x' near x. For $t \in \left[\frac{1}{3}, \frac{2}{3}\right]$, these estimates give some leeway to shift y_t up to distance δ without spoiling the inequality $g_y(x') := -c(x', y) + c(x, y) + u(x) \le u(x')$ on the boundary $x' \in \partial B_\epsilon(x)$. Since $g_y(x) = u(x)$, this inequality does not extend to the interior of the ball $B_\epsilon(x)$, unless we subtract some nonnegative constant from $g_y(\cdot)$. Subtracting the smallest such constant λ yields a function $g_y(\cdot) - \lambda \le u(\cdot)$ on $B_\epsilon(x)$, with equality at some $x_* \in B_\epsilon(x)$. Corollary 4.10 implies $(x_*, y) \in \partial^c u$. For almost every such $y \in S_\delta$ this provides the desired preimage $x_* \in G^{-1}(y)$. □

5. Multidimensional screening: an application to economic theory

We now sketch an application [55] of the mathematics we have developed to one of the central problems in microeconomic theory: making pricing or policy decisions for a monopolist transacting business with a field of anonymous agents whose preferences are known only statistically. Economic keywords associated with problems of this type include "asymmetric information," "mechanism design," "incentive compatibility," "nonlinear pricing," "signalling," "screening," and the "principal/agent" framework.

5.1. Monopolist nonlinear pricing: the principal-agent framework. To describe the problem, imagine we are given: a set of "customer" types $M^+ \subset \mathbf{R}^n$ and "product" types $M^- \subset \mathbf{R}^n$ and

$b(x, y)$ = benefit of product $y \in M^-$ to customer $x \in M^+$;
$a(y)$ = monopolist's cost to manufacture $y \in M^-$;
$d\mu^+(x) \geq 0$ relative frequency of different customer types on M^+.

Knowing all this data, the **Monopolist's problem** is to assign a price to each product, for which she will be willing to manufacture that product and sell it to whichever agents choose to buy it. Her task is to design the price menu $v\colon M^- \to \mathbf{R} \cup \{+\infty\}$ so as to maximize profits. The only constraint that prevents her from raising prices arbitrarily high is the existence of a fixed $y_\varnothing \in M^-$, called the "outside option" or "null product," which she is compelled to sell at cost $v(y_\varnothing) = a(y_\varnothing)$. Though it is not necessary, we can fix the cost of the null product to vanish without loss of generality.

The **Agent's problem** consists in computing

$$u(x) = \max_{y \in M^-} b(x, y) - v(y) \tag{5.1}$$

and choosing to buy that product $y_{b,v}(x)$ for which the maximum is attained. The monopolist is generally called the *principal*, while the customers are called *agents*.

Economists use this framework to model many different types of transactions, including tax policy [111] (where the government wants to decide a tax structure which encourages people both to work and report income), contract theory [132] (where a company wants to decide a salary structure which attracts and rewards effective employees without overpaying them), and the monopolist nonlinear pricing problem described above [114]. In the initial studies, the type spaces $M^\pm$ were assumed one-dimensional, with $x \in M^+$ representing the innate ability or talent of the prospective tax-payer or employee, and $y \in M^-$ the amount of work that he chooses to do or the credentials he chooses to acquire. The basic insight of Mirrlees and Spence was that under condition (A2) (which implies (A1) in a single dimension) the variables $x, y \in \mathbf{R}$ would be monotonically correlated by the optimal solution, reducing the monopolist's problem to an ordinary differential equation. For this reason the one-dimensional versions of (A1)–(A2) are called *Spence–Mirrlees* (or *single-crossing*) conditions in the economics literature; both Mirrlees and Spence were awarded Nobel prizes for exploring the economic implications of their solution.

Of course, many types of products are more realistically modeled using several parameters $y \in \mathbf{R}^n$ — in the case of cars these might include fuel efficiency, size, comfort, safety, reliability, and appearance — while the preferences of customers for such parameters are similarly nuanced. Thus it is natural and desirable to want to solve the multidimensional version $n \geq 2$ of the problem, about which much less is known [12]. Monteiro and Page [113] and independently Carlier [26] showed only that enough compactness remains to conclude that the monopolist's optimal strategy exists. An earlier connection to optimal transportation can be discerned in the work of Rochet [122], who proved a version of Theorem 2.3 (Rockafellar) for general utility functions b ($= -c$ in our earlier notation).

Rochet and Choné [123] studied the special case $b(x,y) = x{\cdot}y$ on $M^{\pm} = [0,\infty[^n$. Taking $a(y) = \frac{1}{2}|y|^2$, $d\mu^+ = \chi_{[0,1]^2}\, d^2x$, and $y_\varnothing = (0,0)$, they deduced that the mapping $y_{b,v}\colon M^+ \to M^-$ was the gradient of a convex function; it sends a positive fraction of the square to the point mass $y_\varnothing$, and a positive fraction to the line segment $y_1 = y_2$, while the remaining positive fraction gets mapped in a bijective manner, so that

$$\mu^- := (y_{b,n})_{\#}\mu^+ = f_0^-\delta_{y_\varnothing} + f_1^-\, d\mathcal{H}^1 + f_2^-\, d\mathcal{H}^2. \tag{5.2}$$

They interpreted this solution to mean that while the top end of the market gets customized vehicles f_2^-, price discrimination alone forces those customers in the next market segment to choose from a more limited set f_1^- of economy vehicles offering a compromise between attributes y_1 and y_2. A fraction $f_0^- > 0$ of consumers will be priced out of the market altogether — which had already been observed by Armstrong [10] to be a hallmark of nonlinear pricing in more than one dimension $n \geq 2$. Economists refer to this general phenomenon (5.2) as "bunching," and to the fact that $f_0^- > 0$ as "the desirability of exclusion."

How robust is this picture? It remains a pressing question to understand whether the bunching phenomena of Rochet and Choné is robust, or merely an accident of the particular example they explored. As we now explain, their results were obtained by reducing the monopolist's problem to the minimization of a Dirichlet energy:

$$\min_{0 \leq u \text{ convex}} \int_{[0,1]^2} \left(\tfrac{1}{2}|Du|^2 - \langle x, Du(x)\rangle + u(x)\right) d\mathcal{H}^2(x). \tag{5.3}$$

The constraint that $u\colon M^+ \to \mathbf{R}$ be *convex* makes this problem nonstandard: its solution satisfies a Poisson type equation only on the set where u is strongly convex ($\operatorname{rank} D^2u = 2 = n$), and there are free boundaries separating the regions where the different constraints $u \geq 0$ and $D^2u \geq 0$ begin to bind ($\operatorname{rank} D^2u = 0$ or 1 respectively).

5.2. Variational formulation using optimal transportation. The principal's problem is to choose $v\colon M^- \to \mathbf{R} \cup \{+\infty\}$ to maximize her profits, or equivalently to minimize her net losses:

$$\min_{\{v \mid v(y_\varnothing) = a(y_\varnothing)\}} \int_{M^+} \left[a\big(y_{b,v}(x)\big) - v\big(y_{b,v}(x)\big)\right] d\mu^+(x). \tag{5.4}$$

Note that the integrand vanishes for all customers x who choose the null product $y_{b,v}(x) = y_\varnothing$.

Where the agent's maximum (5.1) is achieved, we have $Du(x) = D_x b\big(x, y_{b,v}(x)\big)$, so using the twist condition (A1) from our previous lectures we can invert this relation to get $y_{b,v}(x) = Y\big(x, Du(x)\big)$. Moreover, the function $u(x)$ from (5.1) is a b-convex function, called the *surplus* or *indirect utility* $u = u^{\tilde{b}b}$. (In our previous notation, u is a $(-b)$-convex function and $u = u^{(\widetilde{-b})(-b)}$, but we suppress the minus signs hereafter.)

Since $v(Y(x, Du(x)) = b(x, Y(x, Du(x)) - u(x)$, we may reformulate the variational problem (5.4) as the minimization of the principal's losses

$$L(u) := \int_{M^+} \Big[a\Big(Y\big(x, Du(x)\big)\Big) - b\Big(x, Y\big(x, Du(x)\big)\Big) + u(x)\Big] \,\mathrm{d}\mu^+(x). \tag{5.5}$$

over the set $\mathcal{U}_\varnothing = \{u \in \mathcal{U} \mid u \geq u_\varnothing\}$ of b-convex functions $\mathcal{U} = \{u \mid u = u^{\tilde{b}b}\}$ which exceed the reservation utility $u_\varnothing(\cdot) = b(\cdot, y_\varnothing) - a(y_\varnothing)$ associated with the outside option or null product. This strange reformulation due to Carlier [26]

$$\min_{u \in \mathcal{U}_\varnothing} L(u) \tag{5.6}$$

reduces to (5.3) in the case considered by Rochet and Choné.

5.3. When is this optimization problem convex? From [26, 113] we know that a minimizer exists. The contribution of Figalli, Kim and McCann is to give sufficient conditions for the variational problem (5.6) to become convex — in which case it is considerably simpler to analyze, theoretically and computationally. It is very interesting that the Ma, Trudinger and Wang criteria for the regularity of optimal mappings turn out to be related to this question. The following are among the main results of [55]:

Theorem 5.1 (Convexity of the principal's strategy space [55]). *If $b = (-c)$ satisfies* (A0) – (A2) *and* (A4) *then the set $\mathcal{U} = \{u = u^{\tilde{b}b}\}$ is convex if and only if* (B3) *holds, i.e., if and only if* $\mathrm{cross}(p, q) \geq 0$ *for all tangent vectors $(p, x_0) \in TM^+$ and $(q, y_0) \in TM^-$.*

Remark 5.2. It was pointed out subsequently by Brendan Pass [117] that the convexity of $M^-_{x_0}$ assumed in (A4) for each $x_0 \in M^+$ is also necessary for convexity of $\mathcal{U}$.

Sketch of proof. First assume (B3) holds — assuming always (A0) – (A2) and (A4). Given $u_0, u_1 \in \mathcal{U}$ and $t \in [0, 1]$ we claim $u_t := (1 - t)u_0 + tu_1$ is b-convex. This can be established by finding for each $x_0 \in M^+$ a $y_t \in M^-$ such that

$$u_t(\cdot) \geq b(\cdot, y_t) - b(x_0, y_t) + u_t(x_0) \quad \text{throughout } M^+, \tag{5.7}$$

for then $u_t(\cdot)$ is the supremum of such functions. Corresponding to $t = 0, 1$ the desired points $y_0, y_1 \in M^-$ exist, by b-convexity of $u_i = u_i^{\tilde{b}b}$ for $i = 0, 1$. By (A4), we can solve the equation $D_x b(x_0, y_t) = (1 - t)D_x b(x_0, y_0) + tD_x b(x_0, y_t)$; the solution $y_t \in M^-$ makes $f(\cdot, t) := b(\cdot, y_t) - b(x_0, y_t)$ a convex function of $t \in [0, 1]$, according to Remark 4.8. Inequality (5.7) holds at the endpoints $t = 0, 1$; taking a convex combination yields the desired inequality for intermediate values of $t \in [0, 1]$. For the converse direction, we refer to [55]. □

Theorem 5.3 (Convexity of principal's losses and uniqueness of optimal strategy [55]). *If* (A0) – (A4) *and* (B3) *hold and if $a = a^{b\tilde{b}}$, then the functional $u \in \mathcal{U} \mapsto L(u)$ defined by* (5.5) *is convex. Furthermore, it has enough strict convexity to conclude the optimizer $u \in \mathcal{U}_\varnothing$ is uniquely determined (at least μ^+-a.e.) if $\mu^+ \ll \mathcal{H}^n$ and either* (i) *$y_{b,a}\colon M^+ \to M^-$ is continuous or else* (ii) *b has positive cross-curvature* $(\mathrm{B3})_{\mathrm{s}}$.

Sketch of proof. To deduce convexity of $L\colon \mathcal{U} \to \mathbf{R}$, recall $\tilde{b}$-convexity of a implies

$$a\big(Y(x,p)\big) - b\big(x, Y(x,p)\big) = \sup_{x_1 \in M^+} b\big(x_1, Y(x,p)\big) - b\big(x, Y(x,p)\big) - a^b(x_1).$$

For each $x \in M^+$ fixed, the functions under the supremum are convex with respect to $p \in M^-_x$, according to Remark 4.8. Thus the integrand in (5.5) is linear in $u(x)$ and convex with respect to $p = Du(x)$, which establishes the desired convexity of the integral $L(u)$. In case (i) the integrand is strictly convex, while in case (ii) it is strictly convex for all $x \in \operatorname{dom} Da^b$, which is a set of full $\mu^+ \ll \mathcal{H}^n$ measure. We refer to [55] for details. □

Regarding robustness: we may mention that, as in Remark 4.6, the bilinear function $b(x,y)$ lies on the borderline of costs which satisfy (B3). Thus there will be perturbations of this function which destroy convexity of the problem, and we can anticipate that under such perturbations, uniqueness and other properties of its solution may no longer persist. In fact, for $a = 0$ and $b(x,y) = -d^2_M(x,y)$ on a Riemannian ball $M^+ = M^- = B_r(y_\varnothing)$, we arrive at a problem equivalent to a fourfold symmetrized version of Rochet and Choné's in the Euclidean case, but which satisfies or violates (B3) depending on whether the metric is spherical or hyperbolic. This can used to model local delivery of a centralized resource for a town in the mountains [55]; cf. Examples 4.3 – 4.4. On the other hand, under the hypotheses of Theorem 5.3 we are able to show that Armstrong's "desirability of exclusion" [10] continues to hold. We give the statement only and refer to [55] for its proof.

Theorem 5.4 (The desirability of exclusion). *Assume* (A0) – (A4), (B3), $a = a^{b\tilde{b}}$ *and that* $\mathrm{d}\mu^+ = f^+ \,\mathrm{d}\mathcal{H}^n$ *with* $f^+ \in W^{1,1}(M^+)$ *and the convex set* $M^-_{y_\varnothing} = D_y b(M^+, y_\varnothing) \subset \mathbf{R}^n$ *has no* $(n-1)$*-dimensional facets. Then a positive fraction of agents will be priced out of the market by the principal's optimal strategy.*

Remark 5.5. It is interesting to note that $M^+_{y_\varnothing}$ actually has $n-1$ facets if $n = 1$ — where Armstrong noted counterexamples to the desirability of exclusion — and for the example of Rochet-Choné discussed above, where $M^+_{y_\varnothing}$ is the unit square.

5.4. Variant: maximizing social welfare. Idealistic readers may be taken somewhat aback by the model just presented, for it is the very theory which predicts, among other things, just how uncomfortable airlines ought to make their economy seating to ensure — without sacrificing too much economy-class revenue — that passengers with the means to secure a business-class ticket have sufficient incentive to do so. Such readers will doubtless be glad to know that the same mathematics is equally relevant to the more egalitarian question of how to price public services so as to maximize societal benefit.

For example, suppose the welfare $w(x, u(x))$ of agent x is a concave function of the indirect utility $u(x)$ he receives. A public service provider would like to set a price menu for which $u = v^b$ maximizes the total welfare among all agents:

$$\max_{\substack{u \in \mathcal{U}_\varnothing \\ L(u) \le 0}} \int_{M^+} w\big(x, u(x)\big)\, \mathrm{d}\mu^+(x),$$

subject to the constraint $L(u) \le 0$ that the service provider not sustain losses. Introducing a Lagrange multiplier λ for this budget constraint, the problem can be rewritten in the unconstrained form

$$\max_{u \in \mathcal{U}_\varnothing} -\lambda L(u) + \int_{M^+} w\big(x, u(x)\big)\, \mathrm{d}\mu^+(x),$$

(for a suitable $\lambda \geq 0$). Under the same assumptions (A0) – (A2), (A4) and $a = a^{b\tilde{b}}$ as before, we see from the results above that this becomes a concave maximization problem if and only if (B3) holds, in which case existence and uniqueness of solution follow directly, and the problem is rendered amenable to further study, both theoretical and computational.

6. A pseudo-Riemannian and symplectic geometric afterword

The conditions of Ma, Trudinger and Wang for regularity of optimal transport have a differential geometric significance uncovered by Kim and McCann [80], which led to their discovery with Warren [83] of a surprising connection of optimal transport to the theory of volume-maximizing special Lagrangian submanifolds in split geometries.

Indeed, since the smoothness of optimal maps $G\colon M^+ \to M^-$ is a question whose answer is independent of coordinates chosen on M^+ and M^-, it follows that the necessary and sufficient condition (A3) for continuity in Theorem 4.11 should have a geometrically invariant description. We give this description, below, as the positivity of certain sectional curvatures of a metric tensor h induced on the product manifold $N := M^+ \times M^-$ by the cost function $c \in C^4(N)$. This motivates the appellation *cross-curvature*. The rationale for such a description to exist is quite analogous to that underlying general relativity, Einstein's theory of gravity, which can be expressed in the language of pseudo-Riemannian geometry due to the coordinate invariance that results from the equivalence principle (observer independence).

Use the cost function to define the symmetric and antisymmetric tensors

$$h = \sum_{i=1}^{n} \sum_{j=1}^{n} \frac{\partial^2 c}{\partial x^i \partial y^j} (\mathrm{d}x^i \otimes \mathrm{d}y^j + \mathrm{d}y^j \otimes \mathrm{d}x^i) \tag{6.1}$$

$$\omega = \sum_{i=1}^{n} \sum_{j=1}^{n} \frac{\partial^2 c}{\partial x^i \partial y^j} (\mathrm{d}x^i \otimes \mathrm{d}y^j - \mathrm{d}y^j \otimes \mathrm{d}x^i) \tag{6.2}$$

on $N = M^+ \times M^-$. Then condition (A2) is equivalent to nondegeneracy of the metric tensor h, which in turn is equivalent to the assertion that ω is a symplectic form. Note however that h is not positive-definite, but has equal numbers of positive and negative eigenvalues in any chosen coordinates, i.e., signature (n, n). Conditions (A3) and (A4) are conveniently re-expressed in terms of the pseudo-metric h, and its pseudo-Riemannian curvature tensor R_{ijkl} [80]. Indeed, condition (A4) asserts the h-geodesic convexity of $\{x_0\} \times M^-$ and $M^+ \times \{y_0\}$, while the formula

$$\operatorname{cross}(p, q) = R_{ijkl} p^i q^j p^k q^l$$

shows the cross-curvature is simply proportional to the pseudo-Riemannian sectional curvature of the 2-plane $(p \oplus 0) \wedge (0 \oplus q)$. The restriction distinguishing (A3) from (B3) is that $p \oplus q$ be lightlike, which is equivalent to the h-orthogonality of $p \oplus 0$ and $0 \oplus q$.

Kim and McCann went on to point out that the graph of any c-optimal map is h-spacelike and ω-Lagrangian, meaning any tangent vectors $P, Q \in T_{(x,G(x))}N$

to this graph satisfy $h(P,P) \geq 0$ and $\omega(P,Q) = 0$. This is a consequence of Corollary 2.7; it is also very illustrative to check it directly by hand in the case of the quadratic cost in $\mathbb{R}^n$. When (A0)–(A4) hold they also showed the converse to be true: any diffeomorphism whose graph is h-spacelike and ω-Lagrangian is also c-cyclically monotone, hence optimal. With Warren [83], they introduced a pseudo-metric

$$h_c^{f^\pm} = \left(\frac{f^+(x) f^-(y)}{|\det c_{i,j}(x,y)|} \right)^{1/n} h \tag{6.3}$$

conformally equivalent to h, but whose volume element is given by $\det h_c^{f^\pm} = \left(f^+(x)f^-(y)\right)^2$. In this new metric, they show the graph of the c-optimal map pushing $\mathrm{d}\mu^+(x) = f^+(x)\,\mathrm{d}^n x$ forward to $\mu^-(y) = f^-(y)\,\mathrm{d}^n y$ has maximal area with respect to compactly supported perturbations. In particular, Graph(G) has zero mean-curvature as a submanifold (with half the dimension) of $(N, h_c^{f^\pm})$—yielding another unexpected connection of optimal transportation to more classical problems in geometry and geometric measure theory. (Note that the metric (6.3) depends only on the measures $f^\pm$ and the sign of the mixed partial $D^2_{xy}c$ in dimension $n = 1$.) The preprint of Harvey and Lawson [73] contains a wealth of related information concerning special Lagrangian submanifolds in pseudo-Riemannian (= semi-Riemannian) geometry.

References

1. T. Abdellaoui and H. Heinich, *Sur la distance de deux lois dans le cas vectoriel*, C. R. Acad. Sci. Paris Sér. I Math. **319** (1994), no. 4, 397–400.
2. M. Agueh, N. Ghoussoub, and X. Kang, *Geometric inequalities via a general comparison principle for interacting gases*, Geom. Funct. Anal. **14** (2004), no. 1, 215–244.
3. N. Ahmad, H. K. Kim, and R. J. McCann, *Optimal transportation, topology and uniqueness*, Bull. Math. Sci. **1** (2011), no. 1, 13–32.
4. G. Alberti and L. Ambrosio, *A geometrical approach to monotone functions in* $\mathbf{R}^n$, Math. Z. **230** (1999), no. 2, 259–316.
5. S. Alesker, S. Dar, and V. Milman, *A remarkable measure preserving diffeomorphism between two convex bodies in* $\mathbf{R}^n$, Geom. Dedicata **74** (1999), no. 2, 201–212.
6. L. Ambrosio, *Lecture notes on optimal transport problems*, Mathematical Aspects of Evolving Interfaces (Funchal, 2000), Lecture Notes in Math., vol. 1812, Springer, Berlin, 2003, pp. 1–52.
7. L. Ambrosio and N. Gigli, *A user's guide to optimal transport*, Lecture Notes in Math., Springer, Berlin, to appear, available at `http://cvgmt.sns.it/paper/195/`.
8. L. Ambrosio, N. Gigli, and G. Savaré, *Gradient flows in metric spaces and in the space of probability measures*, Lectures Math. ETH Zurich, Birkhäuser, Basel, 2005.
9. E. J. Anderson and P. Nash, *Linear programming in infinite-dimensional spaces*, Wiley-Intersci. Ser. Discrete Math. Optim., Wiley, Chichester, 1987.
10. M. Armstrong, *Multiproduct nonlinear pricing*, Econometrica **64** (1996), no. 1, 51–75.
11. F. Barthe, *On a reverse form of the Brascamp–Lieb inequality*, Invent. Math. **134** (1998), no. 2, 335–361.
12. S. Basov, *Multidimensional screening*, Stud. Econom. Theory, vol. 22, Springer, Berlin, 2005.
13. P. Bernard and B. Buffoni, *Optimal mass transportation and Mather theory*, J. Eur. Math. Soc. (JEMS) **9** (2007), no. 1, 85–121.
14. M. Bernot, V. Caselles, and J.-M. Morel, *Optimal transportation networks: Models and theory*, Lecture Notes in Math., vol. 1955, Springer, Berlin, 2009.
15. S. Bianchini and F. Cavalletti, *The Monge problem in geodesic spaces*, Nonlinear Conservation Laws and Applications (Minneapolis, MN, 2009) (A. Bressan, G.-Q. G. Chen, M. Lewicka, and D. Wang, eds.), IMA Vol. Math. Appl., vol. 153, Springer, New York, 2011, pp. 217–233.

16. G. Bouchitté and G. Buttazzo, *Characterization of optimal shapes and masses through Monge–Kantorovich equation*, J. Eur. Math. Soc. (JEMS) **3** (2001), no. 2, 139–168.
17. G. Bouchitté, W. Gangbo, and P. Seppecher, *Michell trusses and lines of principal action*, Math. Models Methods Appl. Sci. **18** (2008), no. 9, 1571–1603.
18. Y. Brenier, *Décomposition polaire et réarrangement monotone des champs de vecteurs*, C. R. Acad. Sci. Paris Sér. I Math. **305** (1987), no. 19, 805–808.
19. ———, *Polar factorization and monotone rearrangement of vector-valued functions*, Comm. Pure Appl. Math. **44** (1991), no. 4, 375–417.
20. L. A. Caffarelli, *Boundary regularity of maps with convex potentials*, Comm. Pure Appl. Math. **45** (1992), no. 9, 1141–1151.
21. ———, *The regularity of mappings with a convex potential*, J. Amer. Math. Soc. **5** (1992), no. 1, 99–104.
22. ———, *Allocation maps with general cost functions*, Partial Differential Equations and Applications (P. Marcellini, G. G. Talenti, and E. Vesentini, eds.), Lecture Notes in Pure and Appl. Math., vol. 177, Dekker, New York, 1996, pp. 29–35.
23. ———, *Boundary regularity of maps with convex potentials.* II, Ann. of Math. (2) **144** (1996), no. 3, 453–496.
24. L. A. Caffarelli, M. Feldman, and R. J. McCann, *Constructing optimal maps for Monge's transport problem as a limit of strictly convex costs*, J. Amer. Math. Soc. **15** (2002), no. 1, 1–26 (electronic).
25. L. A. Caffarelli and R. J. McCann, *Free boundaries in optimal transport and Monge–Ampère obstacle problems*, Ann. of Math. (2) **171** (2010), no. 2, 673–730.
26. G. Carlier, *A general existence result for the principal-agent problem with adverse selection*, J. Math. Econom. **35** (2001), no. 1, 129–150.
27. G. Carlier and I. Ekeland, *Equilibrium structure of a bidimensional asymmetric city*, Nonlinear Anal. Real World Appl. **8** (2007), no. 3, 725–748.
28. J. A. Carrillo, R. J. McCann, and C. Villani, *Contractions in the 2-Wasserstein length space and thermalization of granular media*, Arch. Ration. Mech. Anal. **179** (2006), no. 2, 217–263.
29. P. Castillon, *Submanifolds, isoperimetric inequalities and optimal transportation*, J. Funct. Anal. **259** (2010), no. 1, 79–103.
30. T. Champion and L. De Pascale, *The Monge problem in* $\mathbb{R}^d$, Duke Math. J. **157** (2011), no. 3, 551–572.
31. T. Champion, L. De Pascale, and P. Juutinen, *The* ∞*-Wasserstein distance: local solutions and existence of optimal transport maps*, SIAM J. Math. Anal. **40** (2008), no. 1, 1–20.
32. S.-Y. A. Chang and Y. Wang, *On Aleksandrov–Fenchel inequalities for k-convex domains*, Milan J. Math. **79** (2011), no. 1, 13–38.
33. P.-A. Chiappori, R. J. McCann, and L. P. Nesheim, *Hedonic price equilibria, stable matching, and optimal transport: equivalence, topology, and uniqueness*, Econom. Theory **42** (2010), no. 2, 317–354.
34. D. Cordero-Erausquin, *Sur le transport de mesures périodiques*, C. R. Acad. Sci. Paris Sér. I Math. **329** (1999), no. 3, 199–202.
35. ———, *Inégalité de Prékopa-Leindler sur la sphère*, C. R. Acad. Sci. Paris Sér. I Math. **329** (1999), no. 9, 789–792.
36. D. Cordero-Erausquin, R. J. McCann, and M. Schmuckenschläger, *A Riemannian interpolation inequality à la Borell, Brascamp and Lieb*, Invent. Math. **146** (2001), no. 2, 219–257.
37. ———, *Prékopa–Leindler type inequalities on Riemannian manifolds, Jacobi fields, and optimal transport*, Ann. Fac. Sci. Toulouse Math. (6) **15** (2006), no. 4, 613–635.
38. D. Cordero-Erausquin, B. Nazaret, and C. Villani, *A mass-transportation approach to sharp Sobolev and Gagliardo–Nirenberg inequalities*, Adv. Math. **182** (2004), no. 2, 307–332.
39. J. A. Cuesta and C. Matrán, *Notes on the Wasserstein metric in Hilbert spaces*, Ann. Probab. **17** (1989), no. 3, 1264–1276.
40. J. A. Cuesta-Albertos and A. Tuero-Díaz, *A characterization for the solution of the Monge–Kantorovich mass transference problem*, Statist. Probab. Lett. **16** (1993), no. 2, 147–152.
41. M. J. P. Cullen, *A mathematical theory of large-scale atmosphere/ocean flows*, Imperial College Press, London, 2006.
42. M. J. P. Cullen, R. J. Douglas, I. Roulstone, and M. J. Sewell, *Generalized semi-geostrophic theory on a sphere*, J. Fluid Mech. **531** (2005), 123–157.

43. M. J. P. Cullen and R. J. Purser, *An extended Lagrangian theory of semigeostrophic frontogenesis*, J. Atmospheric Sci. **41** (1984), no. 9, 1477–1497.
44. ______, *Properties of the Lagrangian semigeostrophic equations*, J. Atmospheric Sci. **46** (1989), no. 17, 2684–2697.
45. P. Delanoë, *Classical solvability in dimension two of the second boundary-value problem associated with the Monge – Ampère operator*, Ann. Inst. H. Poincaré Anal. Non Linéaire **8** (1991), no. 5, 443–457.
46. P. Delanoë and Y. Ge, *Regularity of optimal transport on compact, locally nearly spherical, manifolds*, J. Reine Angew. Math. **646** (2010), 65–115.
47. P. Delanoë and F. Rouvière, *Positively curved Riemannian locally symmetric spaces are positively square distance curved*, Canad. J. Math., to appear, available at `http://hal-unice.archives-ouvertes.fr/hal-00667519`.
48. R. L. Dobrushin, *Definition of a system of random variables by means of conditional distributions*, Teor. Verojatnost. i Primenen. **15** (1970), 469–497 (Russian); English transl., Theor. Probab. Appl. **15** (1970), no. 3, 458–486.
49. R. M. Dudley, *Probabilities and metrics: Convergence of laws on metric spaces, with a view to statistical testing*, Lecture Notes Ser., vol. 45, Aarhus Univ., Matematisk Institut, Aarhus, 1976.
50. I. Ekeland, *An optimal matching problem*, ESAIM Control Optim. Calc. Var. **11** (2005), no. 1, 57–71 (electronic).
51. L. C. Evans, *Partial differential equations*, Grad. Stud. Math., vol. 19, Amer. Math. Soc., Providence, RI, 1998.
52. L. C. Evans and W. Gangbo, *Differential equations methods for the Monge – Kantorovich mass transfer problem*, Mem. Amer. Math. Soc. **137** (1999), no. 653.
53. A. Fathi and A. Figalli, *Optimal transportation on non-compact manifolds*, Israel J. Math. **175** (2010), 1–59.
54. A. Figalli and N. Gigli, *Local semiconvexity of Kantorovich potentials on non-compact manifolds*, ESAIM Control Optim. Calc. Var. **17** (2011), no. 3, 648–653.
55. A. Figalli, Y.-H. Kim, and R. J. McCann, *When is multidimensional screening a convex program?*, J. Econom. Theory **146** (2011), no. 2, 454–478.
56. ______, *Regularity of optimal transport maps on multiple products of spheres*, J. Eur. Math. Soc., to appear, available at `arXiv:1006.1957`.
57. ______, *Hölder continuity and injectivity of optimal maps*, available at `arXiv:1107.1014`.
58. A. Figalli and G. Loeper, *C^1 regularity of solutions of the Monge – Ampère equation for optimal transport in dimension two*, Calc. Var. Partial Differential Equations **35** (2009), no. 4, 537–550.
59. A. Figalli and F. Maggi, *On the shape of liquid drops and crystals in the small mass regime*, Arch. Ration. Mech. Anal. **201** (2011), no. 1, 143–207.
60. A. Figalli, F. Maggi, and A. Pratelli, *A mass transportation approach to quantitative isoperimetric inequalities*, Invent. Math. **182** (2010), no. 1, 167–211.
61. A. Figalli and L. Rifford, *Continuity of optimal transport maps and convexity of injectivity domains on small deformations of $\mathbb{S}^2$*, Comm. Pure Appl. Math. **62** (2009), no. 12, 1670–1706.
62. A. Figalli, L. Rifford, and C. Villani, *On the Ma – Trudinger – Wang curvature on surfaces*, Calc. Var. Partial Differential Equations **39** (2010), no. 3-4, 307–332.
63. ______, *Nearly round spheres look convex*, Amer. J. Math. **134** (2012), no. 1, 109–139.
64. L. Forzani and D. Maldonado, *Properties of the solutions to the Monge – Ampère equation*, Nonlinear Anal. **57** (2004), no. 5-6, 815–829.
65. W. Gangbo, *Quelques problèmes d'analyse non convexe*, Habilitation Thesis, Université de Metz, 1995.
66. W. Gangbo and R. J. McCann, *Optimal maps in Monge's mass transport problem*, C. R. Acad. Sci. Paris Sér. I Math. **321** (1995), no. 12, 1653–1658.
67. ______, *The geometry of optimal transportation*, Acta Math. **177** (1996), no. 2, 113–161.
68. ______, *Shape recognition via Wasserstein distance*, Quart. Appl. Math. **58** (2000), no. 4, 705–737.
69. N. Gigli, *On the inverse implication of Brenier – McCann theorems and the structure of $(\mathcal{P}_2(M), W_2)$*, Methods Appl. Anal. **18** (2011), no. 2, 127–158.

70. D. Gilbarg and N. S. Trudinger, *Elliptic partial differential equations of second order*, 2nd ed., Grundlehren Math. Wiss., vol. 224, Springer, Berlin, 1983.
71. T. Glimm and V. Oliker, *Optical design of single reflector systems and the Monge-Kantorovich mass transfer problem*, J. Math. Sci. (N. Y.) **117** (2003), no. 3, 4096–4108. Nonlinear problems and function theory.
72. S. Haker, L. Zhu, Tannenbaum Allen, and S. Angenent, *Optimal mass transport for registration and warping*, Int. J. Comput. Vis. **60** (2004), no. 3, 225–240.
73. F. R. Harvey and H. B. Lawson Jr., *Split special Lagrangian geometry*, available at `arXiv:1007.0450`.
74. S. Kakutani, *A generalization of Brouwer's fixed point theorem*, Duke Math. J. **8** (1941), 457–459.
75. L. V. Kantorovich, *On the translocation of masses*, C. R. (Doklady) Acad. Sci. URSS (N.S.) **37** (1942), 199–201.
76. ———, *On a problem of Monge*, Uspekhi Mat. Nauk **3** (1948), no. 2, 225–226 (Russian).
77. L. V. Kantorovich and G. Sh. Rubinstein, *On a space of completely additive functions*, Vestnik Leningrad. Univ. **13** (1958), no. 7, 52–59 (Russian,).
78. H. G. Kellerer, *Duality theorems for marginal problems*, Z. Wahrsch. Verw. Gebiete **67** (1984), no. 4, 399–432.
79. Y.-H. Kim, *Counterexamples to continuity of optimal transport maps on positively curved Riemannian manifolds*, Int. Math. Res. Not. IMRN (2008), Art. ID rnn120.
80. Y.-H. Kim and R. J. McCann, *Continuity, curvature, and the general covariance of optimal transportation*, J. Eur. Math. Soc. (JEMS) **12** (2010), no. 4, 1009–1040.
81. ———, *Towards the smoothness of optimal maps on Riemannian submersions and Riemannian products (of round spheres in particular)*, J. Reine Angew. Math., to appear, available at `arXiv:math/0806.0351`.
82. ———, Appendices to original version of *Continuity, curvature, and the general covariance of optimal transportation*, available at `arXiv:math/0712.3077`.
83. Y.-H. Kim, R. J. McCann, and M. Warren, *Pseudo-Riemannian geometry calibrates optimal transportation*, Math. Res. Lett. **17** (2010), no. 6, 1183–1197.
84. T. C. Koopmans, *Optimum utilization of the transportation system*, Econometrica **17** (1949), Suppl., 136–146.
85. P. W. Y. Lee, *New computable necessary conditions for the regularity theory of optimal transportation*, SIAM J. Math. Anal. **42** (2010), no. 6, 3054–3075.
86. P. W. Y. Lee and J. Li, *New examples satisfying Ma – Trudinger – Wang conditions*, SIAM J. Math. Anal. **44** (2012), no. 1, 61–73.
87. P. W. Y. Lee and R. J. McCann, *The Ma – Trudinger – Wang curvature for natural mechanical actions*, Calc. Var. Partial Differential Equations **41** (2011), no. 1-2, 285–299.
88. V. Levin, *Abstract cyclical monotonicity and Monge solutions for the general Monge – Kantorovich problem*, Set-Valued Anal. **7** (1999), no. 1, 7–32.
89. J. Li, *Smooth optimal transportation on hyperbolic space*, Master's Thesis, University of Toronto, 2009.
90. J. Liu, *Hölder regularity of optimal mappings in optimal transportation*, Calc. Var. Partial Differential Equations **34** (2009), no. 4, 435–451.
91. J. Liu, N. S. Trudinger, and X.-J. Wang, *Interior $C^{2,\alpha}$ regularity for potential functions in optimal transportation*, Comm. Partial Differential Equations **35** (2010), no. 1, 165–184.
92. G. Loeper, *On the regularity of solutions of optimal transportation problems*, Acta Math. **202** (2009), no. 2, 241–283.
93. ———, *Regularity of optimal maps on the sphere: the quadratic cost and the reflector antenna*, Arch. Ration. Mech. Anal. **199** (2011), no. 1, 269–289.
94. G. Loeper and C. Villani, *Regularity of optimal transport in curved geometry: the nonfocal case*, Duke Math. J. **151** (2010), no. 3, 431–485.
95. J. Lott, *Optimal transport and Perelman's reduced volume*, Calc. Var. Partial Differential Equations **36** (2009), no. 1, 49–84.
96. J. Lott and C. Villani, *Ricci curvature for metric-measure spaces via optimal transport*, Ann. of Math. (2) **169** (2009), no. 3, 903–991.
97. X.-N. Ma, N. S. Trudinger, and X.-J. Wang, *Regularity of potential functions of the optimal transportation problem*, Arch. Ration. Mech. Anal. **177** (2005), no. 2, 151–183.

98. F. Maggi and C. Villani, *Balls have the worst best Sobolev inequalities.* II: *Variants and extensions*, Calc. Var. Partial Differential Equations **31** (2008), no. 1, 47–74.
99. R. P. McAfee and J. McMillan, *Multidimensional incentive compatibility and mechanism design*, J. Econom. Theory **46** (1988), no. 2, 335–354.
100. R. J. McCann, *A convexity theory for interacting gases and equilibrium crystals*, Ph.D. Thesis, Princeton University, 1994.
101. ———, *Existence and uniqueness of monotone measure-preserving maps*, Duke Math. J. **80** (1995), no. 2, 309–323.
102. ———, *A convexity principle for interacting gases*, Adv. Math. **128** (1997), no. 1, 153–179.
103. ———, *Equilibrium shapes for planar crystals in an external field*, Comm. Math. Phys. **195** (1998), no. 3, 699–723.
104. ———, *Exact solutions to the transportation problem on the line*, R. Soc. Lond. Proc. Ser. A Math. Phys. Eng. Sci. **455** (1999), no. 1984, 1341–1380.
105. ———, *Polar factorization of maps on Riemannian manifolds*, Geom. Funct. Anal. **11** (2001), no. 3, 589–608.
106. ———, *Stable rotating binary stars and fluid in a tube*, Houston J. Math. **32** (2006), no. 2, 603–631.
107. R. J. McCann and P. M. Topping, *Ricci flow, entropy and optimal transportation*, Amer. J. Math. **132** (2010), no. 3, 711–730.
108. R. J. McCann, B. Pass, and M. Warren, *Rectifiability of optimal transportation plans*, Canad. J. Math **64** (2012), no. 4, 924–933.
109. V. D. Milman and G. Schechtman, *Asymptotic theory of finite-dimensional normed spaces*, with an appendix by M. Gromov, Lecture Notes in Math., vol. 1200, Springer, Berlin, 1986.
110. G. J. Minty, *Monotone (nonlinear) operators in Hilbert space*, Duke Math. J. **29** (1962), 341–346.
111. J. A. Mirrlees, *An exploration in the theory of optimum income taxation*, Rev. Econom. Stud. **38** (1971), no. 2, 175–208.
112. G. Monge, *Mémoire sur la théorie des déblais et de remblais*, Histoire de l'Académie Royale des Sciences de Paris, avec les Mémoires de Mathématique et de Physique pour la même année, 1781, pp. 666–704.
113. P. K. Monteiro and F. H. Page Jr., *Optimal selling mechanisms for multiproduct monopolists: incentive compatibility in the presence of budget constraints*, J. Math. Econom. **30** (1998), no. 4, 473–502.
114. M. Mussa and S. Rosen, *Monopoly and product quality*, J. Econom. Theory **18** (1978), no. 2, 301–317.
115. F. Otto, *The geometry of dissipative evolution equations: the porous medium equation*, Comm. Partial Differential Equations **26** (2001), no. 1-2, 101–174.
116. F. Otto and C. Villani, *Generalization of an inequality by Talagrand and links with the logarithmic Sobolev inequality*, J. Funct. Anal. **173** (2000), no. 2, 361–400.
117. B. Pass, *Convexity and multi-dimensional screening between spaces with different dimensions*, J. Econom. Theory, to appear, available at `arXiv:1108.3798`.
118. G. Perelman, *The entropy formula for the Ricci flow and its geometric applications*, available at `arXiv:math/0211159`.
119. A. Yu. Plakhov, *Newton's problem of the body of minimal mean resistance*, Mat. Sb. **195** (2004), no. 7, 105–126 (Russian); English transl., Sb. Math. **195** (2004), no. 7-8, 1017–1037.
120. R. J. Purser and M. J. P. Cullen, *A duality principle in semigeostrophic theory*, J. Atmospheric Sci. **44** (1987), no. 23, 3449–3468.
121. S. T. Rachev and L. Rüschendorf, *Mass transportation problems.* Vols. I & II, Probab. Appl. (N. Y.), Springer, New York, 1998.
122. J.-Ch. Rochet, *A necessary and sufficient condition for rationalizability in a quasilinear context*, J. Math. Econom. **16** (1987), no. 2, 191–200.
123. J.-Ch. Rochet and Ph. Choné, *Ironing, sweeping and multidimensional screening*, Econometrica **66** (1998), no. 4, 783–826.
124. R. T. Rockafellar, *Characterization of the subdifferentials of convex functions*, Pacific J. Math. **17** (1966), 497–510.
125. L. Rüschendorf, *Bounds for distributions with multivariate marginals*, Stochastic Orders and Decision under Risk (Hamburg, 1989) (K. Mosler and M. Scarsini, eds.), IMS Lecture Notes Monogr. Ser., vol. 19, Inst. Math. Statist., Hayward, CA, 1991, pp. 285–310.

126. L. Rüschendorf, *Optimal solutions of multivariate coupling problems*, Appl. Math. (Warsaw) **23** (1995), no. 3, 325–338.
127. L. Rüschendorf, *On c-optimal random variables*, Statist. Probab. Lett. **27** (1996), no. 3, 267–270.
128. L. Rüschendorf and S. T. Rachev, *A characterization of random variables with minimum L^2-distance*, J. Multivariate Anal. **32** (1990), no. 1, 48–54.
129. T. Sei, *A Jacobian inequality for gradient maps on the sphere and its application to directional statistics*, Comm. Statist. Theory Methods, to appear, available at arXiv:0906.0874.
130. C. S. Smith and M. Knott, *Note on the optimal transportation of distributions*, J. Optim. Theory Appl. **52** (1987), no. 2, 323–329.
131. C. S. Smith and M. Knott, *On Hoeffding - Fréchet bounds and cyclic monotone relations*, J. Multivariate Anal. **40** (1992), no. 2, 328–334.
132. M. Spence, *Competitive and optimal responses to signals: An analysis of efficiency and distribution*, J. Econom. Theory **7** (1974), no. 3, 296–332.
133. K.-T. Sturm, *On the geometry of metric measure spaces.* I, Acta Math. **196** (2006), no. 1, 65–131; II, 133–177.
134. V. N. Sudakov, *Geometric problems in the theory of infinite-dimensional probability distributions*, Proc. Steklov Inst. Math. **2** (1979), i–v, 1–178.
135. H. Tanaka, *An inequality for a functional of probability distributions and its application to Kac's one-dimensional model of a Maxwellian gas*, Z. Wahrscheinlichkeitstheorie und Verw. Gebiete **27** (1973), 47–52.
136. P. Topping, *$\mathcal{L}$-optimal transportation for Ricci flow*, J. Reine Angew. Math. **636** (2009), 93–122.
137. N. S. Trudinger, *Isoperimetric inequalities for quermassintegrals*, Ann. Inst. H. Poincaré Anal. Non Linéaire **11** (1994), no. 4, 411–425.
138. N. S. Trudinger and X.-J. Wang, *On the Monge mass transfer problem*, Calc. Var. Partial Differential Equations **13** (2001), no. 1, 19–31.
139. ______, *On the second boundary value problem for Monge - Ampère type equations and optimal transportation*, Ann. Sc. Norm. Super. Pisa Cl. Sci. (5) **8** (2009), no. 1, 143–174.
140. ______, *On strict convexity and continuous differentiability of potential functions in optimal transportation*, Arch. Ration. Mech. Anal. **192** (2009), no. 3, 403–418.
141. J. Urbas, *On the second boundary value problem for equations of Monge - Ampère type*, J. Reine Angew. Math. **487** (1997), 115–124.
142. ______, *Mass transfer problems*, Lecture Notes 41, Universität Bonn, 1998.
143. J. Viaclovsky, *Conformal geometry and fully nonlinear equations*, Inspired by S. S. Chern (Ph. A. Griffiths, ed.), Nankai Tracts Math., vol. 11, World Sci. Publ., Singapore, 2006, pp. 435–460.
144. C. Villani, *Topics in optimal transportation*, Grad. Stud. Math., vol. 58, Amer. Math. Soc., Providence, RI, 2003.
145. ______, *Optimal transport: Old and new*, Grundlehren Math. Wiss., vol. 338, Springer, Berlin, 2009.
146. J. von Neumann, *Zur Theorie der Gesellschaftsspiele*, Math. Ann. **100** (1928), no. 1, 295–320.
147. ______, *Über ein Okonomisches Gleichungssystems und eine Verallgemeinerung die Brouwerschen Fixpunktsatzes*, Ergebuisse eines Mathematischen Seminars **8** (1937), 73–83.
148. M.-K. von Renesse and K.-T. Sturm, *Transport inequalities, gradient estimates, entropy, and Ricci curvature*, Comm. Pure Appl. Math. **58** (2005), no. 7, 923–940.
149. X.-J. Wang, *On the design of a reflector antenna*, Inverse Problems **12** (1996), no. 3, 351–375.
150. ______, *On the design of a reflector antenna.* II, Calc. Var. Partial Differential Equations **20** (2004), no. 3, 329–341.
151. L. N. Wasserstein, *Markov processes over denumerable products of spaces describing large system of automata*, Problemy Peredači Informacii **5** (1969), no. 3, 64–72 (Russian); English transl., Problems of Information Transmission **5** (1969), no. 3, 47–52.
152. Q. Xia, *The formation of a tree leaf*, ESAIM Control Optim. Calc. Var. **13** (2007), no. 2, 359–377 (electronic).

Department of Mathematics, University of Toronto, Toronto ON Canada M5S 2E4
E-mail address: mccann@math.toronto.edu

Department of Mathematics, University of Texas at Austin, Austin TX 78712 USA
Current address: Department of Mathematics, University of California at Los Angeles, Los Angeles CA 90095 USA
E-mail address: nestor@math.ucla.edu

Centre de Recherches Mathématiques
CRM Proceedings and Lecture Notes
Volume **56**, 2013

A Proof of Bobkov's Spectral Bound For Convex Domains via Gaussian Fitting and Free Energy Estimation

Emanuel Milman

Abstract. We obtain a new proof of Bobkov's lower bound on the first positive eigenvalue of the (negative) Neumann Laplacian (or equivalently, the Cheeger constant) on a bounded convex domain K in Euclidean space. Our proof avoids employing the localization method or any of its geometric extensions. Instead, we deduce the lower bound by invoking a spectral transference principle for log-concave measures, comparing the uniform measure on K with an appropriately scaled Gaussian measure which is conditioned on K. The crux of the argument is to establish a good overlap between these two measures (in say the relative-entropy or total-variation distances), which boils down to obtaining sharp lower bounds on the free energy of the conditioned Gaussian measure.

1. Introduction

The following theorem was proved by Sergey Bobkov in [4] (we use the formulation from [37], which is formally stronger but ultimately equivalent in the cases of interest):

Theorem 1.1 (Bobkov). *Let K denote a convex bounded domain in Euclidean space $(\mathbb{R}^n, |\cdot|)$. Let X denote a random vector uniformly distributed in K (with respect to normalized Lebesgue measure). Then:*

$$D_{\mathrm{Che}}(K) \geq \sup_{x_0 \in \mathbb{R}^n} \frac{c}{\sqrt{\mathbb{E}(|X - x_0|)\mathbb{S}(|X - x_0|)}}, \tag{1.1}$$

for some universal constant $c > 0$.

Let us explain the notation used above. We denote the expectation of a random variable Y by $\mathbb{E}(Y)$, and set $\mathbb{S}(Y) := \sqrt{\mathbb{E}\big(\big(Y - \mathbb{E}(Y)\big)^2\big)}$ to denote the square root of the variance. We use $D_{\mathrm{Che}}(\Omega)$ to denote the Cheeger constant of the domain $\Omega \subset (\mathbb{R}^n, |\cdot|)$, defined as:

$$D_{\mathrm{Che}}(\Omega) := \inf_{A \subset \Omega} \frac{\mathcal{H}^{n-1}(\partial A \cap \Omega)}{\min\big(\mathcal{H}^n(A), \mathcal{H}^n(\Omega \setminus A)\big)},$$

2010 *Mathematics Subject Classification.* Primary 52A23, 60E15, 80M22.

Supported by the Israel Science Foundation (grant no. 900/10), the German Israeli Foundation's Young Scientist Program (grant no. I-2228-2040.6/2009), the Binational Science Foundation (grant no. 2010288), and the Taub Foundation (Landau Fellow).

This is the final form of the paper.

where the infimum ranges over all Borel subsets $A \subset \Omega$, and $\mathcal{H}^k$ denotes the k-dimensional Hausdorff measure. When K is a convex domain, it is known by results of Maz′ya [34, 35] and Cheeger [17] on one hand, and Ledoux [31] on the other, that $D_{\mathrm{Che}}(K) \simeq \sqrt{\lambda_1(K)}$, where $\lambda_1(\Omega)$ denotes the first positive eigenvalue of the (negative) Laplacian $-\Delta$ on Ω with Neumann boundary conditions. Here and elsewhere $A \simeq B$ signifies that there exist universal numeric constants $C_1, C_2 > 0$ so that $C_1 A \leq B \leq C_2 A$. Bobkov's theorem may therefore be interpreted as a spectral bound for general convex domains.

Of course, classical lower bounds on $\lambda_1(K)$ are known for convex domains K, and we mention the sharp lower bound in Euclidean space due to Payne and Weinberger [43]:

$$\lambda_1(K) \geq \frac{\pi^2}{\mathrm{diam}(K)^2}. \tag{1.2}$$

This was extended by Li and Yau [32] and Zhong and Yang [45] to locally convex domains with smooth boundary on Riemannian manifolds with nonnegative Ricci curvature. Back to the Euclidean setting, the bound (1.2) was extended (using the equivalent $D_{\mathrm{Che}}(K)$ parameter) by Kannan, Lovász and Simonovits [27] to the bound:

$$D_{\mathrm{Che}}(K) \geq \sup_{x_0 \in \mathbb{R}^n} \frac{\log 2}{\mathbb{E}(|X - x_0|)}, \tag{1.3}$$

requiring control over the average distance to the centroid x_0 rather than control over the diameter $\mathrm{diam}(K)$. To be completely accurate, the bound obtained in [27] was for a slightly different isoperimetric parameter, which is equivalent to within a factor of 2 to the D_{Che} parameter; however, a simple modification of the argument (as in [5, Section 9]) yields the asserted (1.3). The latter bound was extended to the Riemannian setting (with a different numerical constant) in [40].

By testing very elongated n-dimensional cylinders or cones essentially degenerating to one-dimensional densities (and taking the limit as $n \to \infty$ in the latter case), it is easy to verify that π^2 and $\log 2$ are the best possible (dimension independent) constants one can use in (1.2) and (1.3), respectively. However, assuming that such "biased" or "degenerate" domains are prohibited (for instance, a natural nondegeneracy condition is the isotropic condition, requiring that the variance of all unit linear functionals is equal to 1), Bobkov's bound is in fact the best general bound known to date. Indeed, up to a numeric constant it is certainly better than (1.3), since $\mathbb{S}(|X-x_0|)^2 \leq \mathbb{E}(|X-x_0|^2) \leq C(\mathbb{E}|X-x_0|)^2$ for some universal constant $C > 1$, where the last Khinchine-type inequality is a well known consequence of the convexity of K (as follows from Borell's lemma [9], see [41]). Furthermore, Bobkov's bound was used in [25] to deduce the currently best-known estimate on the Cheeger constant for general isotropic convex domains K in $\mathbb{R}^n$, namely $D_{\mathrm{Che}}(K) \geq cn^{-5/12}$; a conjecture of Kannan – Lovász – Simonovits [27] asserts that the bound should be $D_{\mathrm{Che}}(K) \geq c$, for some dimension independent constant $c > 0$.

Bobkov's proof in [4] is based on a localization method, having its origins in the work of Payne and Weinberger [43], rediscovered by Gromov and V. Milman [23], and systematically developed by Kannan – Lovász – Simonovits [27,33]. A more geometric proof of Bobkov's theorem was given in [37, Theorem 5.15], which was based on a general bound on $D_{\mathrm{Che}}(K)$ obtained by Kannan – Lovász – Simonovits

[27] using again localization. The latter general bound was recently given a completely geometric proof in [40], which generalizes to Riemannian manifolds with nonnegative Ricci curvature.

In this work, we propose yet another alternative proof of Bobkov's Theorem 1.1 in the Euclidean setting. However, we believe that the method of proof is of independent interest, and leads to interesting questions on the distribution of mass on convex sets. Furthermore, amongst the known proofs of Theorem 1.1, the approach described in this work is the only one which avoids directly employing the localization method (or its isoperimetric geometric extension, developed in [40]).

Our method is based on a comparison between the uniform probability measure on K, denoted μ_K, and a suitably scaled Gaussian measure which we condition on K:

$$\gamma_K^w := \exp\big(Z_K(w)\big) \exp\Big(-\frac{w}{2}|x|^2\Big)\, \mathrm{d}\mu_K(x),$$

where $Z_K(w) \geq 0$ is a normalization term ensuring that γ_K^w is a probability measure. When K has unit volume, this measure may be thought of as the Gibbs measure at inverse-temperature $w > 0$ associated with the Hamiltonian:

$$H(x) := \begin{cases} |x|^2/2 & x \in K \\ +\infty & \text{otherwise}, \end{cases}$$

and so $\exp\big(-Z_K(w)\big)$ represents the partition function and $Z_K(w)/w$ represents the system's (Helmholtz) free energy. We have learned from Dario Cordero-Erausquin that the idea of adding $wH(x)$ to a given potential and optimizing on $w > 0$ may be traced back to the work of Hörmander (e.g., [26, Theorem 2.2.3]), who derived a generalized version of the Payne–Weinberger estimate (1.2) for pseudo-convex domains in $\mathbb{C}^n$.

Clearly, it is enough to prove the desired bound (1.1) for $x_0 = 0$, since $D_{\mathrm{Che}}(\Omega)$ is invariant under translation of the domain Ω. We consequently denote $E = E_0 = \mathbb{E}(|X|)$ and $S = S_0 = \mathbb{S}(|X|)$. It turns out that when $w \leq \frac{c}{ES}$, for an appropriately chosen small constant $c > 0$, μ_K and γ_K^w overlap rather well in the total-variation or relative entropy senses, and general transference principles from [37,39] imply that μ_K and γ_K^w have comparable Cheeger constants (or equivalently, spectral-gaps). Since γ_K^w is well-known to inherit all the isoperimetric and spectral properties of the (nonconditioned) Gaussian measure $\gamma^w := \gamma_{\mathbb{R}^n}^w$ (e.g., [1, 2, 15] or see below), a delicate estimation of the overlap yields the desired bound (1.1).

Intuitively it is clear that w should be of the order of $1/(ES)$ to get a good overlap. Indeed, since μ_K is mostly concentrated (by Chebyshev's inequality) in the annulus $A := K \cap \big(B(E+2S) \setminus B(E-2S)\big)$ (where $B(R)$ denotes the Euclidean ball of radius R), we would like the density of γ_K^w to be almost uniform in A, yielding the requirement that $w((E+2S)^2 - (E-2S)^2) \simeq 1$, i.e., that $w \simeq 1/(ES)$. However, we also need to control $Z_K(w)$ to push this argument through, which turns out to be a rather delicate task. This constitutes the main part of this work, and leads to questions on the distribution of $|X|$ in K which may be of independent interest. In particular, the crux of the argument relies on controlling from below the free energy $Z_K(w)/w$ for $w \leq c/(ES)$.

The rest of this work is organized as follows. We recall some preliminaries in Section 2, and provide the full justification to the intuitive claims made above in Section 3. We conclude with Section 4, where we demonstrate that our lower bound for the free energy is in fact sharp. Unless otherwise stated, all constants c, c', c'', C, C' etc. appearing in this work are universal numeric constants, whose values may change from one occurrence to the next.

Acknowledgement. I thank Dario Cordero-Erausquin for his interest, discussions and references, and for making me realize that there is something to prove. I also thank the organizers of the 2011 Séminaire de mathématiques supérieures for the invitation to give a mini-course and to present this work.

2. Preliminaries

2.1. Isoperimetry, spectral-gap and concentration. We start by extending our definitions of D_{Che} and λ_1 to a general metric-measure space setting. A standard text-book reference for most of the notions we employ below is the excellent book by Ledoux [30] (see also [39] for a general overview).

Let (Ω, d, μ) denote a measure-metric space, meaning that (Ω, d) is a separable metric space and μ is a Borel probability measure on (Ω, d). There are various known ways of measuring the interaction between the metric d and the measure μ. One of the strongest forms of measuring this interplay is given by isoperimetric inequalities. Recall that Minkowski's (exterior) boundary measure of a Borel set $A \subset \Omega$, denoted $\mu^+(A)$, is defined as $\mu^+(A) := \liminf_{\varepsilon \to 0} \big(\mu(A^d_\varepsilon) - \mu(A)\big)/\varepsilon$, where $A^d_\varepsilon = A^{\Omega,d}_\varepsilon := \{x \in \Omega : \exists y \in A \; d(x,y) < \varepsilon\}$ denotes the ε-neighborhood of A in (Ω, d). The isoperimetric profile $\mathcal{I} = \mathcal{I}_{(\Omega,d,\mu)}$ is then defined as the function $\mathcal{I} \colon [0,1] \to \mathbb{R}_+$ given by $\mathcal{I}(v) = \inf\{\mu^+(A) : \mu(A) = v\}$. An isoperimetric inequality measures the relation between the boundary measure and the measure of a set, by providing a lower bound on $\mathcal{I}_{(\Omega,d,\mu)}$. In this work, we will only be interested in the Cheeger constant D_{Che} of the space (Ω, d, μ), defined as:

$$D_{\mathrm{Che}}(\Omega, d, \mu) := \inf_{v \in [0,1]} \frac{\mathcal{I}_{(\Omega,d,\mu)}(v)}{\min(v, 1-v)} = \inf_{A \subset \Omega} \frac{\mu^+(A)}{\min\big(\mu(A), 1 - \mu(A)\big)},$$

measuring a certain linear isoperimetric property of the space. When d and (or) μ are implied from the context, we simply write $D_{\mathrm{Che}}(\Omega)$ or $D_{\mathrm{Che}}(\mu)$. We will mostly work in Euclidean space $(\mathbb{R}^n, |\cdot|)$, and so given a bounded domain $\Omega \subset \mathbb{R}^n$, we denote $D_{\mathrm{Che}}(\Omega) = D_{\mathrm{Che}}(\Omega, |\cdot|, \mu_\Omega)$, where μ_Ω denotes the uniform probability measure on Ω, and given a Borel probability measure ν, denote $D_{\mathrm{Che}}(\nu) = D_{\mathrm{Che}}(\mathbb{R}^n, |\cdot|, \nu)$. Note that in the former case, when $A \subset \Omega$ has smooth boundary, then $\mu^+_\Omega(A) = \mathcal{H}^{n-1}(\partial A \cap \Omega)/\mathcal{H}^n(\Omega)$. For the standard Gaussian measure γ on $(\mathbb{R}^n, |\cdot|)$, a classical result of Sudakov – Tsirelson [44] and independently Borell [10], asserts that $\mathcal{I}_\gamma = \mathcal{I}_{(\mathbb{R}^n,|\cdot|,\gamma)} = \varphi \circ \Phi^{-1}$, where $\varphi(t) := (1/\sqrt{2\pi}) \exp(-t^2/2)$ and $\Phi(t) := \int_{-\infty}^t \varphi(s)\, ds$, corresponding to the fact that half-planes are isoperimetric minimizers of Gaussian boundary measure. It is elementary to verify that $\mathcal{I}_\gamma$ is a concave function on $[0,1]$ vanishing at the endpoints and symmetric about the point $\frac{1}{2}$, and so consequently $D_{\mathrm{Che}}(\gamma) = 2\mathcal{I}_\gamma(\frac{1}{2}) = \sqrt{2/\pi}$, and by scaling $D_{\mathrm{Che}}(\gamma^w) = \sqrt{2/\pi}\sqrt{w}$ (see [37, 38] for the concavity of the isoperimetric profile in more general spaces).

Another way of measuring the interaction between metric and measure is given by Sobolev inequalities. In this work, we will only be interested in the Poincaré inequality. Let $\mathcal{F} = \mathcal{F}(\Omega, d)$ denote the space of functions which are Lipschitz on every ball in (Ω, d). Given $f \in \mathcal{F}$, define $|\nabla f|$ as the following Borel function:

$$|\nabla f|(x) := \limsup_{d(y,x)\to 0+} \frac{|f(y) - f(x)|}{d(x,y)}.$$

(and we define it as 0 if x is an isolated point — see [7, pp. 184, 189] for more details). In the smooth Euclidean setting, $|\nabla f|$ of course coincides with the Euclidean length of the gradient of f. We say that (Ω, d, μ) satisfies a Poincaré inequality if:

$$\exists \lambda_1 > 0 \quad \forall f \in \mathcal{F} \quad \int |\nabla f|^2 \, d\mu \geq \lambda_1 \Big(\int f^2 \, d\mu - \Big(\int f \, d\mu \Big)^2 \Big).$$

The best possible constant λ_1 above, called the Poincaré or spectral-gap constant, is denoted $\lambda_1(\Omega, d, \mu)$. As usual, we will use $\lambda_1(\Omega)$ or $\lambda_1(\mu)$ when the space is implied from the context. Note that when Ω is a smooth domain in Euclidean space, $\lambda_1(\Omega)$ coincides with the first positive eigenvalue of the (negative) Laplacian $-\Delta$ on Ω with Neumann boundary conditions. For instance, it is well known (e.g. [30]) for the standard Gaussian measure γ on $(\mathbb{R}^n, |\cdot|)$ that $\lambda_1(\gamma) = 1$, and hence by scaling $\lambda_1(\gamma^w) = w$.

As already alluded to in the Introduction, there is an intimate relation between D_{Che} and λ_1. It was shown by Cheeger [17] and independently by Maz′ya [34, 35] that $\sqrt{\lambda_1(\Omega, d, \mu)} \geq \frac{1}{2} D_{\mathrm{Che}}(\Omega, d, \mu)$ (their proof extends from the Euclidean or Riemannian setting to the general metric-measure space one, cf. [37]). The converse inequality is in general false, due to the possible existence of narrow "necks" in the space (see, e.g., [37]). However, it was shown by Buser [14] for the case of a closed Riemannian manifold having *nonnegative* Ricci curvature, and extended by Ledoux [31] to the case of a manifold-with-density having *nonnegative* generalized Ricci curvature, that the converse inequality also holds up to a universal numeric constant. We do not provide unnecessary definitions here and only mention that in particular, in the case of Euclidean space $(\mathbb{R}^n, |\cdot|)$ endowed with a log-concave probability measure μ, it follows that $\sqrt{\lambda_1(\mu)} \simeq D_{\mathrm{Che}}(\mu)$. Recall that a measure μ on $\mathbb{R}^n$ is called log-concave if $\mu = \exp(-V(x))\, dx$ with $V \colon \mathbb{R}^n \to \mathbb{R} \cup \{+\infty\}$ convex (cf. Borell [9]). Note that all of the probability measures mentioned in the Introduction μ_K, γ_K^w and γ^w are log-concave.

A third way of measuring the interaction between metric and measure is given by concentration inequalities. The concentration profile of our space $\mathcal{K} = \mathcal{K}_{(\Omega,d,\mu)} \colon \mathbb{R}_+ \to [0, \frac{1}{2}]$ is defined as the pointwise minimal function so that $1 - \mu(A_r^d) \leq \mathcal{K}(r)$ for all Borel sets $A \subset \Omega$ with $\mu(A) \geq \frac{1}{2}$ $(r \geq 0)$. Clearly $\mathcal{K}$ is a nonincreasing function. Concentration inequalities measure how tightly the measure μ is concentrated around sets having measure $\frac{1}{2}$ as a function of the distance r away from these sets, by providing an upper bound on $\mathcal{K}(r)$ which decays to 0 as $r \to \infty$. To measure this decay, we denote by $D_{\mathrm{Exp}_p} = D_{\mathrm{Exp}_p}(\Omega, d, \mu) \geq 0$ the best constant in the following inequality $(p > 0)$:

$$\forall r \geq 0 \quad \mathcal{K}(r) \leq \exp(-(D_{\mathrm{Exp}_p} r)^p).$$

When $D_{\mathrm{Exp}_1} > 0$ ($D_{\mathrm{Exp}_2} > 0$) we will say that our space has exponential (Gaussian) concentration (respectively). Indeed, it is well-known (see, e.g., [30]) that the standard Gaussian measure γ on $(\mathbb{R}^n, |\cdot|)$ satisfies $D_{\mathrm{Exp}_2}(\gamma) = 1/\sqrt{2}$, and hence by scaling $D_{\mathrm{Exp}_2}(\gamma^w) = \sqrt{w/2}$. Furthermore, it was shown by M. Gromov and V. Milman [22] that $D_{\mathrm{Exp}_1} \geq c\sqrt{\lambda_1}$ for some universal numeric constant $c > 0$, i.e., that having positive spectral-gap implies exponential concentration (their proof applies in a general metric-measure space setting). The converse inequality is again in general false (consider, e.g., a measure with disconnected support). However, for the class of log-concave measures in Euclidean space, it was shown in our previous work [37] that in fact $D_{\mathrm{Exp}_1} \geq c' D_{\mathrm{Che}}$ for some universal constant $c' > 0$, which implies together with the above mentioned results that for this class $D_{\mathrm{Che}} \simeq \sqrt{\lambda_1} \simeq D_{\mathrm{Exp}_1}$ (see [37] for a much stronger result in a more general setting).

2.2. Stability of D_{Che} in the class of log-concave measures. Next, we will require two notions of proximity between two Borel probability measures μ_1 and μ_2 on (Ω, d). The first is given by the total-variation distance, denoted $d_{\mathrm{TV}}(\mu_1, \mu_2)$ and defined as:

$$d_{\mathrm{TV}}(\mu_1, \mu_2) := \sup_{A \subset \Omega} |\mu_1(A) - \mu_2(A)|.$$

The second is given by the relative entropy (or Kullback – Leibler divergence) of μ_2 with respect to μ_1, denoted $H(\mu_2 \mid \mu_1)$ and defined as:

$$H(\mu_2 \mid \mu_1) := \mathrm{Ent}_{\mu_1}\left(\frac{\mathrm{d}\mu_2}{\mathrm{d}\mu_1}\right) = \int \log\left(\frac{d\mu_2}{d\mu_1}\right) \mathrm{d}\mu_2,$$

if $\mu_2 \ll \mu_1$, and ∞ otherwise.

The key ingredient in our proof of Bobkov's Theorem 1.1 is the following theorem, compiled from our previous results from [37, 39]:

Theorem 2.1 ([37,39]). *Let μ_1, μ_2 denote two log-concave probability measures on $(\mathbb{R}^n, |\cdot|)$.*

- *There exists a universal constant $c > 0$ so that for any $\varepsilon > 0$:*

$$d_{\mathrm{TV}}(\mu_1, \mu_2) \leq 1 - \varepsilon \implies D_{\mathrm{Che}}(\mu_2) \geq c \frac{\varepsilon^2}{\log(1/\varepsilon)} D_{\mathrm{Che}}(\mu_1).$$

- *There exists a universal constant $c > 0$ so that for any $p \geq 1$:*

$$H(\mu_2 \mid \mu_1) \leq L \implies D_{\mathrm{Che}}(\mu_2) \geq \frac{c}{1 + L^{1/p}} D_{\mathrm{Exp}_p}(\mu_1).$$

Since $D_{\mathrm{Exp}_1}(\mu_1) \geq c' D_{\mathrm{Che}}(\mu_1)$, in particular:

$$H(\mu_2 \mid \mu_1) \leq L \implies D_{\mathrm{Che}}(\mu_2) \geq \frac{cc'}{1 + L} D_{\mathrm{Che}}(\mu_1).$$

2.3. Conditioning the Gaussian measure onto K. Recall that given $w > 0$, we define γ_K^w by conditioning the Gaussian measure γ^w onto K, namely:

$$\gamma_K^w := \exp\big(Z_K(w)\big) \exp\left(-\frac{w}{2}|x|^2\right) \mathrm{d}\mu_K(x), \tag{2.1}$$

where μ_K denotes the uniform (probability) measure on K, and $Z_K(w) \geq 0$ is a normalization term ensuring that γ_K^w is a probability measure. The following estimates are well-known:

Theorem 2.2.

$$\forall w > 0 \qquad D_{\mathrm{Che}}(\gamma_K^w) \geq \sqrt{\frac{2}{\pi}} \sqrt{w}, \quad \lambda_1(\gamma_K^w) \geq w, \quad D_{\mathrm{Exp}_2}(\gamma_K^w) \geq \sqrt{w/2}.$$

For the sake of completeness, we present (several variants of) the proof.

PROOF. One way to obtain all these estimates simultaneously and with the optimal constants is by applying Caffarelli's contraction theorem. It was shown in [15] that given a source probability measure μ_1 and target probability measure μ_2 on $\mathbb{R}^n$, having the form $\mu_1 = \gamma^w$ and $\mu_2 = f\gamma^w$ with f log-concave, there exists a Borel map $T\colon (\mathbb{R}^n, |\cdot|) \to (\mathbb{R}^n, |\cdot|)$ pushing forward μ_1 onto μ_2 which contracts Euclidean distance: $|T(x) - T(y)| \leq |x - y|$ for all $x, y \in \mathbb{R}^n$. In fact, T is the Brenier optimal-transport map [13], minimizing the L^2-average transport cost $\int |T(x) - x|^2 \, d\mu_1(x)$ among all maps pushing forward μ_1 onto μ_2 (see also [28] for a different construction in a more general setup). We note that Caffarelli's result applies regardless of the smoothness of f, since we can always approximate f by smooth log-concave functions $\{f_n\}$ so that $\{f_n \mu_1\}$ converge in total-variation to $f\mu_1$, and the corresponding contractions $\{T_n\}$ pushing forward μ_1 onto $\{f_n \mu_1\}$ will necessarily converge (by the Arzelà – Ascoli theorem) to our desired contraction T (see [28, Lemma 3.3] for the precise argument). Note that the convexity of K ensures that $f_0 := d\gamma_K^w / d\gamma^w = c_{K,w} \mathbf{1}_K$ is log-concave, and so by Caffarelli's contraction theorem there exists a contraction pushing forward γ^w onto γ_K^w.

It is easy to verify that all of forms of interaction between metric and measure described in Section 2.1 can only improve under contracting maps. Namely, if $T\colon (\Omega_1, d_1, \mu_1) \to (\Omega_2, d_2, \mu_2)$ is a Borel map pushing forward μ_1 onto μ_2 and contracting distances $d_2\big(T(x), T(y)\big) \leq d_1(x, y)$ for all $x, y \in \Omega_1$, then $\mathcal{I}_{(\Omega_2, d_2, \mu_2)} \geq \mathcal{I}_{(\Omega_1, d_1, \mu_2)}$, $|\nabla(f \circ T)|(x) \leq |\nabla f|(Tx)$ and $\mathcal{K}_{(\Omega_2, d_2, \mu_2)} \leq \mathcal{K}_{(\Omega_1, d_1, \mu_2)}$. In particular, it immediately follows that:

$$D_{\mathrm{Che}}(\gamma_K^w) \geq D_{\mathrm{Che}}(\gamma^w), \quad \lambda_1(\gamma_K^w) \geq \lambda_1(\gamma^w), \quad D_{\mathrm{Exp}_2}(\gamma_K^w) \geq D_{\mathrm{Exp}_2}(\gamma^w),$$

and the asserted bounds follow from the well-known bounds for the Gaussian measure γ^w mentioned in Section 2.1.

Another possible way to derive the above mentioned bounds (with perhaps inferior numerical constants) is as follows. The Brascamp – Lieb inequality [12] states that if $\mu = \exp\big(-V(x)\big)\, dx$ is a log-concave measure on $\mathbb{R}^n$ with $V \in C^2(\mathbb{R}^n)$ having positive definite Hessian $\operatorname{Hess} V > 0$, then for any $f \in C^1(\mathbb{R}^n)$:

$$\int \langle (\operatorname{Hess} V)^{-1} \nabla f, \nabla f \rangle \, d\mu \geq \int f^2 \, d\mu - \Big(\int f \, d\mu \Big)^2.$$

In particular, it follows that if $\operatorname{Hess} V \geq \lambda \mathrm{Id}$ with $\lambda > 0$ then $\lambda_1(\mu) \geq \lambda$ (the latter statement may also be attributed to Lichnerowicz, at least in an analogous Riemannian setting, cf. [20, Theorem 4.70]). Moreover, if $\operatorname{Hess} V \geq \lambda Id$ with $\lambda > 0$, then by the Bakry – Émery criterion [1], $(\mathbb{R}^n, |\cdot|, \mu)$ satisfies in fact a stronger log – Sobolev inequality, and the Herbst argument implies that $D_{Exp_2}(\mu) \geq c\sqrt{\lambda}$ for some universal constant $c > 0$ (the interested reader may consult [30, 39] for missing definitions and arguments). The inequality $D_{\mathrm{Che}}(\mu) \geq c'\sqrt{\lambda}$ may be deduced by using that $D_{\mathrm{Che}}(\mu) \simeq \sqrt{\lambda_1(\mu)}$ for log-concave measures μ; alternatively, one may directly deduce that $D_{\mathrm{Che}}(\mu) \geq D_{\mathrm{Che}}(\gamma^\lambda)$ by applying the isoperimetric inequality of Bakry – Ledoux [2]. Finally, an approximation argument as the one described above ensures that all isoperimetric, Sobolev and concentration estimates pass onto $\gamma_K^w = f_0 \gamma_w$ (see [36–38] for more technical approximation results, which in fact apply in the more challenging Riemannian setting). □

3. Proof of Theorem 1.1

In this section we provide a proof of Bobkov's Theorem 1.1 via Gaussian Fitting and estimation of the free energy $Z_K(w)/w$.

We denote $E_2 = \sqrt{\mathbb{E}(|X|^2)}$. Clearly $E \leq E_2$ by Jensen's inequality, but as mentioned in the Introduction, it is also well-known that the convexity of K ensures that $E_2 \leq CE$ for some universal numeric constant $C > 1$. Together with the obvious reduction to the case $x_0 = 0$ described in the Introduction, the proof of Theorem 1.1 thus reduces to proving the following equivalent bound:

$$D_{\mathrm{Che}}(K) \geq \frac{c}{\sqrt{E_2 S}}. \tag{3.1}$$

3.1. Reduction to estimation of the free energy. Note that the normalization term $Z_K(w) \geq 0$ appearing in (2.1) is given by:

$$Z_K(w) := -\log \int \exp\Big(-\frac{w}{2}|x|^2\Big)\, d\mu_K(x).$$

We summarize several useful properties of $Z_K(w)$ below:

Lemma 3.1.

- *The function* $[0,\infty) \ni w \mapsto Z_K(w)$ *is concave increasing.*
- $Z_K(0) = 0$ *and* $(d/dw) Z_K(0) = \frac{1}{2}E_2^2$.
- $Z_K(w) = (n/2)\log(w/(2\pi)) - \log \gamma(\sqrt{w}K)$, *where* γ *denotes the standard Gaussian measure on* $\mathbb{R}^n$.
- $(0,\infty) \ni w \mapsto Z_K(w)/w$ *decreases continuously from* $\frac{1}{2}E_2^2$ *to* 0.

PROOF. The first assertion follows immediately by direct differentiation or by employing Hölder's inequality. The second and third ones are immediate to verify as well. The fourth one is an immediate consequence of the first three. □

Remark 3.2. Some properties and extremal characterizations of $\gamma(K)$ for symmetric convex sets K in $\mathbb{R}^n$ have been studied by Cordero-Erausquin, Fradelizi and Maurey [18] and by Bobkov [6]. In view of the third assertion of the above lemma, it is immediate to translate these properties into statements about $Z_K(w)$. However, we have not found any concrete applications of these connections, and so we only vaguely mention these in passing, and refer the interested reader to the above mentioned references.

To transfer the lower bound on $D_{\mathrm{Che}}(\gamma_K^w)$ given in Theorem 2.2, namely $D_{\mathrm{Che}}(\gamma_K^w) \geq c\sqrt{w}$, onto a lower bound on $D_{\mathrm{Che}}(\mu_K)$, we would like to invoke one of the transference principles given by Theorem 2.1. To this end, we must control the total-variation distance $d_{\mathrm{TV}}(\mu_K, \gamma_K^w)$ or relative entropy $H(\mu_K \mid \gamma_K^w)$. The calculation is a little more concise for the latter option, so we proceed with the task of bounding the relative entropy from above.

Lemma 3.3.

-

$$\begin{aligned} H(\mu_K \mid \gamma_K^w) &= \int \Big(\frac{w}{2}|x|^2 - Z_K(w)\Big)\, d\mu_K(x) \\ &= \frac{1}{2}E_2^2 w - Z_K(w) = w\Big(\frac{1}{2}E_2^2 - \frac{Z_K(w)}{w}\Big). \end{aligned} \tag{3.2}$$

- *The function* $[0,\infty) \ni w \mapsto H(\mu_K \mid \gamma_K^w)$ *is convex and increases from* 0 *to* ∞.

PROOF. The first assertion follows by definition. The second assertion is a direct consequence of the first one and the concavity of $Z_K(w)$. □

Since the measure γ_K^w has Gaussian concentration by Theorem 2.2, namely $D_{\mathrm{Exp}_2}(\gamma_K^2) \geq \sqrt{w/2}$, the transference principle for the relative entropy given by Theorem 2.1 guarantees that if:

$$H(\mu_K \mid \gamma_K^w) \leq wL_w,$$

then for some universal constant $c' > 0$:

$$D_{\mathrm{Che}}(\mu_K) \geq c' \frac{D_{\mathrm{Exp}_2}(\gamma_K^w)}{1+\sqrt{wL_w}} \geq \frac{c'/\sqrt{2}}{1/\sqrt{w}+\sqrt{L_w}}. \tag{3.3}$$

Optimizing on $w > 0$ in (3.3), we see that we would like to choose $w_0 > 0$ as large as possible so that $H(\mu_K \mid \gamma_K^{w_0}) \leq w_0 L_{w_0} \simeq 1$, resulting in the bound $D_{\mathrm{Che}}(\mu_K) \geq c''\sqrt{w_0}$. Recalling (3.2), this boils down to choosing $w_0 > 0$ as large as possible so that:

$$\frac{Z_K(w_0)}{w_0} \geq \frac{1}{2}E_2^2 - \frac{C}{w_0}, \tag{3.4}$$

for some numeric constant $C > 0$. We will show that (3.4) is indeed satisfied with $w_0 = c/E_2S$, for some appropriately small numeric constant $c > 0$, from whence Bobkov's bound (3.1) follows.

Theorem 3.4 (Free energy lower bound). *There exists universal constants* $c, C > 0$ *so that for any convex bounded domain* K *in* $(\mathbb{R}^n, |\cdot|)$, *we have the following estimate for the free energy*:

$$w \leq \frac{c}{E_2S} \implies \frac{Z_K(w)}{w} \geq \frac{1}{2}E_2^2 - CE_2S. \tag{3.5}$$

Before proceeding with the proof of Theorem 3.4, we summarize below its consequences as described in the preceding discussion:

Corollary 3.5. *There exist two numeric constants* $c', c'' > 0$ *so that for any convex bounded domain* K *in* $(\mathbb{R}^n, |\cdot|)$:

- *Setting* $w_0 = c'/(E_2S)$, *we have*:
$$H(\mu_K \mid \gamma_K^{w_0}) \leq \tfrac{1}{2}, \quad d_{\mathrm{TV}}(\mu_K, \gamma_K^{w_0}) \leq \tfrac{1}{2}.$$
- *Bobkov's bound*: $D_{\mathrm{Che}}(K) \geq c''/\sqrt{E_2S}$.

PROOF. Theorem 3.4 and (3.2) guarantee that setting $c' = \min(c, 1/(2C))$, where $c, C > 0$ are given in Theorem 3.4, the bound on $H(\mu_K \mid \gamma_K^{w_0})$ follows. The bound on $d_{\mathrm{TV}}(\mu_K, \gamma_K^{w_0})$ follows by the Csiszár – Kullback – Pinsker inequality (e.g., [30]), stating that for any two Borel probability measures μ_1, μ_2:

$$d_{\mathrm{TV}}(\mu_1, \mu_2) \leq \sqrt{H(\mu_2 \mid \mu_1)/2}.$$

Bobkov's lower bound on $D_{\mathrm{Che}}(K)$ follows by the discussion preceding the statement of Theorem 3.4, employing the transference principle given by Theorem 2.1 and the well-known bound $D_{\mathrm{Che}}(\gamma_K^w) \geq c'''\sqrt{w}$ stated in Theorem 2.2. □

3.2. Main calculation. For the proof of Theorem 3.4, we require the following:

Lemma 3.6. *There exists a universal constant* $c_1 := \log(3)/4 > 0$ *so that for any convex bounded domain* K *in* $(\mathbb{R}^n, |\cdot|)$*:*

$$\mu_K\{|x| \leq r\} \leq \exp\left(-c_1 \frac{E_2 - r}{S}\right) \quad \forall r \in [0, E_2 - 3S].$$

PROOF. Note that by the Brunn–Minkowski inequality (e.g., [21]), the function $r \mapsto \log \mu_K\{|x| \leq r\}$ is concave. Indeed, denoting $K_\infty := \mathbb{R} \times K$ and $L_\infty = \{(r,x) \in \mathbb{R} \times \mathbb{R}^n : r \geq 0, |x| \leq r\}$, then $\mu_K\{|x| \leq r\} = \mathrm{Vol}(K_+ \cap H_r)/\mathrm{Vol}(K)$, where K_+ is the convex set $K_\infty \cap L_\infty$ and H_r is the hyperplane $\{r\} \times \mathbb{R}^n$ in $\mathbb{R}^{n+1}$. Consequently, the function $r \mapsto (\mu_K\{|x| \leq r\})^{1/n}$ is concave, and in particular, the asserted log-concavity of $r \mapsto \mu_K\{|x| \leq r\}$ follows.

Next, note that by Chebyshev's inequality:

$$\mu_K\{|x| \leq E - 2S\} \leq \tfrac{1}{4} \quad \mu_K\{|x| \leq E + 2S\} \geq \tfrac{3}{4}.$$

The log-concavity of $r \mapsto \mu_K\{|x| \leq r\}$ consequently implies:

$$\mu_K\{|x| \leq E - 2S - t\} \leq \frac{1}{4} \exp\left(-\frac{\ln(3)}{4} \frac{t}{S}\right) \quad \forall t \geq 0,$$

or equivalently:

$$\mu_K\{|x| \leq r\} \leq \frac{1}{4} \exp\left(-\frac{\ln(3)}{4}\left(\frac{E - r}{S} - 2\right)\right) \quad \forall r \leq E - 2S.$$

Observing that $E_2^2 = E^2 + S^2$ and hence $E_2 \leq E + S$, it follows that:

$$\mu_K\{|x| \leq r\} \leq \frac{1}{4} \exp\left(3\frac{\ln(3)}{4}\right) \exp\left(-\frac{\ln(3)}{4} \frac{E_2 - r}{S}\right) \quad \forall r \leq E_2 - 3S,$$

and the assertion of the lemma follows by direct numerical inspection. □

PROOF OF THEOREM 3.4. First, we may assume that $E_2 \geq 3S$, since otherwise the assertion follows trivially with $C = \frac{3}{2}$. Next, since $Z_K(w)/w$ is decreasing, it is enough to prove the assertion for $w_0 = c_1/(\alpha E_2 S)$, where c_1 is the constant from the above lemma and $\alpha := 2$. Integrating by parts, evaluating separately the integral over the ranges $[0, E_2 - 3S]$ and $[E_2 - 3S, \infty)$, and using the Lemma, we have:

$$\begin{aligned}
&\int \exp\left(-\frac{w_0}{2}|x|^2\right) d\mu_K(x) \\
&\quad = \int_0^\infty w_0 r \exp\left(-\frac{w_0}{2} r^2\right) \mu_K\{|x| \leq r\}\, dr \\
&\quad \leq \exp\left(-\frac{w_0}{2}(E_2 - 3S)^2\right) + \int_0^{E_2 - 3S} w_0 r \exp\left(-\frac{w_0}{2} r^2\right) \exp\left(-c_1 \frac{E_2 - r}{S}\right) dr.
\end{aligned}$$

To evaluate the second term above, denoted T_2, we use the change of variables $r = c_1/(w_0 S) - t = \alpha E_2 - t$:

$$\begin{aligned} T_2 &= \exp\Big(-c_1\frac{E_2}{S} + \frac{c_1^2}{2w_0S^2}\Big)\int_0^{E_2-3S} w_0 r\exp\Big(-\frac{1}{2}\Big(\sqrt{w_0}r - \frac{c_1}{\sqrt{w_0}S}\Big)^2\Big)\,\mathrm{d}r \\ &\le \exp\Big(-c_1\frac{E_2}{S} + \frac{c_1^2}{2w_0S^2}\Big)\int_{E_2(\alpha-1)+3S}^{\infty}(\alpha w_0 E_2 - w_0 t)\exp\Big(-\frac{w_0}{2}t^2\Big)\,\mathrm{d}t. \end{aligned}$$

Since $\alpha \ge 1$, we estimate the latter expression by:

$$\begin{aligned} &\le \exp\Big(-c_1\frac{E_2}{S} + \frac{c_1^2}{2w_0S^2}\Big)\Big(\frac{\alpha E_2}{E_2(\alpha-1)+3S} - 1\Big)\int_{E_2(\alpha-1)+3S}^{\infty} w_0 t\exp\Big(-\frac{w_0}{2}t^2\Big)\,\mathrm{d}t \\ &\le \exp\Big(-c_1\frac{E_2}{S} + \frac{c_1^2}{2w_0S^2}\Big)\frac{1}{\alpha-1}\exp\Big(-\frac{w_0}{2}(E_2(\alpha-1)+3S)^2\Big). \end{aligned}$$

Recalling that $w_0 = c_1/(\alpha E_2 S)$, this is the same as:

$$= \exp\Big(c_1\frac{E_2}{S}\Big(\frac{\alpha}{2} - 1\Big)\Big)\frac{1}{\alpha-1}\exp\Big(-\frac{w_0}{2}(E_2(\alpha-1)+3S)^2\Big).$$

It is easy to check that, in fact, for any $\alpha > 0$:

$$\exp\Big(c_1\frac{E_2}{S}\Big(\frac{\alpha}{2} - 1\Big)\Big)\exp\Big(-\frac{w_0}{2}(E_2(\alpha-1))^2\Big) = \exp\Big(-\frac{w_0}{2}E_2^2\Big),$$

but this is most apparent for $\alpha = 2$ (which satisfies the requirement $\alpha \ge 1$ used above). Consequently:

$$T_2 \le \frac{1}{\alpha-1}\exp\Big(-\frac{w_0}{2}E_2^2\Big) \le \frac{1}{\alpha-1}\exp\Big(-\frac{w_0}{2}(E_2-3S)^2\Big),$$

and we conclude that for $w_0 = c_1/(2E_2S)$:

$$\int \exp\Big(-\frac{w_0}{2}|x|^2\Big)\,\mathrm{d}\mu_K(x) \le 2\exp\Big(-\frac{w_0}{2}(E_2-3S)^2\Big).$$

It follows that:

$$\begin{aligned} \frac{Z_K(w_0)}{w_0} &= -\frac{1}{w_0}\log\Big(\int\exp\Big(-\frac{w_0}{2}|x|^2\Big)\,\mathrm{d}\mu_K(x)\Big) \\ &\ge \frac{1}{2}(E_2-3S)^2 - \frac{2\log(2)}{c_1}E_2S \ge \frac{1}{2}E_2^2 - \Big(3 + \frac{\log(4)}{c_1}\Big)E_2S. \end{aligned}$$

The assertion now follows from the monotonicity of $w \mapsto Z_K(w)/w$ with $c = c_1/2$ and $C = 3 + \log(4)/c_1$. □

Remark 3.7. To appreciate the delicate nature of the bound (3.5) obtained in Theorem 3.4, note that the requirement that $\alpha > 1$ was crucially used in the proof. In particular, we do *not* know how to extend (3.5) to the following statement, which is perhaps suggested by the heuristic argument outlined in the Introduction:

$$\forall c > 0 \quad \exists C > 0 \quad \text{such that} \quad w \le \frac{c}{E_2S} \implies \frac{Z_K(w)}{w} \ge \frac{1}{2}E_2^2 - CE_2S.$$

Furthermore, note that our proof above crucially relied on the estimate

$$\int_a^\infty \exp(-t^2/2)\,\mathrm{d}t \le (1/a)\exp(-a^2/2)$$

valid for positive $a > 0$, and that the slightly rougher bound $\int_a^\infty \exp(-t^2/2)\,\mathrm{d}t \le C\exp(-a^2/2)$ would have incurred an extraneous logarithmic term in the final result: $Z_K(w_0)/w_0 \ge \frac{1}{2}E_2^2 - CE_2 S\log(E_2/S)$.

Remark 3.8. A more delicate analysis of the proof above (e.g., using $\alpha = 5$) reveals that we may in fact assert the existence of a constant $c > 0$ so that:

$$w \le \frac{c}{E_2 S} \implies \frac{Z_K(w)}{w} \ge \frac{1}{2}(E_2 - 3S)^2.$$

4. Sharpness of free energy estimate

Before concluding, we observe that the lower bound on the free energy we obtained in Theorem 3.4 is in fact sharp in the following sense:

Proposition 4.1 (Free energy upper bound). *There exist two universal constants $c, C > 0$ so that for any convex bounded domain K in $\mathbb{R}^n$:*

$$w \ge \frac{C}{E_2 S} \implies \frac{Z_K(w)}{w} \le \frac{1}{2}E_2^2 - cE_2 S.$$

The proof is immediate given the following lemma, asserting a reverse Chebyshev-type inequality for log-concave measures, which may be of independent (technical) interest:

Lemma 4.2. *Given a log-concave probability measure μ on Euclidean space $(\mathbb{R}^n, |\cdot|)$, let X be a random vector distributed according to μ and set $E = \mathbb{E}(|X|)$ and $S = \mathbb{S}(|X|)$. There exists a universal constant $c_0 > 0$ so that:*

$$\mathbb{P}(|X| \le E - c_0 S) \ge c_0. \tag{4.1}$$

Proof of Proposition 4.1. Since μ_K is log-concave and $E \le E_2$, we know by the previous lemma that:

$$\mu_K\{|x| \le E_2 - c_0 S\} \ge c_0.$$

Consequently, we may estimate:

$$\int \exp\left(-\frac{w}{2}|x|^2\right)\mathrm{d}\mu_K(x) \ge c_0 \exp\left(-\frac{w}{2}(E_2 - c_0 S)^2\right),$$

and hence:

$$\frac{Z_K(w)}{w} \le \frac{1}{2}(E_2 - c_0 S)^2 + \frac{\log(1/c_0)}{w} \quad \forall w > 0.$$

It follows that if $C > 0$ is chosen large enough, then:

$$w \ge \frac{C}{E_2 S} \implies \frac{Z_K(w)}{w} \le \frac{1}{2}E_2^2 - cE_2 S,$$

where:

$$c := 2c_0 - c_0^2 - \frac{\log(1/c_0)}{C} > 0. \qquad \square$$

It remains to establish Lemma 4.2. Surprisingly, we have not found a reference for it in the literature, perhaps due to the fact that the inequality (4.1) goes in the opposite direction to the standard large-deviation or small-ball estimates for $|X|$, and so for completeness we provide a proof. It is clear for experts that the statement of the lemma basically reduces to the one-dimensional case using the localization technique, and so our task is ultimately one-dimensional. However, to make the argument as short as possible, we prefer to rely on the available tools in the literature, and do not make any attempts to obtain good numerical constants. We therefore turn to the known reverse-Hölder Khinchine-type inequalities between positive and negative moments of $f(X) = |X|^2 - E^2$. Such inequalities between positive moments of *semi-norms* are classical and readily follow by Borell's Lemma [9] (see, e.g., [41]); these have been extended to pth order moments for $p = 0$ and $p \in (-1, 0)$ in [24, 29]. However, we require such moment inequalities for the more general function f, and thus turn to the reverse-Hölder inequalities initiated by Bourgain in [11] for polynomials, and extended using the localization technique by Bobkov [3], Carbery and Wright [16] and Nazarov, Sodin and Volberg [42]. For more on this topic, we refer to the excellent survey paper of Fradelizi [19], which also extends these results even further.

Theorem 4.3 ([8,19,42]). *Let P denote a degree d polynomial on $\mathbb{R}^n$. Then for any $-1/d < q \leq p < \infty$ and random-vector X distributed according to a log-concave measure on $\mathbb{R}^n$:*

$$\|P(X)\|_p \leq C(p, q, d)\|P(X)\|_q,$$

for some (explicit) constant $C(p, q, d) > 0$ depending solely on its arguments.

Here and below we use $\|Y\|_p$ to denote $\mathbb{E}(|Y|^p)^{1/p}$ (with the usual interpretation when $p = 0$, which in any case we will not require).

Proof of Lemma 4.2. By the union bound and Markov – Chebyshev inequality:

$$\begin{aligned}\mathbb{P}\big(\big||X| - E\big| \leq c_1 S\big) &\leq \mathbb{P}(|X| \geq 8E) + \mathbb{P}\big(\big||X| - E\big| \leq c_1 S \wedge |X| \leq 8E\big)\\ &\leq \tfrac{1}{8} + \mathbb{P}\big(\big||X|^2 - E^2\big| \leq 9c_1 SE\big).\end{aligned}$$

The function $f(x) = |x|^2 - E^2$ is a degree two polynomial, and so applying Theorem 4.3 with $q = -\frac{1}{4}$ and $p = 2$, preceded by Chebyshev's inequality, we have:

$$\mathbb{P}\big(\big||X|^2 - E^2\big| \leq 9c_1 SE\big) \leq (9c_1 SE)^{1/4}\|f(X)\|_{-1/4}^{-1/4} \leq C'(9c_1 SE)^{1/4}\|f(X)\|_2^{-1/4}.$$

Since:

$$\|f(X)\|_2 \geq E\||X| - E\|_2 = ES,$$

we see that choosing the constant $c_1 > 0$ small enough, we can ensure that:

$$\mathbb{P}\big(\big||X| - E\big| \leq c_1 S\big) \leq \tfrac{1}{4}.$$

Now set $\varepsilon := \mathbb{P}(|X| \leq E - c_1 S)$, and observe that:

$$\begin{aligned}(4.2)\quad 0 &= \mathbb{E}(|X| - E)\\ &\geq \big(\mathbb{P}(|X| \geq E + c_1 S) - P(|X| \leq E + c_1 S)\big)c_1 S - \int_{c_1 S}^{E} \mathbb{P}(|X| \leq E - r)\,\mathrm{d}r\\ &\geq S\Big((\tfrac{1}{2} - 2\varepsilon)c_1 - \int_0^{\infty} \mathbb{P}(|X| \leq E - (c_1 + t)S)\,\mathrm{d}t\Big).\end{aligned}$$

Since $\mathbb{P}(|X| \leq E + 2S) \geq \frac{3}{4}$ by Chebyshev's inequality, we may argue exactly as in the proof of Lemma 3.6 that the log-concavity of the function $r \mapsto \mathbb{P}(|X| \leq r)$ ensures that:

$$\mathbb{P}(|X| \leq E - (c_1 + t)S) \leq \varepsilon \exp\left(-\frac{\log \frac{3}{4} - \log \varepsilon}{2 + c_1} t\right) \quad \forall t \geq 0.$$

Plugging this into (4.2), we obtain:

$$0 \geq (\tfrac{1}{2} - 2\varepsilon)c_1 - \varepsilon \frac{2 + c_1}{\log \frac{3}{4} - \log \varepsilon}.$$

However, this is clearly impossible if $\varepsilon \geq 0$ is too small, and so ε must be bounded from below by some universal constant $c_2 > 0$. Setting $c_0 = \min(c_1, c_2)$, the asserted claim follows. □

References

1. D. Bakry and M. Émery, *Diffusions hypercontractives*, Séminaire de probabilités, XIX, 1983/84 (J. Azéma and M. Yor, eds.), Lecture Notes in Math., vol. 1123, Springer, Berlin, 1985, pp. 177–206.
2. D. Bakry and M. Ledoux, *Lévy – Gromov's isoperimetric inequality for an infinite-dimensional diffusion generator*, Invent. Math. **123** (1996), no. 2, 259–281.
3. S. G. Bobkov, *Remarks on the growth of L^p-norms of polynomials*, Geometric aspects of functional analysis, Lecture Notes in Math., vol. 1745, Springer, Berlin, 2000, pp. 27–35.
4. ______, *On isoperimetric constants for log-concave probability distributions*, Geometric Aspects of Functional Analysis (V. D. Milman and G. Schechtman, eds.), Lecture Notes in Math., vol. 1910, Springer, Berlin, 2007, pp. 81–88.
5. ______, *Large deviations and isoperimetry over convex probability measures with heavy tails*, Electron. J. Probab. **12** (2007), 1072–1100.
6. ______, *On Milman's ellipsoids and M-position of convex bodies*, Concentration, Functional Inequalities and Isoperimetry (Boca Raton, FL, 2009) (Ch. Houdré, M. Ledoux, E. Milman, and M. Milman, eds.), Contemp. Math., vol. 545, Amer. Math. Soc., Providence, RI, 2011, pp. 23–33.
7. S. G. Bobkov and Ch. Houdré, *Isoperimetric constants for product probability measures*, Ann. Probab. **25** (1997), no. 1, 184–205.
8. S. G. Bobkov and F. L. Nazarov, *Sharp dilation-type inequalities with fixed parameter of convexity*, Zap. Nauchn. Sem. S.-Peterburg. Otdel. Mat. Inst. Steklov. (POMI) **351** (2007), Veroyatnost i Statistika. 12, 54–78, 299.
9. Ch. Borell, *Convex measures on locally convex spaces*, Ark. Mat. **12** (1974), 239–252.
10. ______, *The Brunn – Minkowski inequality in Gauss space*, Invent. Math. **30** (1975), no. 2, 207–216.
11. J. Bourgain, *On the distribution of polynomials on high-dimensional convex sets*, Geometric Aspects of Functional Analysis (J. Lindenstrauss and V. D. Milman, eds.), Lecture Notes in Math., vol. 1469, Springer, Berlin, 1991, pp. 127–137.
12. H. J. Brascamp and E. H. Lieb, *On extensions of the Brunn – Minkowski and Prékopa – Leindler theorems, including inequalities for log concave functions, and with an application to the diffusion equation*, J. Functional Analysis **22** (1976), no. 4, 366–389.
13. Y. Brenier, *Polar factorization and monotone rearrangement of vector-valued functions*, Comm. Pure Appl. Math. **44** (1991), no. 4, 375–417.
14. P. Buser, *A note on the isoperimetric constant*, Ann. Sci. École Norm. Sup. (4) **15** (1982), no. 2, 213–230.
15. L. A. Caffarelli, *Monotonicity properties of optimal transportation and the FKG and related inequalities*, Comm. Math. Phys. **214** (2000), no. 3, 547–563.
16. A. Carbery and J. Wright, *Distributional and L^q norm inequalities for polynomials over convex bodies in $\mathbb{R}^n$*, Math. Res. Lett. **8** (2001), no. 3, 233–248.
17. J. Cheeger, *A lower bound for the smallest eigenvalue of the Laplacian*, Problems in Analysis (Princeton, NJ, 1969) (R. Gunning C., ed.), Princeton Univ. Press, Princeton, NJ, 1970, pp. 195–199.

18. D. Cordero-Erausquin, M. Fradelizi, and B. Maurey, *The* (B) *conjecture for the Gaussian measure of dilates of symmetric convex sets and related problems*, J. Funct. Anal. **214** (2004), no. 2, 410–427.
19. M. Fradelizi, *Concentration inequalities for s-concave measures of dilations of Borel sets and applications*, Electron. J. Probab. **14** (2009), no. 71, 2068–2090.
20. S. Gallot, D. Hulin, and J. Lafontaine, *Riemannian geometry*, Universitext, Springer, Berlin, 1987.
21. R. J. Gardner, *The Brunn–Minkowski inequality*, Bull. Amer. Math. Soc. (N.S.) **39** (2002), no. 3, 355–405.
22. M. Gromov and V. D. Milman, *A topological application of the isoperimetric inequality*, Amer. J. Math. **105** (1983), no. 4, 843–854.
23. ———, *Generalization of the spherical isoperimetric inequality to uniformly convex Banach spaces*, Compositio Math. **62** (1987), no. 3, 263–282.
24. O. Guédon, *Kahane–Khinchine type inequalities for negative exponent*, Mathematika **46** (1999), no. 1, 165–173.
25. O. Guédon and E. Milman, *Interpolating thin-shell and sharp large-deviation estimates for isotropic log-concave measures*, Geom. Funct. Anal. **21** (2011), no. 5, 1043–1068.
26. L. Hörmander, *L^2 estimates and existence theorems for the $\bar{\partial}$ operator*, Acta Math. **113** (1965), 89–152.
27. R. Kannan, L. Lovász, and M. Simonovits, *Isoperimetric problems for convex bodies and a localization lemma*, Discrete Comput. Geom. **13** (1995), no. 3-4, 541–559.
28. Y.-H. Kim and E. Milman, *A generalization of Caffarelli's contraction theorem via (reverse) heat flow*, Math. Ann., to appear, available at `arXiv:1002.0373`.
29. R. Latała, *On the equivalence between geometric and arithmetic means for log-concave measures*, Convex Geometric Analysis (Berkeley, CA, 1996) (K. M. Ball and V. Milman, eds.), Math. Sci. Res. Inst. Publ., vol. 34, Cambridge Univ. Press, Cambridge, 1999, pp. 123–127.
30. M. Ledoux, *The concentration of measure phenomenon*, Math. Surveys Monogr., vol. 89, Amer. Math. Soc., Providence, RI, 2001.
31. ———, *Spectral gap, logarithmic Sobolev constant, and geometric bounds*, Surveys in Differential Geometry. Vol. IX (A. Grigor′yan and S. T. Yau, eds.), Surv. Differ. Geom., vol. 9, Int. Press, Somerville, MA, 2004, pp. 219–240.
32. P. Li and S. T. Yau, *Estimates of eigenvalues of a compact Riemannian manifold*, Geometry of the Laplace Operator (Honolulu, HI, 1979) (R. Osserman and A. Weinstein, eds.), Proc. Sympos. Pure Math., vol. 36, Amer. Math. Soc., Providence, RI, 1980, pp. 205–239.
33. L. Lovász and M. Simonovits, *Random walks in a convex body and an improved volume algorithm*, Random Structures Algorithms **4** (1993), no. 4, 359–412.
34. V. G. Maz′ja, *Classes of domains and imbedding theorems for function spaces*, ZDokl. Akad. Nauk SSSR **133** (1960), 527–530 (Russian); English transl., Soviet Math. Dokl. **1** (1960), 882–885.
35. ———, *The negative spectrum of the higher-dimensional Schrödinger operator*, Dokl. Akad. Nauk SSSR **144** (1962), 721–722 (Russian); English transl., Soviet Math. Dokl. **3** (1962), 808–810.
36. E. Milman, *On the role of convexity in functional and isoperimetric inequalities*, Proc. Lond. Math. Soc. (3) **99** (2009), no. 1, 32–66.
37. ———, *On the role of convexity in isoperimetry, spectral gap and concentration*, Invent. Math. **177** (2009), no. 1, 1–43.
38. ———, *Isoperimetric and concentration inequalities: equivalence under curvature lower bound*, Duke Math. J. **154** (2010), no. 2, 207–239.
39. ———, *Properties of isoperimetric, functional and transport-entropy inequalities via concentration*, Probab. Theory Related Fields **152** (2012), no. 3-4, 475–507.
40. ———, *Isoperimetric bounds on convex manifolds*, Concentration, Functional Inequalities and Isoperimetry (Boca Raton, FL, 2009) (Ch. Houdré, M. Ledoux, E. Milman, and M. Milman, eds.), Contemp. Math., vol. 545, Amer. Math. Soc., Providence, RI, 2011, pp. 195–208.
41. V. D. Milman and G. Schechtman, *Asymptotic theory of finite-dimensional normed spaces*, with an appendix by M. Gromov, Lecture Notes in Math., vol. 1200, Springer, Berlin, 1986.
42. F. Nazarov, M. Sodin, and A. Vol′berg, *The geometric Kannan–Lovász–Simonovits lemma, dimension-free estimates for the distribution of the values of polynomials, and the distribution*

of the zeros of random analytic functions, Algebra i Analiz **14** (2002), no. 2, 214–234 (Russian); English transl., St. Petersburg Math. J. **14** (2003), no. 2, 351–366.
43. L. E. Payne and H. F. Weinberger, *An optimal Poincaré inequality for convex domains*, Arch. Rational Mech. Anal. **5** (1960), 286–292 (1960).
44. V. N. Sudakov and B. S. Cirel′son, *Extremal properties of half-spaces for spherically invariant measures*, Zap. Naučn. Sem. Leningrad. Otdel. Mat. Inst. Steklov. (LOMI) **41** (1974), 14–24, 165 (Russian).
45. J. Q. Zhong and H. C. Yang, *On the estimate of the first eigenvalue of a compact Riemannian manifold*, Sci. Sinica Ser. A **27** (1984), no. 12, 1265–1273.

Department of Mathematics, Technion — Israel Institute of Technology, Haifa 32000, Israel
E-mail address: emilman@tx.technion.ac.il

Centre de Recherches Mathématiques
CRM Proceedings and Lecture Notes
Volume **56**, 2013

A Visual Introduction to Riemannian Curvatures and Some Discrete Generalizations

Yann Ollivier

ABSTRACT. We try to provide a visual introduction to some objects used in Riemannian geometry: parallel transport, sectional curvature, Ricci curvature, Bianchi identities... We then explain some of the strategies used to define analogues of curvature in nonsmooth or discrete spaces, beginning with Alexandrov curvature and δ-hyperbolic spaces, and insisting on various notions of generalized Ricci curvature, which we briefly compare.

The first part of this text covers in a hopefully intuitive and visual way some of the usual objects of Riemannian geometry: parallel transport, sectional curvature, Ricci curvature, the Riemann tensor, Bianchi identities... For each of those we try to provide one or several pictures that convey the meaning (or at least one possible interpretation) of the formal definition.

Next, we consider the problem of defining analogues of curvature for nonsmooth or discrete objets. This is in the spirit of "synthetic" or "coarse" geometry, in which large-scale properties of metric spaces are investigated instead of fine small-scale properties; one of the aims being to be impervious to small perturbations of the underlying space. Motivation for these nonsmooth or discrete extensions often comes from various other fields of mathematics, such as group theory or optimal transport. From this viewpoint is emerging a theory of *metric measure spaces*, see for instance [35].

Thus we cover the definitions and mention some applications of δ-hyperbolic spaces and Alexandrov curvature (generalizing sectional curvature), and of two notions of generalized Ricci curvature: *displacement convexity of entropy* introduced by Sturm and Lott – Villani (following von Renesse – Sturm and others), and *coarse Ricci curvature* as used by the author. We mention new theorems obtained for the original Riemannian case thanks to this more general viewpoint. We also briefly discuss the differences between these two approaches to Ricci curvature.

Note that scalar curvature is not covered, for lack of the author's competence about recent developments and applications in this field.

2010 *Mathematics Subject Classification.* 53B20, 51K10.
This is the final form of the paper.

1. Riemannian curvatures

The purpose of this section is not to give a formal course in Riemannian geometry. We will therefore not reproduce formal definitions of the basic objects, but try to convey some pictorial intuition behind them. For a more formal introduction we refer, for instance, to [22], or to [8] for a survey of numerous developments in Riemannian geometry.

Riemannian manifolds. The most basic example of a Riemannian manifold is a smooth surface embedded in three-dimensional Euclidean space. More generally, any smooth manifold can be seen as a set $X \subset \mathbb{R}^p$ for some integer p, such that, at any point $x \in X$, there exists an N-dimensional affine subspace of $\mathbb{R}^p$ which coincides with X at first order around x. This subspace is the *tangent space* $T_x X$ at $x \in X$, and N is the *dimension* of X.

Note that if $t \mapsto c(t)$ is a smooth curve inside X, then its derivative $\mathrm{d}c(t)/\,\mathrm{d}t$ is a vector tangent to X at $c(t)$.

A *Riemannian* manifold is a manifold equipped with a *Riemannian metric*, that is, for each $x \in X$, a definite positive quadratic form defined on $T_x X$. For instance, if X is included in $\mathbb{R}^p$, such a quadratic form might be the restriction to $T_x X$ of the canonical Euclidean structure on $\mathbb{R}^p$. We will assume that the quadratic form depends smoothly on $x \in X$.

Such a quadratic form, applied to any tangent vector, allows to define the norm of this vector. By integration, one can then define the length of a curve in X. The distance (inside X) between two points of X is then defined as the infimum of the lengths of all curves between these two points. This turns X into a metric space.

We will always assume that X is connected and complete for this metric.

A *geodesic* is a curve γ in X such that, for any two close enough points on γ, the distance in X between those two points is obtained by travelling along γ. Locally, such curves always exist. Moreover, given a tangent vector $x \in X$ and a tangent vector $v \in T_x X$ at x, there exists exactly one geodesic starting at x such that its initial velocity is v and which has constant speed. This will be the *geodesic starting along* v.

The *endpoint* of v will be the point, denoted $\exp_x v$, obtained after following the geodesic starting along v for a unit time.

Parallel transport. Say we are given two very close points x and y in a Riemannian manifold. Is there a way to compare a tangent vector at x and a tangent vector at y, even though they live, a priori, in different vector spaces? This is done via *parallel transport.*

So let w_x be a tangent vector at x; we are looking for a tangent vector w_y at y which would be "the same" as w_x. Since x and y are very close, we may assume that y is the endpoint of a small tangent vector v at x. For simplicity, we will assume that w_x is orthogonal to v, and that the norm of w_x is very small. Then, there exists a particular tangent vector w_y at y: it is the one whose endpoint is closest to the endpoint of w_x, given the restriction that w_y be orthogonal to v (if this orthogonality condition is lifted, then of course there is a tangent vector at y whose endpoint is exactly the endpoint of w_x, but this vector "turns back towards x"). The vector w_y is the best candidate to be "the same" as w_x, translated to y (Figure 1).

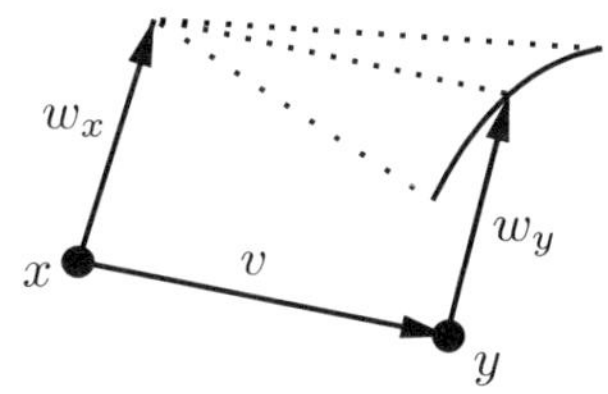

FIGURE 1. Parallel transport of w_x along v.

Parallel transport is the operation mapping w_x to w_y. (More exactly, since we have assumed that w_x is very small, we take the linear component of the map $w_x \mapsto w_y$ for small w_x and then extend by linearity to a map between the whole spaces T_xX and T_yX. We have also assumed that w_x is orthogonal to v; by definition, we define the parallel transport of v from x to y as the velocity at y of the geodesic starting along v.)

More generally, parallel transport of a vector w along any smooth curve starting at x can be defined by decomposing the curve into small intervals and performing successive parallel transports along these subintervals.

Sectional curvature and Ricci curvature. Let us now define various curvatures. The first one we consider is *sectional* curvature.

Let again x be a point in X, v a small tangent vector at x, y the endpoint of v, w_x a small tangent vector at x, and w_y the parallel transport of w_x from x to y along v. If, instead of a Riemannian manifold, we were working in ordinary Euclidean space, the endpoints x' and y' of w_x and w_y would constitute a rectangle with x and y. But in a manifold, generally these four points do not constitute a rectangle any more.

Indeed, because of curvature, the two geodesics starting along w_x and w_y may diverge from or converge towards each other. Thus, on a sphere (positive curvature), two meridians starting at two points on the equator have parallel initial velocities, yet they converge at the North (and South) pole. Since the initial velocities w_x and w_y are parallel to each other, this effect is at second order in the distance along the geodesics (Figure 2).

Thus, let us consider the points lying at distance ε from x and y on the geodesics starting along w_x and w_y, respectively. In a Euclidean setting, the distance between

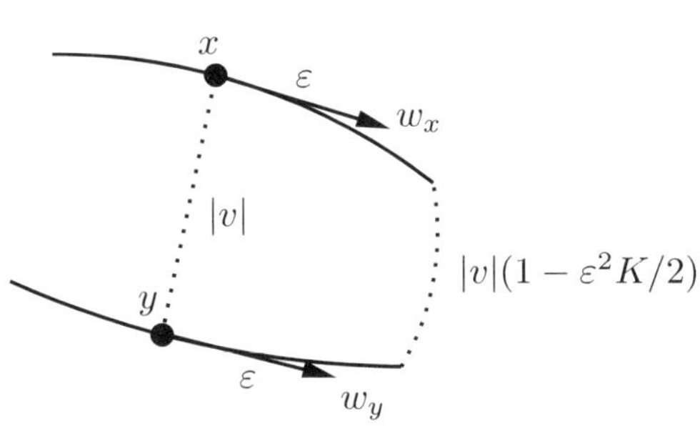

FIGURE 2. The sectional curvature K.

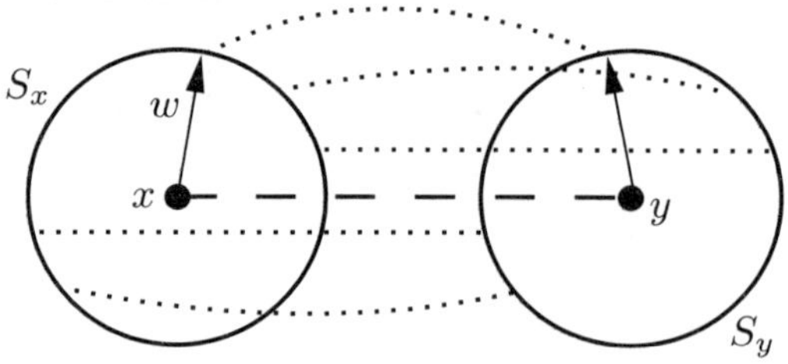

FIGURE 3. The Ricci curvature Ric.

those two points would be $|v|$, the same as the distance between x and y. The discrepancy from this Euclidean case is used as a definition of a curvature.

Definition 1.1 (Sectional curvature). Let (X,d) be a Riemannian manifold. Let v and w_x be two unit-length tangent vectors at some point $x \in X$. Let $\varepsilon, \delta > 0$. Let y be the endpoint of δv and let w_y be obtained by parallel transport of w_x from x to y. Then

$$d(\exp_x \varepsilon w_x, \exp_y \varepsilon w_y) = \delta\left(1 - \frac{\varepsilon^2}{2} K(v,w) + O(\varepsilon^3 + \varepsilon^2\delta)\right)$$

when $(\varepsilon,\delta) \to 0$. This defines a quantity $K(v,w)$, which is the *sectional curvature* at x in the directions (v,w).

In contrast, *Ricci curvature* only depends on one tangent vector v; it is obtained by averaging $K(v,w)$ over all the directions w (Figure 3).

Definition 1.2 (Ricci curvature). Let x be a point in an N-dimensional Riemannian manifold. Let v be a unit tangent vector at x. The *Ricci curvature along* v is the quantity $\mathrm{Ric}(v)$ defined as N times the average of $K(v,w)$, where the average is taken over w running over the unit sphere in the tangent space T_xX.

The scaling factor N comes from the traditional definition of Ricci curvature as the trace of a linear map, which results in a sum over a basis rather than an average over the unit sphere. Actually $\mathrm{Ric}(v)$ is a quadratic form in v and is usually denoted $\mathrm{Ric}(v,v)$; the Ricci tensor is the corresponding bilinear form $\mathrm{Ric}(v,v')$.

This definition can be rephrased as follows, due to the fact that we used parallel transport $w_x \mapsto w_y$ in the definition of curvature.

Corollary 1.3. *Let v be a unit tangent vector at a point x in a Riemannian manifold. Let $\varepsilon,\delta > 0$ and let y be the endpoint of δv.*

Let S_x be the set of endpoints of all tangent vectors at x with norm ε, and likewise let S_y be the set of endpoints of the sphere of radius ε in the tangent space at y. Then, if S_x is mapped to S_y using parallel transport, the average distance between a point of S_x and its image is

$$\delta\left(1 - \frac{\varepsilon^2}{2N}\,\mathrm{Ric}(v,v) + O(\varepsilon^3 + \varepsilon^2\delta)\right)$$

when $(\varepsilon,\delta) \to 0$.

If balls are used instead of spheres, the scaling factor is $\varepsilon^2/\big(2(N+2)\big)$ instead of $\varepsilon^2/(2N)$.

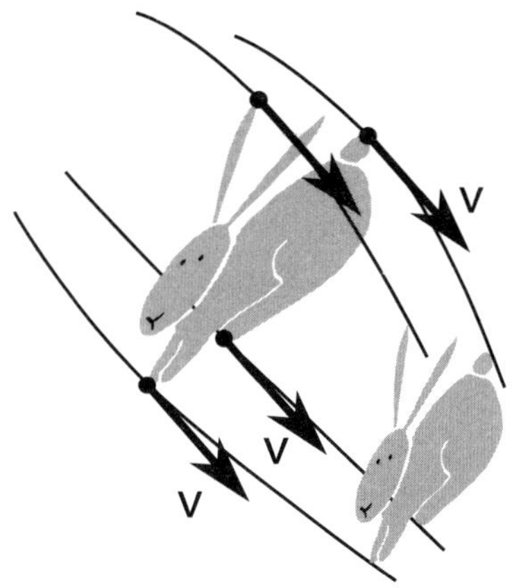

FIGURE 4. Curvature and volume change under geodesic flow.

Thus, Ricci curvature is positive when "balls are closer than their centers are." We will see that this property is well suited as a definition in more general, non-smooth spaces.

Let us mention another, more dynamical way to visualize Ricci curvature. Let again v be a unit tangent vector at some point x in the Riemannian manifold X. Let C be any small neighborhood of X, having an arbitrary shape. For each point z in C, let us consider the geodesic $t \mapsto z_t$ starting at z and whose initial velocity is v (we implicitly use parallel transport to identify v with a tangent vector at z). On average, the geodesics starting from various choices of z get closer or further apart depending on the sign of curvature. Now let us have C "slide along" these geodesics; more exactly, let C_t be the set $\{z_t,\ z \in C\}$. In particular $C_0 = C$ (Figure 4).

Then we have

$$\operatorname{vol} C_t = \operatorname{vol} C \left(1 - \frac{t^2}{2} \operatorname{Ric}(v) + \text{smaller terms} \right)$$

(note that the derivative of $\operatorname{vol} C_t$ is 0 for $t = 0$ because we choose geodesics with parallel initial speeds). Thus Ricci curvature controls the evolution of volumes under the geodesic flow.

Signs of curvature. Positive curvature and negative curvature are very different worlds; we refer to [34] for a thorough discussion of the "sign and geometric meaning of curvature." It is often the case, in Riemannian geometry, that theorems in negative curvature assume that all sectional curvatures $K(v, w)$ are negative, whereas theorems in positive curvature assume that the Ricci curvatures $\operatorname{Ric}(v, v)$ is positive for any v (which is weaker than having all sectional curvatures $K(v, w) \geqslant 0$).

We give here two basic theorems of Riemannian geometry illustrating each of these situations. When Ricci curvature is positive, the Bonnet–Myers theorem states that if a Riemannian manifold is more positively curved than a sphere, then its diameter is smaller.

Theorem 1.4 (Bonnet – Myers). *Let X be an N-dimensional Riemannian manifold. Let* $\inf \mathrm{Ric}(X)$ *be the infimum of the Ricci curvature* $\mathrm{Ric}(v, v)$ *over all unit tangent vectors v.*

Let $S^N \subset \mathbb{R}^{N+1}$ be the unit sphere of the same dimension as X. Then, if $\inf \mathrm{Ric}(X) \geqslant \inf \mathrm{Ric}(S^N)$ *then* $\operatorname{diam} X \leqslant \operatorname{diam} S^N$.

Several other theorems about positive Ricci curvature follow this pattern of a comparison with the sphere of the same dimension.

As for negative curvature, let us mention the Cartan – Hadamard theorem, which implies that all topological information about a negatively curved manifold is contained in its fundamental group. This is obviously false in positive curvature, since, for instance, a sphere is simply connected.

Theorem 1.5 (Cartan – Hadamard). *Let X be an N-dimensional Riemannian manifold and assume that $K(v, w) \leqslant 0$ for any pair of tangent vectors v, w at the same point. Then the universal cover of X is homeomorphic to $\mathbb{R}^N$.*

The Riemann curvature tensor. The next object in our journey through curvatures is the *Riemann curvature tensor*. It depends on three tangent vectors u, v, w at a given point x, and its output is another tangent vector at x, denoted $R(v, w)u$. Before we define this tensor, we need to say more about parallel transport.

Given two tangent vectors v and w at point x, we may parallel-transport w along v; this yields a tangent vector w' at the endpoint of v. Likewise, we may also parallel-transport v along w, getting a tangent vector v' at the endpoint of w. In Euclidean space, the vectors v, w, v', w' make up a parallelogram; in particular, the endpoints of v' and of w' coincide. But we could expect that, in a general Riemannian manifold, the endpoints of v' and w' may not coincide anymore. Actually things are not so bad: for small v and w, the endpoints of v' and w' do coincide at the first order at which this is nontrivially true (more precisely, up to $o(|v||w|)$). This property of parallel transport is called *torsion-freeness* and expresses the fact that "parallelograms close up."

Actually this property was used implicitly before: when we defined curvature, we had to look at a phenomenon of order $O(|v||w|^2)$.

Let us now consider three vectors u, v, w instead of two. Let us keep the parallelogram built from u and v, and let us parallel-transport u along the path vw, or along the path wv. This time, we obtain two different results; their difference is the *Riemann curvature* $R(v, w)u$, which is again a tangent vector (Figure 5).

This proves that, in contrast to parallelograms, cubes don't close up in Riemannian geometry.

Let us take a further look at this. Given three directions u, v, w for the sides of a cube whose base corner is x, there are several ways to build the farthest corner of the cube: we saw that the paths vwu and wvu do not end up at the same point, and that the difference is $R(v, w)u$. (When we say "the path vwu", it means: follow the tangent vector v to its endpoint, parallel-transport w along this path, follow the obtained vector to its endpoint, parallel-transport u along the whole path, and follow the obtained vector to its endpoint.)

All in all, given u, v and w, there are six possible paths to the farthest corner, hence six possible ways to define this farthest corner. However, these six paths define only three distinct endpoints: indeed, torsion-freeness guarantees that $vw =$

wv, but torsion-freeness at the endpoint of u also guarantees that $uvw = uwv$. The latter equality just expresses that "lateral faces of the cube close up."

These three distinct endpoints form a triangle. We saw that, by definition, one of the sides of this triangle is $R(v,w)u$; by symmetry the other sides are $R(w,u)v$ and $R(u,v)w$. Since the three sides of any triangle add up to 0 we get

$$R(u,v)w + R(v,w)u + R(w,u)v = 0$$

which is the *first Bianchi identity* and was discovered by Ricci.

With Bruno Sévennec (unpublished work), we turned this heuristic argument into a very short formal proof along the same lines, by interpreting expressions such as uvw in a suitable quotient of a jet space of curves in X starting at the base point x.

Riemannian volume measure. Another notion we will use from Riemannian geometry is that of *Riemannian volume measure.* This is a measure on an N-dimensional Riemannian manifolds which gives volume ε^N to a small cube made of N tangent vectors at the same point, each of length ε and orthogonal to each other.

This informal definition yields a volume n-form associated with a Riemannian manifold. This volume form can be denoted $\sqrt{\det g}$ where g is the bilinear form defining the metric. The reason is as follows: At a given point x, the metric g is a bilinear form on the tangent space $E = T_x X$. A bilinear form is an application from E to its dual E^*. Thus, from $g\colon E \to E^*$ we can apply the textbook definition

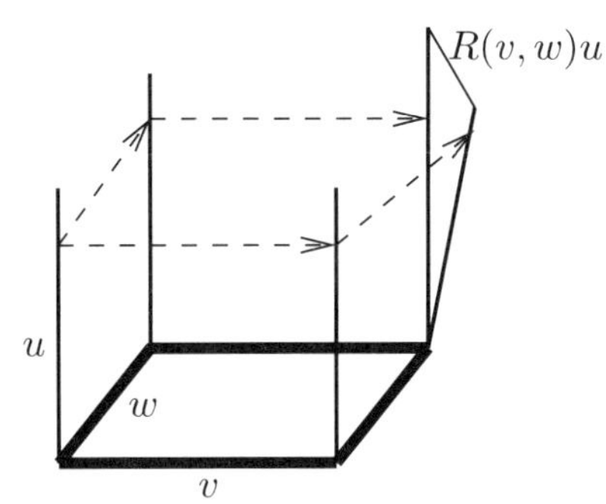

FIGURE 5. The Riemann curvature $R(v,w)u$.

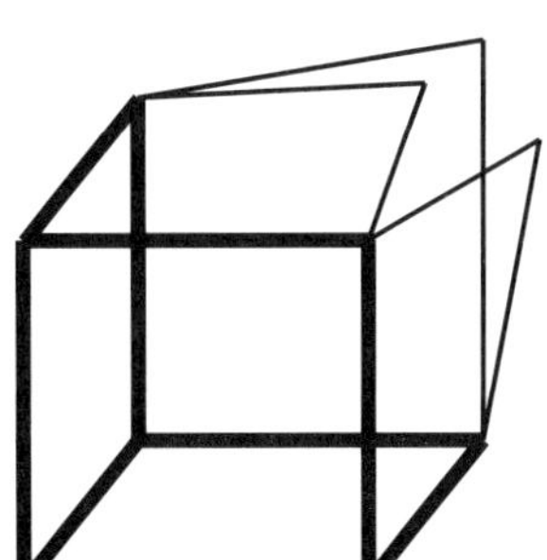

FIGURE 6. The first Bianchi identity.

of the determinant of a linear map, passing to the exterior product: $\det g = \bigwedge^N g$ is a map from $\bigwedge^N E$ to $\bigwedge^N E^* \simeq \left(\bigwedge^N E\right)^*$. That is to say, $\det g$ is a bilinear form on $\bigwedge^N E$ (which is a 1-dimensional space since $N = \dim E$). Moreover $\det g$ is definite positive whenever g is. The square root $\sqrt{\det g}$ is thus a *norm* on $\bigwedge^N E$, that is, a volume form.

2. Discrete curvatures

Let us now consider some generalizations of these notions to more general spaces than smooth manifolds. This is motivated by the consideration of "synthetic" geometry: a geometry that focuses on "large-scale" instead of "small-scale" properties, and uses mainly metric comparisons, avoiding local properties such as differentiability, with the goal of being invariant under small perturbations of the underlying space. We refer for instance to Gromov's influential book [35].

Two types of spaces can be considered. *Geodesic spaces* are continuous spaces in which the distance between points is always realized by the length of a continuous curve; important examples are piecewise smooth objects, or objects obtained as limits of smooth manifolds, which can be thought of as manifolds with singularities.

Just as important are truly *discrete spaces*; typical examples are discrete groups equipped with a Cayley graph metric, or spaces of configurations of discrete statistical physics systems, for instance.

We will briefly mention sectional curvature, the generalizations of which are now well understood especially in negative curvature, with deep connections for instance with group theory. We will focus on generalizations of Ricci curvature, which were developed more recently. In passing we will mention new results obtained thanks to a discrete viewpoint on Ricci curvature, both in the original setting of Riemannian manifolds and in a surprisingly diverse array of discrete settings.

2.1. Nonsmooth and discrete sectional curvature: Alexandrov curvature and Gromov-hyperbolicity. There are at least two notions of metric spaces having bounds on their sectional curvature. The first one is restricted to geodesic spaces and is that of *spaces with curvature bounds in the sense of Alexandrov* (also termed *locally* CAT(k) *spaces* for curvature bounded above, *locally* CAT$^+(k)$ *spaces* for curvature bounded below). The second is that of *δ-hyperbolic spaces* (or *Gromov-hyperbolic spaces*) and applies to any metric space, but only for negative curvature "at large scale."

Both notions have been extremely successful, in particular in negative curvature, as δ-hyperbolic spaces and CAT(0) spaces have deep connections to geometric group theory, for instance. We cannot list all the relevant work and we refer the reader to [11] or [13] for a more extensive treatment.

Curvature bounds in the sense of Alexandrov go as follows. Let X be a geodesic space. A *triangle* in X is a triplet of points $(a, b, c) \in X^3$, together with three curves from a to b, b to c and c to a, respectively (the *sides* of the triangle), such that the lengths of these curves realize the distances $d(a, b)$, $d(b, c)$ and $d(c, a)$, respectively.

The curvature criterion of Alexandrov (also called CAT criterion, for Cartan–Alexandrov–Topogonov), states that triangles become "thinner" in negative curvature and "fatter" in positive curvature. This will be measured by the distance

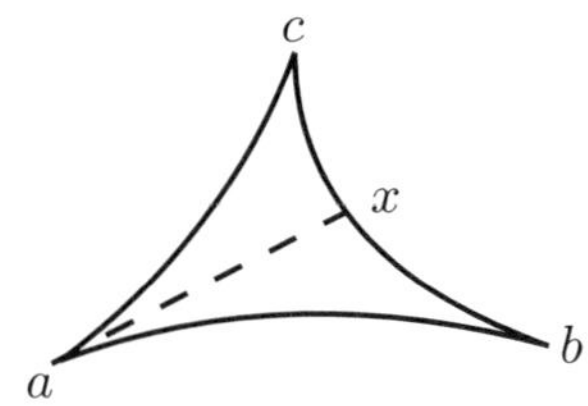

FIGURE 7. Comparison triangles and Alexandrov curvature.

between a vertex of the triangle, and a point on the opposite side, compared to that expected in the Euclidean situation (Figure 7).

For any triangle in X, there exists a *comparison triangle* (A, B, C) in the Euclidean plane whose sides have the same lengths. More generally, let S_k is the standard 2-dimensional space of constant curvature k (Euclidean plane if $k = 0$, Euclidean sphere of radius $1/\sqrt{k}$ with its intrinsic metric when $k > 0$, and hyperbolic plane with curvature k if $k < 0$), and let r_k be the diameter of diam S_k (which is infinite if $k \leqslant 0$). Then, for any triangle in X with perimeter at most $2r_k$, there exists a comparison triangle (A_k, B_k, C_k) in S_k, whose sides have the same lengths as (a, b, c). Such a comparison triangle is unique up to isometry, and will be called *the* comparison triangle of (a, b, c) in curvature k.

Definition 2.1. Let X be a geodesic space and k a real number. X is a *space of curvature* $\leqslant k$ *in the sense of Alexandrov* if, for any small enough triangle (a, b, c) in X, and for any point x on the side bc of this triangle, the following holds:

$$d(a, x) \leqslant d_{S_k}(A_k, X_k)$$

where (A_k, B_k, C_k) is the comparison triangle of (a, b, c) in curvature k, and X_k is the point on the side B_kC_k corresponding to x, i.e., such that $d(X_k, B_k) = d(x, b)$. Likewise, X is a *space of curvature* $\geqslant k$ *in the sense of Alexandrov* if, in the same situation, the reverse inequality

$$d(a, x) \geqslant d_{S_k}(A_k, X_k)$$

holds.

This definition is compatible with the Riemannian notion of sectional curvature. For instance, we have the following ([11, Theorem II.1A.6]).

Theorem 2.2. *Let X be a complete smooth Riemannian manifold and $k \in \mathbb{R}$. Then X has curvature at most k in the sense of Alexandrov if and only if for any two tangent vectors v, w at the same point, with unit length and orthogonal to each other, one has $K(v, w) \leqslant k$.*

The terminology CAT(k) *space*, for curvature $\leqslant k$, applies to the case when *all* triangles, not only small ones, satisfy the comparison criterion. For complete smooth Riemannian manifolds this is equivalent to negative sectional curvature and being simply connected. On the other hand, for spaces of curvature bounded *below*, imposing the condition on small triangles implies the same for all triangles (Topogonov's theorem, [13, Theorem 10.5]).

Spaces with curvature bounds in the sense of Alexandrov keep a number of properties of Riemannian manifolds with controlled sectional curvature. For instance, the Cartan – Hadamard theorem (Theorem 1.5) in this context states that any CAT(0) space is contractible ([11, Corollary II.1.5]).

The Bonnet – Myers theorem (Theorem 1.4) applies as well: a space with curvature at least $k > 0$ in the sense of Alexandrov has diameter at most $\pi/\sqrt{k}$ ([13, Theorem 10.4.1]); otherwise, triangles would get too big with respect to their comparison triangle in the sphere. Note that here, contrary to Theorem 1.4, we do not work with a Ricci curvature bound but with a (stronger) sectional curvature bound instead.

Spaces with curvature bounds in the sense of Alexandrov do not stray too far from smooth manifolds. For instance ([13, Theorem 10.8.3]), an Alexandrov space with Hausdorff dimension n contains an open dense subset which is an n-dimensional manifold.

For *discrete* spaces, there is a very successful notion of negative sectional curvature, namely *δ-hyperbolicity* (also called Gromov-hyperbolicity). This notion is usually attributed to Rips and was thoroughly developed by Gromov [33], especially in the context of geometric group theory, where it led to the notion of hyperbolic groups.

δ-hyperbolicity states that, at large scales, a space behaves metrically almost like a tree. It disregards any phenomena occurring at small scales; in particular, a bounded metric space X is always δ-hyperbolic with $\delta = \operatorname{diam} X$.

The simplest definition of δ-hyperbolicity involves triangles, and thus assumes that the space is geodesic. For nonnecessarily geodesic metric spaces, the condition is expressed in terms of quadrilaterals, as follows.

The underlying idea is that large n-gons in the hyperbolic plane look almost like trees, their sides clinging to each other. Let us pass to the ideal limit case and consider four points w, x, y, z in a tree. Assume the general situation where no branch of the subtree joining these points is degenerate: this subtree is made of five segments, one of which lies in the middle.

Let us consider the three pairwise sums of distances $d(w,x)+d(y,z)$, $d(w,y)+d(x,z)$ and $d(w,z)+d(x,y)$. Of these three sums, only one (the smallest one) does not involve the length of the middle segment of the tree; the other two are equal. This can be summed up by the ultrametric-like inequality

$$d(w,x)+d(y,z) \leqslant \max\bigl(d(w,y)+d(x,z), d(w,z)+d(x,y)\bigr)$$

for all points w, x, y, z in a metric tree. By permuting the roles of the points, we get that, among the three sums involved, the largest two must be equal (because the largest is bounded by the max of the smallest and second-largest).

Now we just perturb this inequality by an additive term. This means that, in a hyperbolic space, the metric relations between any four points are the same as in a tree up to a small error (Figure 8).

Definition 2.3. Let $\delta \geqslant 0$. A metric space X is *δ-hyperbolic* if

$$d(w,x)+d(y,z) \leqslant \max\bigl(d(w,y)+d(x,z), d(w,z)+d(x,y)\bigr) + 2\delta$$

for any points w, x, y, z in X.

This does not imply any local topological property when $\delta > 0$: indeed, anything can happen at scales smaller than δ. One key feature of this definition is its invariance under *quasi-isometries* [11], which makes it very robust.

This notion is an extension of negative sectional curvature "at large scales" in the sense that any Riemannian manifold with negative sectional curvature and which is simply connected, is δ-hyperbolic. Moreover CAT$(-k)$ spaces for $k > 0$ are δ-hyperbolic, so that δ-hyperbolicity is compatible with global negative curvature in the sense of Alexandrov. Finally, δ-hyperbolic groups (groups whose Cayley graph is δ-hyperbolic) are important objects and keep many of the features of fundamental groups of negatively curved manifolds. Interestingly, these groups turn out to be, in some sense, "generic" among discrete groups [33, 58].

2.2. Discrete Ricci curvature: the coarse Ricci curvature approach. Let us now turn to generalizations of Ricci curvature. The survey [46] contains a discussion of the geometric interest of bounds on Ricci curvature, and the need for a generalized notion of positive Ricci curvature for metric measure spaces.

This time, let us begin with the discrete case, and the notion of *coarse Ricci curvature* used by the author [59, 60]. Another approach introduced by Sturm [65] and Lott–Villani [47], based on *displacement convexity of entropy*, is also presented below. Coarse Ricci curvature keeps fewer properties from Riemannian manifolds, but is somewhat simpler to present and has a wider range of examples.

Consider the following motivation: From a statistical physics viewpoint, at a macroscopic level there should not be much difference between a system of n particles whose total energy is exactly E, and a system of n particles whose energies are chosen independently with average E/n, when n is large. For free particles, the energy E is (proportional to) the sum of square velocities; so in the first situation the configuration space is a sphere of radius $\sqrt{E}$ in speed space $\mathbb{R}^{3n}$, while for the case of independent particles, the configuration space for speeds is the whole of $\mathbb{R}^{3n}$, but equipped with a Gaussian measure (the Gaussian measure being the Maxwell–Boltzmann distribution $\exp(-E/kT)$ for this energy E). If we pick a large number of points according to this latter Gaussian distribution, a set will emerge which is, in some sense, close to a sphere. A third model for our particles of average energy E would be, for each particle, to choose at random an energy either 0 or $2E/3n$ along each axis; in this third model, the configuration space would be the discrete hypercube $\{0, 2E/3n\}^{3n}$.

One can wonder whether these three macroscopically similar viewpoints have anything in common from a geometric point of view. Since the sphere (in the first example) is the archetypical space of positive curvature, one can wonder whether

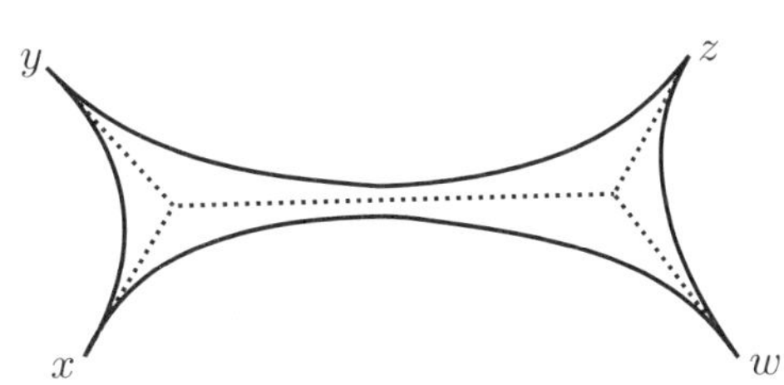

FIGURE 8. A δ-hyperbolic quadrilateral.

the last two examples (Euclidean space with a Gaussian measure, and discrete hypercube) somehow have positive curvature. For instance, Gromov argues in [35] that the discrete hypercube shares a lot of geometric properties with a sphere.

A successful approach in that direction, with a rich progeny over the years, is Bakry – Émery theory [6, 7]. It allows to define a notion of Ricci curvature for a manifold equipped with a measure (more exactly, for a manifold equipped with a diffusion process whose invariant measure is the measure considered). This way, the Euclidean space endowed with a Gaussian measure acquires positive Ricci curvature. However, the use of diffusions make its adaptation to a discrete setting delicate, and this theory is not stable if we replace the space considered with a "close enough" space.

Thus it would be nice to have at hand a notion of Ricci curvature, which would be valid "at a certain scale" and thus robust and insensitive to the local structure, discrete or continuous. Ideally, this should be compatible with the Bakry – Émery approach, be easy to check on examples, capture some common features for our three examples above, and allow to generalize some of the properties of Riemannian manifolds with positive Ricci curvature.

Coarse Ricci curvature directly transposes Definition 1.2, or more precisely the Corollary 1.3. The point is to compare the distances between two small balls, with the distance between their centers. Ricci curvature will be positive if *small balls are closer than their centers are.*

For this we need to define what the distance between balls is. It will be practical to think of a ball as a measure of mass 1 around some point (in line with Definition 1.2 which involves an average over the sphere). Then we can use the well-known *transportation distance* (also called Wasserstein distance, Monge – Kantorovich – Rubinstein distance, or earth-mover distance), defined as follows [69].

Definition 2.4 (Transportation distance). Let μ_1, μ_2 be two measures with mass 1 in a metric space (X, d). A *transference plan* (or *coupling*) from μ_1 to μ_2 is a measure ξ on $X \times X$ such that $\int_y \mathrm{d}\xi(x, y) = \mathrm{d}\mu_1(x)$ and $\int_x \mathrm{d}\xi(x, y) = \mathrm{d}\mu_2(y)$. (Thus $\mathrm{d}\xi(x, y)$ represents the amount of mass travelling from x to y.)

The L^1 *transportation distance* between μ_1 and μ_2, denoted $W_1(\mu_1, \mu_2)$, is the best average travelling distance that can be achieved:

$$W_1(\mu_1, \mu_2) = \inf_{\xi \in \Pi(\mu_1, \mu_2)} \iint d(x, y)\, \mathrm{d}\xi(x, y)$$

where $\Pi(\mu_1, \mu_2)$ is the set of transference plans from μ_1 to μ_2.

The transportation distance between small balls can then be used to define a Ricci curvature at some scale. The relevant notion of a "small ball" depends on the situation at hand: for instance, in a graph it is natural to use balls of radius 1, whereas in manifolds, arbitrarily small balls are used.

Let us assume that, for each point x in space X, we have chosen a measure b_x to play the role of "small ball around x." The basic example is when X itself comes with a global measure μ (such as the Riemannian volume measure on a manifold, or the counting measure on a graph) and we have chosen a scale parameter $\varepsilon > 0$; then we can define b_x by restricting the measure to the ε-ball around x and scaling:

$$b_x^{\varepsilon,\mu} = \frac{\mu_{|B(x,\varepsilon)}}{\mu\big(B(x,\varepsilon)\big)}$$

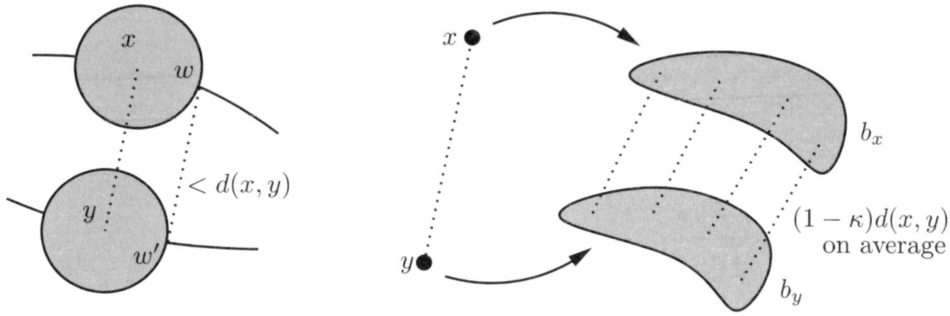

FIGURE 9. In positive curvature, balls are closer than their centers are.

where $B(x,\varepsilon)$ is the closed ball of radius ε around x. This is a very common choice for b_x but is not the only one; thus, we will keep full generality in the definition.

We are now ready to transpose Corollary 1.3.

Definition 2.5 (Coarse Ricci curvature). Let (X,d) be a metric space. We assume that, for each $x \in X$, a probability measure b_x on X is given. Let x and y be two distinct points in X. The *coarse Ricci curvature* along xy is the quantity $\kappa(x,y)$ defined by the relation

$$W_1(b_x,b_y) = \big(1-\kappa(x,y)\big)d(x,y)$$

(see Figure 9).

Compared to the Riemannian case, note that we chose not to apply the $\varepsilon^2/(2N)$ scaling factor. Also, in the Riemannian case Ricci curvature was defined along a tangent vector. Here, in a metric setting, the best we can do for tangent vectors is to have a pair of close points. A simple but key feature of coarse Ricci curvature is that it is enough to compute it for pairs of points which are close enough [59].

The technical assumptions implicit in this definition are basically the ones needed for good behavior of the Wasserstein distance: the space X should be Polish (metric, separable, complete), and the measures b_x should all have a finite first moment [69]. Moreover the map $x \mapsto b_x$ should be measurable.

This kind of condition appears in [71] as one among several ways to characterize a lower Ricci bound on manifolds. It was proposed in [59, 60] (see also [39]) as a possible definition of Ricci curvature extending to metric spaces several results known for positively curved manifolds.

The data $(b_x)_{x\in X}$ exactly define a *Markov chain* on X: the point x is sent to a point chosen at random according to the measure b_x. The condition $W_1(b_x,b_y) \leqslant (1-\kappa)d(x,y)$ actually has quite a long history in the Markov chain literature, under various names: *Dobrushin–Shlosman criterion* [23–25], or [16], or *path coupling* [12]; though, without the link to curvature. It appears in some recent Markov chain textbooks [15, 45]. It is most often used to prove fast convergence of the measure of the Markov chain to its invariant distribution. It is operational in a number of settings: to name just a few, the Ising model of ferromagnetism [25], graph coloring [12], fast generation of random orthogonal matrices [57], sampling of paths with constraints [29] (see also [72]), simulations of waiting queues [40, 41]...

The definition also works well if one replaces the Markov chain b_x with a continuous-time Markov process (e.g., [16]). In this setting, a lower bound K on

Ricci curvature for a Markov process (P^t) states that $W_1(P^t_x, P^t_y) \leqslant \exp^{-Kt} d(x,y)$, and it is enough to check this for close enough points x, y and small enough time t (see [68] for details). Note an interesting connection: in the manifold case, if the distance d happens to change with time according to the backwards Ricci flow, then the transportation distance between P^t_x and P^t_y (measured using the distance at time t) will stay bounded, because the two effects compensate [50]. This could lead to notions of Ricci flow in metric spaces [50, 61].

As examples, let us illustrate how coarse Ricci curvature applies, first, to the original case of Riemannian manifolds, and, second, to the discrete hypercube $\{0,1\}^N$ mentioned above.

Example 2.6 (Riemannian manifolds). Let (X,d) be a smooth Riemannian manifold and let vol be the Riemannian volume measure. Let $\varepsilon > 0$ small enough and consider the ball of radius ε around each point x, i.e., set $b_x = b_x^{\varepsilon,\mathrm{vol}}$ as above. Let $x, y \in X$ be two close enough points. Let v be the unit tangent vector at x directed towards y. Then the coarse Ricci curvature along xy is

$$\kappa(x,y) = \frac{\varepsilon^2 \operatorname{Ric}(v,v)}{2(N+2)} + O\big(\varepsilon^3 + \varepsilon^2 d(x,y)\big)$$

This is basically a reformulation of Corollary 1.3. (Actually the latter only yields an inequality, as there might be better transference plans than parallel transport. But parallel transport gives the exact value of the Wasserstein distance at this order [60].)

The next example is the discrete hypercube.

Example 2.7 (Discrete hypercube). Let $X = \{0,1\}^N$ be the N-dimensional discrete hypercube, equipped with its graph (L^1) metric, and let μ be the uniform measure on X. Let $b_x = b_x^{1,\mu}$ be the uniform probability measure on the 1-ball around x (i.e., on x and its N neighbors). Then the coarse Ricci curvature along two adjacent points x, y is $\kappa(x,y) = 2/(N+1)$.

This is quite clear on Figure 10: indeed, to move b_x to b_y, out of the $N+1$ points in b_x, $N-1$ of them have to move by one edge, and 2 of them do not have to move at all.

Note that we have computed $\kappa(x,y)$ only for adjacent points x and y; but, as mentioned before, this is enough to imply the same lower bound on curvature for *all* pairs of points, by the triangle inequality for W_1 [59, 60].

The inclusion of x in b_x, i.e., the use of balls rather than spheres, is identical to the well-known need to use *lazy* random walks on bipartite graphs: otherwise,

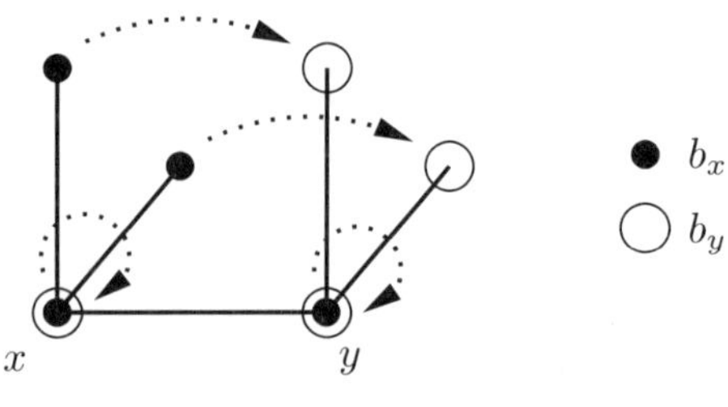

FIGURE 10. Coarse Ricci curvature of the cube.

points located at an odd distance never meet each other. In general metric spaces, spheres may be empty and balls are preferable.

Other very simple examples include the grid $\mathbb{Z}^n$ in $\mathbb{R}^n$ with its graph metric, or $\mathbb{R}^n$ itself equipped with any arbitrary Euclidean or non-Euclidean norm: these have 0 curvature as expected. We mentioned above numerous more sophisticated examples from Markov chains or statistical physics; see also the examples in [60].

Since Ricci curvature is an average of sectional curvature, bounds on sectional curvature imply the same bounds on Ricci curvature, at least for ordinary curvature. One can wonder whether the same holds for the generalized notions used here.

This is indeed the case. For δ-hyperbolic groups, for instance, with b_x the uniform measure on a ball of radius $\gg \delta$ around x, one can check [60] that coarse Ricci curvature is indeed negative (and actually takes the smallest possible value). For spaces with curvature bounded below in the sense of Alexandrov, equipped with their (continuous-time) canonical diffusion semigroup (see, e.g., [31]), a result of Gigli, Kuwada and Ohta [31,53] (combined with [64,73]) implies that if Alexandrov curvature is bounded below, then coarse Ricci curvature is bounded below by the corresponding value.

Let us review some of the theorems obtained thanks to this approach. They are of two kinds: extensions of known theorems about Ricci curvature to a more general setting, and, more interestingly, new theorems proven thanks to the insight provided by the coarse viewpoint. We will say a few words about the former and insist on the latter.

Coarse Ricci curvature does not keep too many topological or metric properties from the manifold case: this can be seen from the discrete examples listed above. For instance, the Bonnet – Myers theorem is true only in a very weak sense [60]. The kind of properties kept from manifolds relate mostly to functional inequalities or *concentration of measure* inequalities, i.e., various inequalities involving measures.

A metric space satisfies a concentration of measure property if every real-valued 1-Lipschitz function on the space is "almost constant", that is to say, if there exists a value such that the function is close to that value on most of the space. The basic example is that of the function "proportion of tails" on the space of length-N sequences of heads and tails: it is very close to $\frac{1}{2}$ on most sequences, with fluctuations of order $1/\sqrt{N}$. Concentration of measure asserts that this is not a property of this particular function, but of the underlying space. In this case the underlying space is the hypercube $\{H,T\}^N$, and the Lipschitz functions are those for which the influence of any single symbol in the sequence on the value of the function is at most $1/N$; any such function will be almost constant. We refer to [44] for more on this topic, or to [35] for a geometric viewpoint.

Thus the hypercube exhibits this concentration phenomenon. A theorem of Gromov [32] implies that manifolds with positive Ricci curvature also share this property, using a method very different from that used for the hypercube. It is shown in [60] that concentration follows from positive coarse Ricci curvature. This allows to put both examples in a common framework with a unified and simple proof (though with some small loss in the numerical constants).

Last, let us give examples of new theorems proven using the technique or intuition of coarse Ricci curvature. Some deal with non-Riemannian settings such as waiting queues or statistical physics [41], but it is interesting that new results have been obtained even in the original Riemannian case, as follows.

First, let us give a result for the empirical average of a function over a trajectory of the Brownian motion. When time grows, this empirical average converges to the average of the function over the manifold, since the invariant law of Brownian motion is uniform. The following is a concentration result for the variations of this quantity around its mean.

Theorem 2.8 ([41]). *Let $(Y_t)_{t\geqslant 0}$ be a trajectory of the Brownian motion starting at a fixed point x_0 on a smooth, compact N-dimensional Riemannian manifold X with Ricci curvature at least $K > 0$. Let*

$$\bar{f}_t = \frac{1}{t}\int_{s=0}^{t} f(Y_s)\,\mathrm{d}s$$

be the empirical mean of the 1-Lipschitz function $f\colon X \to \mathbb{R}$ under the diffusion process (Y_t) starting at some point $x \in X$. Then, for any fixed x_0 one has

$$\Pr(|\bar{f}_t - \mathbb{E}\bar{f}_t| > r) \leqslant 2\mathrm{e}^{-tK^2r^2/32}.$$

[36] contains related results and a beautiful characterization of this kind of property in terms of transportation-information inequalities.

Our second example deals with the spectral gap of a manifold. Let Δ be the Laplace(–Beltrami) operator on the compact Riemannian manifold X. It is a negative operator on L^2; its spectral gap λ_1 is the smallest nonzero eigenvalue of $-\Delta$. Alternatively, λ_1 can be defined as the best constant in either one of those inequalities: the norm of the Poisson equation operator $\|\Delta^{-1}f\|_{L^2} \leqslant (1/\lambda_1)\|f\|_{L^2}$, the rate of smoothing by the heat equation $\|f_t\|_{L^2} \leqslant C_f\mathrm{e}^{-\lambda_1 t}\|f_0\|_{L^2}$ with f_t solving $\partial f_t/\partial t = \Delta f_t$, or the Poincaré(–Sobolev) inequality $\|f\|_{L^2}^2 \leqslant (1/\lambda_1)\|\nabla f\|_{L^2}^2$, all these inequalities applying for functions in L^2 whose integral over X is 0. Thus, for instance, $1/\lambda_1$ is the characteristic decay time for the heat equation.

A classical theorem of Lichnerowicz (see, e.g., [8]) asserts that if X is an N-dimensional manifold with positive Ricci curvature, then $1/\lambda_1$ is controlled by the largest radius of curvature:

$$\frac{1}{\lambda_1} \leqslant \frac{N-1}{N}\sup_{x\in X}\frac{1}{\mathrm{Ric}(x)}$$

where $\mathrm{Ric}(x)$ is the infimum of $\mathrm{Ric}(v,v)$ for v a unit tangent vector at $x \in X$. This was improved by Veysseire as follows.

Theorem 2.9 ([66, 67]). *Let X be a smooth, compact N-dimensional manifold with positive Ricci curvature and let λ_1 be the spectral gap of the Laplace operator on X. Then*

$$\frac{1}{\lambda_1} \leqslant \int_{x\in X}\frac{1}{\mathrm{Ric}(x)}\frac{\mathrm{d}\,\mathrm{vol}(x)}{\mathrm{vol}(X)}$$

Thus we get a harmonic mean of Ricci curvature over the manifold, instead of the worst value. A slightly more complicated formula in [66, 67] allows to recover the Lichnerowicz theorem with the $(N-1)/N$ factor.

The proof given in [66] is very short and purely computational; it was found afterwards. The original intuition [67] was to couple two Brownian motions starting at very close points and using curvature, an approach used in [43] (see [3] for a new elegant viewpoint). If Brownian motion is seen as a large number of small steps, using parallel transport of the trajectory (x_t) starting at x provides a Brownian trajectory (y_t) starting at y close to x; since, at each step, the distance between x_t and y_t decreases by a factor depending on Ricci curvature, the distance after time t decreases exponentially fast: Ricci curvature is a kind of "Lyapunov exponent along Brownian trajectories." This is turned into a spectral gap estimate using the correspondence between Brownian motion and the heat kernel. The new idea here is that the exponential rate at which the trajectories (x_t) and (y_t) get closer only depends on curvature *at the current point* (x_t), and that, since x_t covers the whole manifold with uniform measure, an averaging effect over the value of curvature should occur, instead of using only the worst curvature. This proof is longer than the short computational proof in [66], but may be more adapted to generalization, e.g., for discrete settings.

Finally, let us mention that [61] lists a number of questions about coarse Ricci curvature. Some have been solved (e.g., the relationship with Alexandrov curvature), but most remain open.

2.3. Nonsmooth Ricci curvature: the displacement convexity of entropy approach. The other main approach to Ricci curvature for nonsmooth spaces is *displacement convexity of entropy*. It is introduced in [65] and [47], following a characterization of Ricci curvature for manifolds presented in [71] (elaborating on [38, 49, 63]). It is also based on the use of Wasserstein distances.

This approach allows to define a property $\mathrm{CD}(K, N)$ which, for Riemannian manifolds, is equivalent to having Ricci curvature at least K and dimension at most N. We will only present here the case $N = \infty$; the case $N < \infty$ involves very precise metric comparisons with N-dimensional manifolds. The property $\mathrm{CD}(K, N)$ in this sense will be an extension of Bakry – Émery's $\mathrm{CD}(K, N)$ property [6, 7], though from a different viewpoint.

The idea is that, in positive Ricci curvature, "midpoints spread out": if we take two sets or measures in the manifold X, and consider the set of points that lie "halfway" between the two sets (in a sense to be made precise below), then the set of midpoints is wider than expected from the Euclidean case. For instance, on a sphere, for any reasonable notion of what a midpoint is, the set of midpoints of the two poles will be the whole equator. The reverse is true in negative Ricci curvature. This intuition, illustrated in Figures 11 and 12, was formalized in various ways in [18, 19, 71].

Once more, we will use probability measures instead of sets. So let A_0 and A_1 be two probability measures in the geodesic space X. The extent to which they are spread will be evaluated using the *relative entropy* (or *Kullback – Leibler divergence*) with respect to the uniform measure μ on X (μ can actually be any reference measure of particular interest; in general there is no notion of "uniform measure" on arbitrary spaces). Namely, for a probability measure A define

$$H(A \mid \mu) = \int A(\mathrm{d}x) \log \frac{A(\mathrm{d}x)}{\mu(\mathrm{d}x)}$$

assuming the density $A(\mathrm{d}x)/\mu(\mathrm{d}x)$ exists and the integral makes sense (and $+\infty$ otherwise). This quantity is well-known in information theory [20]; the lower it is, the more "uniform" or "spread out" the measure A is (w.r.t. the reference measure μ).

Next, given two probability measures A_0 and A_1, we need to define the "measure of midpoints" $A_{1/2}$ of A_0 and A_1. We will define $A_{1/2}$ as a measure that lies midway between A_0 and A_1 in L^2 Wasserstein distance. The latter is defined like the L^1 Wasserstein distance introduced above [69].

Definition 2.10. Let μ_1, μ_2 be two measures with mass 1 in a metric space (X, d). The L^2 *transportation distance* (or L^2 *Wasserstein distance*) between μ_1 and μ_2, denoted $W_2(\mu_1, \mu_2)$, is the best average travelling distance that can be achieved:

$$W_2(\mu_1, \mu_2) = \inf_{\xi \in \Pi(\mu_1, \mu_2)} \sqrt{\iint d(x,y)^2 \, \mathrm{d}\xi(x,y)}$$

where $\Pi(\mu_1, \mu_2)$ is the set of transference plans from μ_1 to μ_2, as introduced in Definition 2.4.

Assume for simplicity that the geodesic space X is compact. Then, given any two probability measures A_0 and A_1, there exists a (not necessarily unique) probability measure $A_{1/2}$ such that

$$W_2(A_0, A_{1/2}) = W_2(A_{1/2}, A_1) = \tfrac{1}{2}\, W_2(A_0, A_1)$$

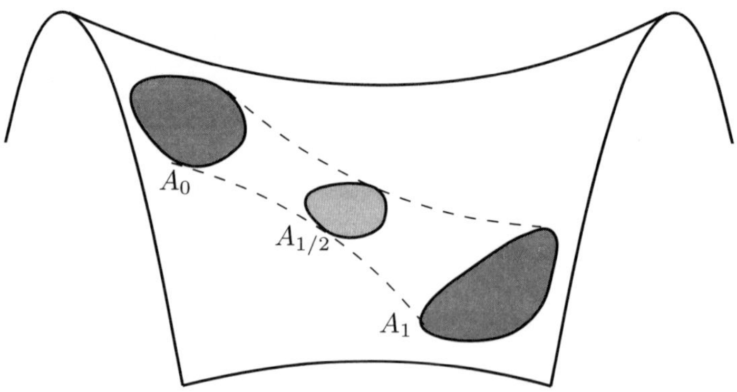

FIGURE 11. In negative curvature, midpoints contract.

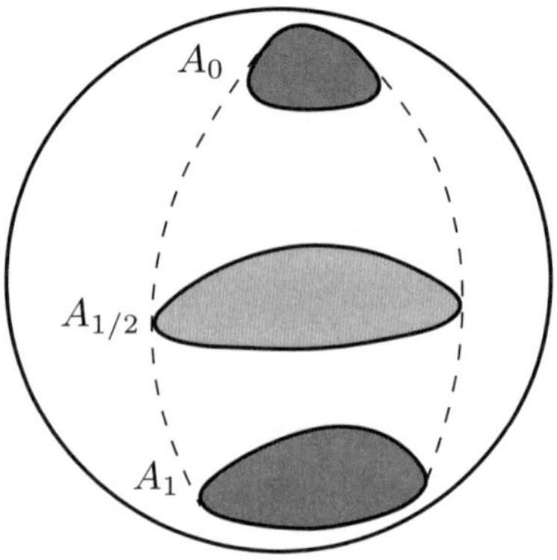

FIGURE 12. In positive curvature, midpoints spread out.

where W_2 is the L^2 Wasserstein distance. We will call such an $A_{1/2}$ a *midpoint measure* of A_0 and A_1.

Curvature will be positive when the relative entropy of a midpoint measure $A_{1/2}$ is smaller than the average of the relatives entropies of A_0 and A_1. The difference is used to define (a lower bound on) curvature.

Definition 2.11. Let (X, d) be a metric space and μ be a reference measure on X, with $\mu(X)$ finite. Then (X, d, μ) satisfies the $\mathrm{CD}(K, \infty)$ if, for any two probability measures A_0 and A_1 with finite entropy with respect to μ, there exists a midpoint measure $A_{1/2}$ of A_0 and A_1 such that

$$H(A_{1/2} \mid \mu) \leqslant \frac{1}{2}\big(H(A_0 \mid \mu) + H(A_1 \mid \mu)\big) - \frac{K}{8} W_2(A_0, A_1)^2$$

The reason for the name *displacement convexity of entropy* [49] is clear: the inequality above states that entropy is a convex function on the space of probability measures, where the convex interpolation between two measures is obtained by transportation. (Actually, we used convexity only for *one* midpoint measure, not for all of them, hence the qualification of *weak* convexity sometimes encountered.)

The restriction that $\mu(X)$ is finite is not essential; it only makes for a slightly less technical definition. The $K/8$ scaling factor reproduces what happens in Riemannian manifolds with Ricci curvature equal to K.

This defines only a global lower bound for curvature; but this definition can be localized by taking A_0 and A_1 in the neighborhood of some point. Then, under various technical assumptions, the global lower bound on curvature is the infimum of the local lower bounds (Theorem I.4.17 in [65], extended in [5]). This is analogous to the property mentioned above, that it is enough to check coarse Ricci curvature $\kappa(x, y)$ for close pairs of points x and y.

Let us describe examples of spaces known to have Ricci curvature bounds in this sense.

First, this definition is a consistent extension of Ricci curvature: For Riemannian manifolds with μ the Riemannian volume measure, the $\mathrm{CD}(K, \infty)$ condition is satisfied exactly when Ricci curvature is at least K ([65, Theorem I.4.9]), and likewise for the refined $\mathrm{CD}(K, N)$ curvature-dimension condition.

Second, spaces with a lower bound on sectional curvature in the sense of Alexandrov (Definition 2.1) have a corresponding lower bound on Ricci curvature in the present sense [64, 73].

Another source of examples are Gromov – Hausdorff limits [13, 35] of Riemannian manifolds with positive Ricci curvature. These limits can have singularities or exhibit collapsing (reduction of dimension) of some parts. One of the main results here is that $\mathrm{CD}(K, \infty)$ and $\mathrm{CD}(K, N)$ lower bounds on curvature are preserved when passing to these limits [47, 65]. Actually, the study of the phenomena arising when taking Gromov–Hausdorff limits of manifolds was a strong motivation for the search of generalized notions of Ricci curvature [46], especially in view of Gromov's precompactness theorem (see, e.g., [35]) which states that the set of all Riemannian manifolds of a given dimension, with diameter bounded above and Ricci curvature bounded below, is precompact in Gromov – Hausdorff topology. This line of thought led to sustained research, for example a series of results by Cheeger and Colding [14]. An elegant way to incorporate the measured aspect in these limits is provided by a "Gromov – Wasserstein" distance [65].

Displacement convexity of entropy has also been successfully applied to manifolds equipped with non-Riemannian metrics (where an arbitrary non-Euclidean norm is used on each tangent space), i.e., Finsler manifolds [54–56], allowing to prove new results such as Poincaré-like inequalities on some of these spaces.

This approach has also been conducted on the Heisenberg group [42]: no Ricci curvature lower bound $\mathrm{CD}(K, N)$ with $K > -\infty$ holds in this space. On the other hand, in [42] it is shown that a related but weaker property holds, the *measure contraction property* [52, 65].

Let us now quickly mention some of the theorems from Riemannian manifolds that carry out to spaces with positive Ricci curvature in the present sense. We refer to [47, 65, 70] for more details.

The condition $\mathrm{CD}(K, \infty)$ entails several powerful functional inequalities, in particular the logarithmic Sobolev inequality, which in turn implies other properties such as Gaussian concentration of measure (as discussed above) or spectral gap (Poincaré) inequalities.

The condition $\mathrm{CD}(K, N)$ is stronger and closer to the Riemannian situation. It implies, for instance, the Bishop–Gromov volume growth inequality, which is essentially a restatement of the definition of $\mathrm{CD}(K, N)$ with the sets A_0 and A_1 taken to be concentric balls. The Bonnet–Myers theorem also holds (keep in mind, though, that the definition of $\mathrm{CD}(K, N)$ is precisely modeled after the Riemannian situation, and explicitly uses functions such as $\sin\left(d(x, y)\sqrt{K/(N-1)}\right)$). The Brunn–Minkowski inequality (similar to the pictures above, but with a larger set of midpoints $A_{1/2}$) is another consequence. This illustrates the power implicit in the $\mathrm{CD}(K, N)$ condition, for various choices of A_0 and A_1; the difficult part, of course, being to prove that the condition holds.

The constants obtained in these various inequalities are often sharp compared to the Riemannian case; for comparison, using coarse Ricci curvature leads to a numerical loss in the constants (even if the order of magnitude stays the same).

It is impossible to mention the displacement convexity approach without hinting at the deep context from which it arose, especially, the important relationship between the heat equation and entropy on metric spaces. The fundamental idea is that "the heat equation is the best way to increase entropy while moving the particles as least as possible" [38]. More precisely, if a measure μ_t on the base space X evolves according to the heat equation $\partial\mu_t/\partial t = \Delta\mu_t$, this evolution can be thought of in the space of all measures on X in the following way: start with μ_0, consider the entropy function around μ_0, and follow the gradient of this entropy function. Here, to define gradients, the (infinite-dimensional) space of measures is equipped with a formally Riemannian metric for which the distance between two very close measures is defined as their W_2 transportation distance [38, 63]. This approach can be used to *define* a heat equation in metric measured spaces [1, 30] and has been very successful.

Let us finally discuss displacement convexity of entropy in relation with coarse Ricci curvature.

First, there is no analogue to $\mathrm{CD}(K, N)$ with $N < \infty$ for coarse Ricci curvature; the latter appears more comparable to $\mathrm{CD}(K, \infty)$.

Second, the CD condition was not designed for discrete spaces; moreover good technical behavior of the CD condition requires that the geodesic space be *non-branching*, i.e., that two geodesics cannot coincide for some time then part away, as typically happens in graphs. On the other hand, coarse Ricci curvature bypasses any local topological structure. An easy way to discretize displacement convexity of entropy is to use *approximate* midpoints for the set $A_{1/2}$, instead of exact midpoints. This approach yields partial results: in [5] it is applied to homogeneous planar graphs, and in [62] to the discrete hypercube. The case of the discrete hypercube is not fully solved: only a weaker property, the Brunn – Minkowski inequality, is obtained in [62] (though intriguing possible connections with combinatorics appear).

Another recent and promising approach to displacement convexity of entropy in discrete spaces has been proposed independently in [17, 26, 48, 51]: these authors introduce a new metric on the space of probability measures on a discrete space, such that the [38] interpretation of the heat equation holds, namely, such that a given Markov kernel on the underlying space coincides with the gradient of the entropy functional on the space of probability measures. Such an approach gives positive curvature to the discrete hypercube [26], as expected. The functional inequalities proven in [26] under discrete positive curvature in this sense are comparable to those obtained from $\mathrm{CD}(K,\infty)$: logarithmic Sobolev inequalities and Gaussian concentration of measure, for instance, with optimal numerical constants recovered for the hypercube.

Coarse Ricci curvature, on the other hand, is based on a simpler criterion and may apply in a wider range of situations. For instance, no reversibility assumption is needed. More, it allows for non-Gaussian concentration: discrete examples in significant applications exhibit, e.g., Poissonian behavior. This is not compatible with $\mathrm{CD}(K,\infty)$.

Another situation in which $\mathrm{CD}(K,\infty)$ and coarse Ricci curvature differ is Finsler spaces. For instance, consider $\mathbb{R}^n$ equipped with a non-Euclidean norm, e.g., the ℓ^1 norm. It is immediate to check that such a space has vanishing coarse Ricci curvature, by translation invariance. These spaces are also known to satisfy the $\mathrm{CD}(0,\infty)$ property; yet the natural heat equation defined by the [38] interpretation is nonlinear (hence non-Markovian, so that the setting of coarse Ricci curvature does not apply to it), and does not satisfy the contraction in transportation distance appearing in the definition of coarse Ricci curvature [55, 56].

Finally, let us mention three variants around the $\mathrm{CD}(K,\infty)$ condition, each with independent interest. The nineteenth-century-old *Brunn – Minkowski inequality* (see, e.g., [28] for an introduction) can be seen in geodesic spaces as a weaker version of $\mathrm{CD}(K,N)$, using a larger set of midpoints $A_{1/2}$; the equivalence with Ricci curvature in the Riemannian case is proven in [18, 19]. As a general property of geodesic spaces, it is weaker than $\mathrm{CD}(K,N)$ but already has a number of the same geometrical consequences [65]. It has been studied in, e.g., Finsler spaces [54], discretizes well [9] and is known to hold in the discrete hypercube [62]. See also [4, 37] for a functional inequality version. The property $\mathrm{RCD}(K,\infty)$, introduced in [2], is a reinforcement both of $\mathrm{CD}(K,\infty)$ and of coarse Ricci curvature. It allows for an elegant control of fine properties of the heat equation and related problems. The *measure contraction property* [52, 65] is weaker than the CD condition and can be thought of as the particular case when the measure A_0 in the pictures above is a Dirac mass and A_1 is the uniform measure on the whole space. Any of those properties is yet another reasonable contender for the title of generalized Ricci curvature.

This closes this tour of some possible generalizations of sectional and Ricci curvature, and their connections to various other areas of mathematics. The field is still very active.

References

1. L. Ambrosio, N. Gigli, and G. Savaré, *Gradient flows in metric spaces and in the space of probability measures*, Lectures Math. ETH Zurich, Birkhäuser, Basel, 2005.
2. ———, *Metric measure spaces with Riemannian Ricci curvature bounded from below*, available at arXiv:1109.0222.
3. M. Arnaudon, K. A. Coulibaly, and A. Thalmaier, *Horizontal diffusion in C^1 path space*, Séminaire de Probabilités XLIII (Poitiers, 2009) (C. Donati-Martin, A. Lejay, and A. Rouault, eds.), Lecture Notes in Math., vol. 2006, Springer, Berlin, 2011, pp. 73–94.
4. K. Bacher, *On Borell – Brascamp – Lieb inequalities on metric measure spaces*, Potential Anal. **33** (2010), no. 1, 1–15.
5. K. Bacher and K.-T. Sturm, *Localization and tensorization properties of the curvature-dimension condition for metric measure spaces*, J. Funct. Anal. **259** (2010), no. 1, 28–56.
6. D. Bakry and M. Émery, *Hypercontractivité de semi-groupes de diffusion*, C. R. Acad. Sci. Paris Sér. I Math. **299** (1984), no. 15, 775–778.
7. D. Bakry and M. Émery, *Diffusions hypercontractives*, Séminaire de probabilités, XIX, 1983/84 (J. Azéma and M. Yor, eds.), Lecture Notes in Math., vol. 1123, Springer, Berlin, 1985, pp. 177–206.
8. M. Berger, *A panoramic view of Riemannian geometry*, Springer, Berlin, 2003.
9. M. Bonnefont, *A discrete version of the Brunn – Minkowski inequality and its stability*, Ann. Math. Blaise Pascal **16** (2009), no. 2, 245–257.
10. A.-I. Bonciocat and K.-T. Sturm, *Mass transportation and rough curvature bounds for discrete spaces*, J. Funct. Anal. **256** (2009), no. 9, 2944–2966.
11. M. R. Bridson and A. Haefliger, *Metric spaces of non-positive curvature*, Grundlehren Math. Wiss., vol. 319, Springer, Berlin, 1999.
12. R. Bubley and M. Dyer, *Path coupling: a technique for proving rapid mixing in Markov chains*, 38th Annual Symposium on Foundations of Computer Science (Miami Beach, FL, 1997), IEEE Computer Society Press, Los Alamitos, CA, 1997, pp. 223–231.
13. D. Burago, Y. Burago, and S. Ivanov, *A course in metric geometry*, Grad. Stud. Math., vol. 33, Amer. Math. Soc., Providence, RI, 2001.
14. J. Cheeger and T. H. Colding, *On the structure of spaces with Ricci curvature bounded below.* I, J. Differential Geom. **46** (1997), no. 3, 406–480; II, **54** (2000), no. 1, 13–35; III, 37–74.
15. M. F. Chen, *From Markov chains to non-equilibrium particle systems*, 2nd ed., World Sci. Publ., Singapore, 2004.
16. M. F. Chen and F. Y. Wang, *Application of coupling method to the first eigenvalue on manifold*, Sci. China Ser. A **37** (1994), no. 1, 1–14.
17. S.-N. Chow, W. Huang, Y. Li, and H. Zhou, *Fokker – Planck equations for a free energy functional or Markov process on a graph*, Arch. Ration. Mech. Anal. **203** (2012), no. 3, 969–1008.
18. D. Cordero-Erausquin, R. J. McCann, and M. Schmuckenschläger, *A Riemannian interpolation inequality à la Borell, Brascamp and Lieb*, Invent. Math. **146** (2001), no. 2, 219–257.
19. ———, *Prékopa – Leindler type inequalities on Riemannian manifolds, Jacobi fields, and optimal transport*, Ann. Fac. Sci. Toulouse Math. (6) **15** (2006), no. 4, 613–635.
20. T. M. Cover and J. A. Thomas, *Elements of information theory*, Wiley Ser. Telecom., Wiley, New York, 1991.
21. H. Djellout, A. Guillin, and L. Wu, *Transportation cost-information inequalities and applications to random dynamical systems and diffusions*, Ann. Probab. **32** (2004), no. 3B, 2702–2732.
22. M. P. do Carmo, *Riemannian geometry*, Math. Theory Appl., Birkhäuser, Boston, MA, 1992.

23. R. L. Dobrushin, *Definition of a system of random variables by means of conditional distributions*, Teor. Verojatnost. i Primenen. **15** (1970), 469–497 (Russian); English transl., Theor. Probability Appl. **15** (1970), no. 3, 458–486.
24. ———, *Perturbation methods of the theory of Gibbsian fields*, Lectures on Probability Theory and Statistics (Saint-Flour, 1994) (P. Bernard, ed.), Lecture Notes in Math., vol. 1648, Springer, Berlin, 1996, pp. 1–66.
25. R. L. Dobrushin and S. B. Shlosman, *Constructive criterion for the uniqueness of Gibbs field*, Statistical Physics and Dynamical Systems (Köszeg, 1984) (J. Fritz, A. Jaffe, and D. Szász, eds.), Progr. Phys., vol. 10, Birkhäuser, Boston, MA, 1985, pp. 347–370.
26. M. Erbar and J. Maas, *Ricci curvature of finite Markov chains via convexity of the entropy*, available at `arXiv:1111.2687`.
27. S. Fang, J. Shao, and K.-T. Sturm, *Wasserstein space over the Wiener space*, Probab. Theory Related Fields **146** (2010), no. 3-4, 535–565.
28. R. J. Gardner, *The Brunn – Minkowski inequality*, Bull. Amer. Math. Soc. (N.S.) **39** (2002), no. 3, 355–405.
29. L. Gerin, *Random sampling of lattice paths with constraints, via transportation*, 21st International Meeting on Probabilistic, Combinatorial, and Asymptotic Methods in the Analysis of Algorithms (AofA'10) (Vienna, 2010) (M. Drmota and B. Gittenberger, eds.), Discrete Math. Theor. Comput. Sci. Proc., AM, Assoc. Discrete Math. Theor. Comput. Sci., Nancy, 2010, pp. 317–328.
30. N. Gigli, *On the heat flow on metric measure spaces: existence, uniqueness and stability*, Calc. Var. Partial Differential Equations **39** (2010), no. 1-2, 101–120.
31. N. Gigli, K. Kuwada, and S.-i. Ohta, *Heat flow on Alexandrov spaces*, available at `arXiv:1008.1319`.
32. M. Gromov, *Large Riemannian manifolds*, Curvature and Topology of Riemannian Manifolds (Katata, 1985) (K. Shiohama, T. Sakai, and T. Sunada, eds.), Lecture Notes in Math., vol. 1201, Springer, Berlin, 1986, pp. 108–121.
33. ———, *Hyperbolic groups*, Essays in Group Theory (S. M. Gersten, ed.), Math. Sci. Res. Inst. Publ., vol. 8, Springer, New York, 1987, pp. 75–263.
34. ———, *Sign and geometric meaning of curvature*, Rend. Sem. Mat. Fis. Milano **61** (1991), 9–123 (1994).
35. M. Gromov, *Metric structures for Riemannian and non-Riemannian spaces*, with appendices by M. Katz, P. Pansu, and S. Semmes, Progr. Math., vol. 152, Birkhäuser, Boston, MA, 1999.
36. A. Guillin, C. Léonard, L. Wu, and N. Yao, *Transportation-information inequalities for Markov processes*, Probab. Theory Related Fields **144** (2009), no. 3-4, 669–695.
37. E. Hillion, *On Prekopa – Leindler inequalities on metric-measure spaces*, available at `arXiv:0912.3593`.
38. R. Jordan, D. Kinderlehrer, and F. Otto, *The variational formulation of the Fokker – Planck equation*, SIAM J. Math. Anal. **29** (1998), no. 1, 1–17.
39. A. Joulin, *Poisson-type deviation inequalities for curved continuous-time Markov chains*, Bernoulli **13** (2007), no. 3, 782–798.
40. ———, *A new Poisson-type deviation inequality for Markov jump processes with positive Wasserstein curvature*, Bernoulli **15** (2009), no. 2, 532–549.
41. A. Joulin and Y. Ollivier, *Curvature, concentration and error estimates for Markov chain Monte Carlo*, Ann. Probab. **38** (2010), no. 6, 2418–2442.
42. N. Juillet, *Geometric inequalities and generalized Ricci bounds in the Heisenberg group*, Int. Math. Res. Not. IMRN **13** (2009), 2347–2373.
43. W. S. Kendall, *Nonnegative Ricci curvature and the Brownian coupling property*, Stochastics **19** (1986), no. 1-2, 111–129.
44. M. Ledoux, *The concentration of measure phenomenon*, Math. Surveys Monogr., vol. 89, Amer. Math. Soc., Providence, RI, 2001.
45. D. A. Levin, Y. Peres, and E. L. Wilmer, *Markov chains and mixing times*, with a chapter by J. G. Propp and D. B. Wilson, Amer. Math. Soc., Providence, RI, 2009.
46. J. Lott, *Optimal transport and Ricci curvature for metric-measure spaces*, Surveys in Differential Geometry. Vol. XI (J. Cheeger and K. Grove, eds.), Surv. Differ. Geom., vol. 11, Int. Press, Somerville, MA, 2007, pp. 229–257.
47. J. Lott and C. Villani, *Ricci curvature for metric-measure spaces via optimal transport*, Ann. of Math. (2) **169** (2009), no. 3, 903–991.

48. J. Maas, *Gradient flows of the entropy for finite Markov chains*, J. Funct. Anal. **261** (2011), no. 8, 2250–2292.
49. R. J. McCann, *A convexity principle for interacting gases*, Adv. Math. **128** (1997), no. 1, 153–179.
50. R. J. McCann and P. M. Topping, *Ricci flow, entropy and optimal transportation*, Amer. J. Math. **132** (2010), no. 3, 711–730.
51. A. Mielke, *Geodesic convexity of the relative entropy in reversible Markov chains*, Calc. Var. Partial Differential Equations, to appear.
52. S.-i. Ohta, *On the measure contraction property of metric measure spaces*, Comment. Math. Helv. **82** (2007), no. 4, 805–828.
53. ———, *Gradient flows on Wasserstein spaces over compact Alexandrov spaces*, Amer. J. Math. **131** (2009), no. 2, 475–516.
54. ———, *Finsler interpolation inequalities*, Calc. Var. Partial Differential Equations **36** (2009), no. 2, 211–249.
55. S.-i. Ohta and K.-T. Sturm, *Heat flow on Finsler manifolds*, Comm. Pure Appl. Math. **62** (2009), no. 10, 1386–1433.
56. ———, *Non-contraction of heat flow on Minkowski spaces*, Arch. Ration. Mech. Anal., available at `arXiv:1008.1319`.
57. R. I. Oliveira, *On the convergence to equilibrium of Kac's random walk on matrices*, Ann. Appl. Probab. **19** (2009), no. 3, 1200–1231.
58. Y. Ollivier, *A January* 2005 *invitation to random groups*, Ensaios Mat., vol. 10, Soc. Brasil. Mat., Rio de Janeiro, 2005.
59. ———, *Ricci curvature of metric spaces*, C. R. Math. Acad. Sci. Paris **345** (2007), no. 11, 643–646.
60. ———, *Ricci curvature of Markov chains on metric spaces*, J. Funct. Anal. **256** (2009), no. 3, 810–864.
61. ———, *A survey of Ricci curvature for metric spaces and Markov chains*, Probabilistic Approach to Geometry (Kyoto, 2008) (M. Kotani, M. Hino, and T. Kumagai, eds.), Adv. Stud. Pure Math., vol. 57, Math. Soc. Japan, Tokyo, 2010, pp. 343–381.
62. Y. Ollivier and C. Villani, *A curved Brunn – Minkowski inequality on the discrete hypercube*, available at `arXiv:1011.4779`.
63. F. Otto and C. Villani, *Generalization of an inequality by Talagrand and links with the logarithmic Sobolev inequality*, J. Funct. Anal. **173** (2000), no. 2, 361–400.
64. A. Petrunin, *Alexandrov meets Lott – Villani – Sturm*, available at `arXiv:1003.5948`.
65. K.-T. Sturm, *On the geometry of metric measure spaces.* I, Acta Math. **196** (2006), no. 1, 65–131; II, 133–177.
66. L. Veysseire, *A harmonic mean bound for the spectral gap of the Laplacian on Riemannian manifolds*, C. R. Math. Acad. Sci. Paris **348** (2010), no. 23-24, 1319–1322.
67. ———, *Improved spectral gap bounds on positively curved manifolds*, available at `arXiv:1105.6080`.
68. ———, *Coarse Ricci curvature for continuous-time Markov processes*, available at `arXiv:1202.0420`.
69. C. Villani, *Topics in optimal transportation*, Grad. Stud. Math., vol. 58, Amer. Math. Soc., Providence, RI, 2003.
70. ———, *Optimal transport: Old and new*, Grundlehren Math. Wiss., vol. 338, Springer, Berlin, 2009.
71. M.-K. von Renesse and K.-T. Sturm, *Transport inequalities, gradient estimates, entropy, and Ricci curvature*, Comm. Pure Appl. Math. **58** (2005), no. 7, 923–940.
72. D. B. Wilson, *Mixing times of Lozenge tiling and card shuffling Markov chains*, Ann. Appl. Probab. **14** (2004), no. 1, 274–325.
73. H.-C. Zhang and X.-P. Zhu, *Ricci curvature on Alexandrov spaces and rigidity theorems*, Comm. Anal. Geom. **18** (2010), no. 3, 503–553.

Laboratoire de recherche en informatique, Université Paris-Sud/CNRS, Bât. 490, 91405 Orsay Cedex, France

E-mail address: `Yann.Ollivier@lri.fr`

Published Titles in This Series

56 **Galia Dafni, Robert John McCann, and Alina Stancu, Editors,** Analysis and Geometry of Metric Measure Spaces, 2013

55 **André Boivin and Javad Mashreghi, Editors,** Complex Analysis and Potential Theory, 2012

54 **Daniel Daigle, Richard Ganong, and Mariusz Koras,** Affine Algebraic Geometry, 2011

53 **Bradd Hart, Thomas G. Kucera, Anand Pillay, Philip J. Scott, and Robert A. G. Seely, Editors,** Models, Logics, and Higher-Dimensional Categories, 2011

52 **Dmitry Jakobson, Stéphane Nonnenmacher, and Iosif Polterovich, Editors,** Spectrum and Dynamics, 2010

51 **Javad Mashreghi, Thomas Ransford, and Kristian Seip, Editors,** Hilbert Spaces of Analytic Functions, 2010

50 **P. Robert Kotiuga, Editor,** A Celebration of the Mathematical Legacy of Raoul Bott, 2010

49 **Miguel Abreu, François Lalonde, and Leonid Polterovich, Editors,** New Perspectives and Challenges in Symplectic Field Theory, 2009

48 **David Avis, David Bremner, and Antoine Deza, Editors,** Polyhedral Computation, 2009

47 **John Harnad and Pavel Winternitz, Editors,** Groups and Symmetries, 2009

46 **Jean-Marie De Koninck, Andrew Granville, and Florian Luca, Editors,** Anatomy of Integers, 2008

45 **Panos M. Pardalos and Pierre Hansen, Editors,** Data Mining and Mathematical Programming, 2008

44 **Stanley Alama, Lia Bronsard, and Peter J. Sternberg, Editors,** Singularities in PDE and the Calculus of Variations, 2008

43 **Andrew Granville, Melvyn B. Nathanson, and József Solymosi, Editors,** Additive Combinatorics, 2007

42 **Donald A. Dawson, Vojkan Jakšić, and Boris Vainberg, Editors,** Probability and Mathematical Physics, 2007

41 **André Bandrauk, Michel C. Delfour, and Claude Le Bris, Editors,** High-Dimensional Partial Differential Equations in Science and Engineering, 2007

40 **Vestislav Apostolov, Andrew Dancer, Nigel Hitchin, and McKenzie Wang, Editors,** Perspectives in Riemannian Geometry, 2006

39 **P. Winternitz, D. Gomez-Ullate, A. Iserles, D. Levi, P. J. Olver, R. Quispel, and P. Tempesta, Editors,** Group Theory and Numerical Analysis, 2005

38 **Jacques Hurtubise and Eyal Markman, Editors,** Algebraic Structures and Moduli Spaces, 2004

37 **P. Tempesta, P. Winternitz, J. Harnad, W. Miller, Jr., G. Pogosyan, and M. Rodriguez, Editors,** Superintegrability in Classical and Quantum Systems, 2004

36 **Hershy Kisilevsky and Eyal Z. Goren, Editors,** Number Theory, 2004

35 **H. E. A. Eddy Campbell and David L. Wehlau, Editors,** Invariant Theory in All Characteristics, 2004

34 **P. Winternitz, J. Harnad, C. S. Lam, and J. Patera, Editors,** Symmetry in Physics, 2004

33 **André D. Bandrauk, Michel C. Delfour, and Claude Le Bris, Editors,** Quantum Control: Mathematical and Numerical Challenges, 2003

32 **Vadim B. Kuznetsov, Editor,** The Kowalevski Property, 2002

31 **John Harnad and Alexander Its, Editors,** Isomonodromic Deformations and Applications in Physics, 2002

30 **John McKay and Abdellah Sebbar, Editors,** Proceedings on Moonshine and Related Topics, 2001

29 **Alan Coley, Decio Levi, Robert Milson, Colin Rogers, and Pavel Winternitz, Editors,** Bäcklund and Darboux Transformations. The Geometry of Solitons, 2001

28 **J. C. Taylor, Editor,** Topics in Probability and Lie Groups: Boundary Theory, 2001

27 **I. M. Sigal and C. Sulem, Editors,** Nonlinear Dynamics and Renormalization Group, 2001

26 **J. Harnad, G. Sabidussi, and P. Winternitz, Editors,** Integrable Systems: From Classical to Quantum, 2000

25 **Decio Levi and Orlando Ragnisco, Editors,** SIDE III—Symmetries and Integrability of Difference Equations, 2000

24 **B. Brent Gordon, James D. Lewis, Stefan Müller-Stach, Shuji Saito, and Noriko Yui, Editors,** The Arithmetic and Geometry of Algebraic Cycles, 2000

23 **Pierre Hansen and Odile Marcotte, Editors,** Graph Colouring and Applications, 1999

22 **Jan Felipe van Diejen and Luc Vinet, Editors,** Algebraic Methods and q-Special Functions, 1999

21 **Michel Fortin, Editor,** Plates and Shells, 1999

20 **Katie Coughlin, Editor,** Semi-Analytic Methods for the Navier-Stokes Equations, 1999

19 **Rajiv Gupta and Kenneth S. Williams, Editors,** Number Theory, 1999

18 **Serge Dubuc and Gilles Deslauriers, Editors,** Spline Functions and the Theory of Wavelets, 1999

17 **Olga Kharlampovich, Editor,** Summer School in Group Theory in Banff, 1996, 1999

16 **Alain Vincent, Editor,** Numerical Methods in Fluid Mechanics, 1998

15 **François Lalonde, Editor,** Geometry, Topology, and Dynamics, 1998

14 **John Harnad and Alex Kasman, Editors,** The Bispectral Problem, 1998

13 **Michel C. Delfour, Editor,** Boundaries, Interfaces, and Transitions, 1998

12 **Peter C. Greiner, Victor Ivrii, Luis A. Seco, and Catherine Sulem, Editors,** Partial Differential Equations and Their Applications, 1997

11 **Luc Vinet, Editor,** Advances in Mathematical Sciences: CRM's 25 Years, 1997

10 **Donald E. Knuth,** Stable Marriage and Its Relation to Other Combinatorial Problems, 1997

9 **Decio Levi, Luc Vinet, and Pavel Winternitz, Editors,** Symmetries and Integrability of Difference Equations, 1996

8 **J. Feldman, R. Froese, and L. M. Rosen, Editors,** Mathematical Quantum Theory II: Schrödinger Operators, 1995

7 **J. Feldman, R. Froese, and L. M. Rosen, Editors,** Mathematical Quantum Theory I: Field Theory and Many-Body Theory, 1994

6 **Guido Mislin, Editor,** The Hilton Symposium 1993, 1994

5 **D. A. Dawson, Editor,** Measure-Valued Processes, Stochastic Partial Differential Equations, and Interacting Systems, 1994

4 **Hershy Kisilevsky and M. Ram Murty, Editors,** Elliptic Curves and Related Topics, 1994

3 **Rémi Vaillancourt and Andrei L. Smirnov, Editors,** Asymptotic Methods in Mechanics, 1993

2 **Philip D. Loewen,** Optimal Control via Nonsmooth Analysis, 1993

1 **M. Ram Murty, Editor,** Theta Functions, 1993